Numerical Determination of the Electronic Structure of Atoms, Diatomic and Polyatomic Molecules

NATO ASI Series

Advanced Science Institutes Series

A Series presenting the results of activities sponsored by the NATO Science Committee, which aims at the dissemination of advanced scientific and technological knowledge, with a view to strengthening links between scientific communities.

The Series is published by an international board of publishers in conjunction with the NATO Scientific Affairs Division

A Life Sciences **B Physics**	Plenum Publishing Corporation London and New York
C Mathematical and Physical Sciences **D Behavioural and Social Sciences** **E Applied Sciences**	Kluwer Academic Publishers Dordrecht, Boston and London
F Computer and Systems Sciences **G Ecological Sciences** **H Cell Biology**	Springer-Verlag Berlin, Heidelberg, New York, London, Paris and Tokyo

Series C: Mathematical and Physical Sciences - Vol. 271

Numerical Determination of the Electronic Structure of Atoms, Diatomic and Polyatomic Molecules

edited by

Mireille Defranceschi

Laboratoire de Chimie,
Ecole Normale Supérieure de Paris,
France

and

Joseph Delhalle

Département de Chimie,
Facultés Universitaires Notre-Dame de la Paix,
Namur, Belgium

Kluwer Academic Publishers

Dordrecht / Boston / London

Published in cooperation with NATO Scientific Affairs Division

Proceedings of the NATO Advanced Research Workshop on
Numerical Determination of the Electronic Structure
of Atoms, Diatomic and Polyatomic Molecules
Versailles, France
17–22 April 1988

Library of Congress Cataloging in Publication Data

NATO Advanced Research Workshop on 'Numerical Determination of the
Electronic Structure of Atoms, Diatomic and Polyatomic Molecules'
(1988 : Versailles, France)
Numerical determination of the electronic structure of atoms,
diatomic and polyatomic molecules / edited by Mireille Defranceschi
and Joseph Delhalle.
p. cm. -- (NATO ASI series. Series C, Mathematical and
physical sciences, vol. 271)
"Proceedings of the NATO Advanced Research Workshop on 'Numerical
Determination of the Electronic Structure of Atoms, Diatomic and
Polyatomic Molecules' held in Versailles, France, 17-22 April 1988"-
-T.p. verso.
"Published in cooperation with NATO Scientific Affairs Division."
Includes index.
ISBN 0-7923-0170-6
1. Atomic structure--Measurement--Congresses. 2. Molecular
structure--Measurement--Congresses. 3. Electronic structure-
-Measurement--Congresses. I. Defranceschi, Mireille, 1955- .
II. Delhalle, Joseph, 1945- . III. Title. IV. Series: NATO ASI
series. Series C, Mathematical and physical sciences ; no. 271.
QC172.N36 1989
539'.1--dc19 89-31057
CIP

ISBN 0–7923–0170–6

Published by Kluwer Academic Publishers,
P.O. Box 17, 3300 AA Dordrecht, The Netherlands.

Kluwer Academic Publishers incorporates the publishing programmes of
D. Reidel, Martinus Nijhoff, Dr W. Junk and MTP Press.

Sold and distributed in the U.S.A. and Canada
by Kluwer Academic Publishers,
101 Philip Drive, Norwell, MA 02061, U.S.A.

In all other countries, sold and distributed
by Kluwer Academic Publishers Group,
P.O. Box 322, 3300 AH Dordrecht, The Netherlands.

Printed in the Netherlands on acid free paper

CONTENTS

Short Contributions

INTRODUCTION

Quantum mechanical calculations in physics, chemistry and biology are widely recognized as useful interpretative and predictive tools. Unfortunately, they are plagued by unfavorable convergence limitations due to the use of finite linear combinations of basis functions. With the current computer technologies, there is a possible way out to the situation by solving numerically the corresponding wave equations.

The present interest and need for numerical determination of electronic structure of atoms, diatomic and polyatomic molecules led us to organize a NATO-ARW devoted to these questions. The aim of the meeting was to provide a review of the state of the art about techniques and applications.

The organizing committee consisted of Drs. G. Berthier, P. Claverie, M. Defranceschi, J. Delhalle, H.J. Monkhorst and P. Pyykkö. It was a great sorrow for us to be informed in January 88 of the death of Professor P. Claverie who supported so enthusiastically the idea of having such a meeting organized.

The NATO Advanced Research Worshop on : " Numerical Determination of the Electronic Structure of Atoms, Diatomic and Polyatomic Molecules" was held at Versailles (France) from April 17th till April 22th, 1988.

The ARW was organized around three related major themes. Three speakers concentrated on the properties of electronic wavefunctions and on the limitations of the current practice using basis functions. Three other speakers presented the stage reached by the numerical approach in configuration space, on diatomic molecules at Hartree-Fock level and beyond, while two speakers covered the recent achievements of the numerical approach in momentum space. Finally, two main lecturers compared the basis and computational strategies of Monte Carlo methods with those already mentionned. The meeting was concluded by a roundtable moderated by Professor W. Kutzelnigg.

This volume contains the ten main lectures of the workshop. Besides the lectures are included the short contributions of those participants who presented either a short communication or a poster during the meeting. At the end, a subject index gives an overview of chemical systems already treated by numerical methods.

We would like to express our deep gratitude to the NATO Scientific Affairs Division not only for its generous support which made this workshop possible, but also for its continuous interest during the preparation of the meeting. We would also like to thank Dr. J.G. Fripiat for his help and effective collaboration as scientific secretary during the meeting.

IN MEMORY of Pierre CLAVERIE

Our friend and colleague Pierre Claverie died January 24,1988 after having faced courageously the cruel disease that had thrust itself upon him.

In the name of all those who have shared with him the joys and sorrows of a path which he sought to enlighten with the knowledge of Nature and with a better understanding of Man, I should like to remind you, in a few words, of his personality and of his remarkable destiny. He was a man of many facets often misunderstood by those who approached him. He attracted people of contrasting attitudes.

Those who felt irritated by his obstination and his zeal, those who were fascinated by the extension and the depth of his knowledge, those whom he exasperated by his awkwardness and his sometimes hopeless precautions, those whom he reassured by the accuracy of his unfailing judgements, those who upbraided him for the obsessional aspects of certain intellectual claims, those who admired his insatiable curiosity - a sign of the profound respect for human nature which drove him forward -, all felt for him an undeniable fondness. All felt attached to him for his immense honesty which was on the borderline of innocence. Even those who disliked his tendency to speak in the name of intellectual and moral rigour, secretly admired him for being able to assume calmly the candour of knowing in a world of calculations and ambitions, competition and interests. Many have misunderstood his real nature. Where some have been able to see only an implacable debater, an inexorable calculator, a fearful annihilator of obstacles, an intellectual bulldozer, there was a great dreamer, always curious, ready to let his mind drift provided a great perspective of knowledge seemed promising at the end of the road. What he valued more than anything else was not so much the explanation of a particular phenomenon as the establishment of a coherent vision that was to extend itself over multiple fields of reality. He tried to make the various pieces of the puzzle of science fit together. He wanted to find the paths between the different valleys of science, not in a reductive but in a mediatory sense. He pleased himself wanting to create a subtle and continuous vision of reality, often repudiating the excesses of mathematical structuralism.

He had graduated from the Ecole Polytechnique and as a researcher he was attracted by the perspectives which had been opened up for theoretical biology by the development of molecular biology and the progress of theoretical molecular physics. It is here that he quickly realized the great difficulty in going from the microscopic (quantum) description to significant macroscopic phenomena. The problem of complexity obsessed him and he tried to improve the techniques which would permit him to curb it.

Thus one can explain his preference for perturbation methods and stochastic processes. These are avenues where one tries never to loose sight of physical meaning, to keep in mind the frames of reference where one can exert fully adaptative strategies close to natural cognitive strategies. It is in this spirit that one should consider all his work on intermolecular forces, on the electronic structure of molecules by means of localised perturbation theory (the PCILO method) or by adjusting empirical methods to ab initio calculations (the SIBFA method), on solvent models accounting for the role of water in molecular phenomena, on quantum Monte Carlo calculations which he developed during the last years by using simultaneously a perturbative formulation and a stochastic calculation of this perturbation expressed in the form of a functional integral.

His passionate search of mediatory roads led him to areas as diverse as biology, ethology, psychology, and the history of religions. He saw everywhere the same manifestations of certain master ideas on the relations between determinism and freedom, between simplicity and complexity, between structuring and evolution.

With a profound conviction he continued his search for an alternative theory of microphysics in the form of stochastic electrodynamics. He refused to abandon microphysics to the theologians and to consider quantum mechanics as a dogma of which one could say: " Credo quia absurdum".

He very much wanted to believe, but only provided he understood. He was ready to accept the singularity, but only if he could place it in the context of a complete picture of the Universe.

S. Diner

Basic Mathematical Properties of Electronic Wave Functions in Configuration Space.

Reinhart Ahlrichs
Inst. f. Physikalische Chemie u. Elektrochemie
Lehrstuhl f. Theoretische Chemie
Universität Karlsruhe
Kaiserstr. 12
D-7500 Karlsruhe
West Germany (FRG)

I. Introduction

The mathematical structure underlying Quantum Mechanics is determined by the following requirements.
(i) The superposition principle - typical for wave theories - implies that admissible wavefunctions form a linear space over complex numbers $\boldsymbol{C}$
(ii) The probability interpretation leads to the consideration of square integrable functions, f, which have finite norm, $\|f\|$,

$$\| f \| = [\int_{B} | f(x)^2 | \, d\tau]^{1/2} < \infty \quad , \tag{1}$$

where $\boldsymbol{B} \subset \boldsymbol{R}^{3n}$ denotes the space to which the n particles under consideration are confined.

We are thus lead to the vector space of square integrable functions with a scalar product

$$\langle f | g \rangle = \int f^*(x) \, g(x) \, d\tau \tag{2}$$

which is well defined if f and g obey (1). Fortunately, the set of functions (1) even forms a complete space provided integration is in the Lebesgue sense. The (complete) vector space of square integrable functions, denoted $\boldsymbol{L}^2(\boldsymbol{B})$, is a Hilbert space $\boldsymbol{H}$. In $\boldsymbol{L}^2$ one is really dealing with equivalence classes: f is equivalent to g if $\| f - g \| = 0$, which does not imply $f(x) = g(x) \; \forall \; x \in \boldsymbol{B}$.
(iii) In Quantum Mechanics we are especially concerned with linear selfadjoint operators, which correspond to observable quantities. Measurements imply a resolution of the identity with a real spectrum (results of measurments), which is characteristic for selfadjoint operators.

M. Defranceschi and J. Delhalle (eds.), Numerical Determination of the Electronic Structure of Atoms, Diatomic and Polyatomic Molecules, 1–15.

This talk is concerned with properties of exact stationary state wavefunctions ψ. Since these are eigenfunctions of Schrödinger operators we have to study their properties. In section II we recall briefly some basic concepts of functional analysis required subsequently. Local properties of ψ are discussed in section III. The main topic of this paper is the asymptotic behaviour of ψ and reduced densities ρ_k (one or more electrons having large distance from nuclei) which is treated in section IV.

II Schrödinger operators in Hilbertspace [1,2]

We are concerned with the electronic Schrödinger equation of atomic or molecular n-electron systems in free space, $\boldsymbol{H} = L^2(\boldsymbol{R}^{3n})$. If no confusion is possible we use the following notation without further comment: $x \in \boldsymbol{R}^{3n}$, $x_i \in \boldsymbol{R}^3$ denotes the coordinates of the i-th electron, $r_i = |x_i|$, $\boldsymbol{\Delta}$ is the Laplacien in 3n dimensions, $\boldsymbol{\Delta}_i$ in $\boldsymbol{R}^3$.

A linear operator $\mathbf{A}$ maps a linear subset $\boldsymbol{D}_\mathbf{A} \subset \boldsymbol{H}$, the domain of $\mathbf{A}$, into $\boldsymbol{H}$. It is important to specify $\boldsymbol{D}_\mathbf{A}$ since we are dealing with unbounded operators (for which no $M < \infty$ exists such that $\| \mathbf{A} f \| \leq M \| f \|$, $f \in \boldsymbol{D}_\mathbf{A}$), which excludes $\boldsymbol{D}_\mathbf{A} = \boldsymbol{H}$. We will always assume that $\boldsymbol{D}_\mathbf{A}$ is dense in $\boldsymbol{H}$.

A first difficulty concerns the definition of selfadjoint operators. $\mathbf{A}$ is called selfadjoint if every operator $\mathbf{B}$ for which (3) holds

$$\langle \mathbf{B} f \mid g \rangle = \langle f \mid \mathbf{A} g \rangle , \quad f \in \boldsymbol{D}_\mathbf{B},\ g \in \boldsymbol{D}_\mathbf{A} \tag{3}$$

is a restriction of $\mathbf{A}$, $\mathbf{B} \subseteq \mathbf{A}$, i.e.

$$\mathbf{A} f = \mathbf{B} f , \quad f \in \boldsymbol{D}_\mathbf{B} \subseteq \boldsymbol{D}_\mathbf{A} . \tag{4}$$

This is often expressed in saying that $\mathbf{A}$ is hypermaximal.

We will frequently use operator inequalities and write, for selfadjoint operators $\mathbf{A}$ and $\mathbf{B}$,

$$\mathbf{A} \geq \mathbf{B} \quad \text{if } \langle f \mid \mathbf{A} f \rangle \geq \langle f \mid \mathbf{B} f \rangle \tag{5}$$

in a suitably chosen subset of $\boldsymbol{H}$.

We will now consider the electronic Hamiltonian $\mathbf{H}$ of a molecular system, in the usual nomenclature,

$$\mathbf{H} = \mathbf{T} + \mathbf{V} \tag{6}$$

$$\mathbf{T} = -\frac{\boldsymbol{\Delta}}{2} = -\sum_{i=1,n} \frac{\boldsymbol{\Delta}_i}{2} \tag{7}$$

$$\mathbf{V} = -\sum_{a,i} Z_a \boldsymbol{r}_{ai}^{-1} + \sum_{i<j} \boldsymbol{r}_{ij}^{-1} \tag{8}$$

H is a differential operator with singular potentials and it is not trivial to see how this defines a selfadjoint operator. [2]

Let us first consider the kinetic energy **T**. **T** is well defined in the space ***S***

$$\boldsymbol{S} = \{ f \ / \ f(x) = P(x) \ exp(-|x|^2 / 2) \} \tag{9}$$

where $P(x)$ denotes a polynomial. **T** restricted to ***S*** is not maximal and thus not selfadjoint. ***S*** has some nice properties. If $f \in \boldsymbol{S}$ then $\mathbf{T} f \in \boldsymbol{S}$ and the Fouriertransform of f is also of the form (9). The momentum space representation of **T** is just multiplication by $|K^2|/2$. Since ***S*** is dense in ***H***, this clearly defines a selfadjoint operator in a unique way and ***S*** is called a core of **T**. In coordinate space this means $\boldsymbol{D}_T$ includes all functions $f(x)$ whose Fouriertransform if multiplied by $|K|^2$ is square integrable. Since ***S*** is a core of **T** there is another useful definition: $\boldsymbol{D}_T$ includes all f for which a sequence $f_n \in \boldsymbol{S}$ exists with $f_n \to f$ such that $\mathbf{T} f_n$ converges, $\| \mathbf{T} f_n - g \| \to 0$, and one puts $\mathbf{T} f = g$. **T** is a differential operator in the generalized (distributional) sense and not in the ususal meaning of standard analysis. The domain $\boldsymbol{D}_T$ includes in fact rather weird functions.

The inclusion of Coulomb potentials **V** causes little problems, since **V** is bounded with respect to **T**. In order to get acquainted with this reasoning consider $f \in \boldsymbol{S} \subset \boldsymbol{L}^2(\boldsymbol{R}^3)$, i.e. the single particle case. Let x denote one of the cartesion coordiantes and consider

$$\langle \left(a \frac{\partial}{\partial x} + xr^{-1} \right) f \,|\, \left(a \frac{\partial}{\partial x} + xr^{-1} \right) f \rangle \geq 0 \tag{10}$$

Standard manipulations - partial integration, allowed for $f \in \boldsymbol{S}$ - then yield

$$\langle f \,|\, \mathbf{r}^{-1} \,|\, f \rangle \leq a \langle f \,|\, -\frac{\Delta}{2} \,|\, f \rangle + (2a)^{-1} \langle f \,|\, f \rangle \tag{11}$$

for arbitrary $a > 0$ and, since ***S*** is dense in ***H***, for all f for which the r.h.s. of (11) is defined. If we put r^{-2} instead of r^{-1} in (10) we get (after optimizing a)

$$\langle f \,|\, \mathbf{r}^{-2} \,|\, f \rangle \leq 8 \langle f \,|\, -\frac{\Delta}{2} \,|\, f \rangle \tag{12}$$

and further, since $\mathbf{A} \geq 0$ implies $\mathbf{A} \leq \alpha \mathbf{A}^2 + (4\alpha)^{-1}$, $\alpha > 0$

$$\| \mathbf{r}^{-1} f \| \leq a \| -\frac{\Delta}{2} f \| + 16 \, a^{-1} \| f \| , \; f \in \boldsymbol{D}_\Delta , \; a > 0 \tag{13}$$

With the aid of (11) and (13) one sees immediately that V is form-bounded

$$| \langle f \,|\, \mathbf{V} \,|\, f \rangle | \leq a \langle f \,|\, \mathbf{T} \,|\, f \rangle + b \langle f \,|\, f \rangle \tag{14}$$

and norm bounded (which is "stronger")

$$\| \mathbf{V} f \| \leq a \, \| \mathbf{T} f \| + b \, \| f \| \tag{15}$$

with respect to **T**. (14) and (15) hold for suitably chosen constants a and b whenever the r.h.s. is well defined. $a > 0$ can be arbitrarily small. This implies that **V** is a "small" addition to **T** and $D_H = D_T$.

The bound (11) cannot be improved since for the hydrogen atom it leads to $\mathbf{H} \geq Z^2/2$. For n-electron atomic or molecular systems one gets $\mathbf{H} \geq -n Z^2/2$, $Z = \sum Z_a$, which is usually poor but proves that H is bounded from below.

III Local properties of eigenfunctions of H [3]

We are now ready to study properties of exact wavefunctions ψ

$$\mathbf{H} \psi = E \psi \, , \; \| \psi \| = 1 \tag{16}$$

where we drop the index which labels different solutions. We will neglect spin - which enters only in connection with the Pauli principle and permutational symmetry- and consider $\psi(x)$, $x \in \mathbf{R}^{3n}$. [4-7]

A first problem concerns the interpretation of eq.(16). Since **H** must be selfadjoint - otherwise there would be no spectral decomposition - it is a generalized differential operator. The proof for the existence of (global) solutions of (16) is simple only in the sense of Hilbert space theory [2]. Fortunately it has been shown that (16) also holds in the sense of usual analysis. The following rigorous results have been proved for ψ.

(i) ψ is bounded and continuous in $\mathbf{R}^{3n}$ (even Hölder continuous)

(ii) ψ is analytical for $x \in \mathbf{R}^{3n}$ for which **V** is regular

Singular points of Coulomb potentials require detailed considerations. Rigorous results are only available for simple singularities where just a single interparticle distance vanishes. Let r denote the vanishing distance, i.e. $r = r_{ia}$ (electron i near nucleus a) or $r = r_{ij}$ (electrons i and j close).

Now choose a coordinate system (x, y), $x \in \mathbf{R}^3$, $y \in \mathbf{R}^{3n-3}$, such that $r = |x|$ and $\mathbf{V}$ is a regular function of y (locally). ψ can then be written as

$$\psi(x, y) = f(x, y) + g(r, y) \tag{17}$$

and for $r \to 0$, $0 < \lambda < 1$

$$\frac{\partial \psi}{\partial x} = \frac{\partial f}{\partial x} + \left(\frac{\partial g}{\partial r} \right)_{r=0} \frac{\partial r}{\partial x} + O(r^{\lambda}) \tag{18}$$

where all terms on the rhs are well defined. (18) provides a measure for the discontinuity of $\frac{\partial \psi}{\partial x}$ at $r \to 0$.

Starting from (18) it is easy to derive the cusp conditions which relate $\frac{\partial g}{\partial r}$ to ψ

$$\left(\frac{\partial g}{\partial r} \right)_{r=0} = -Z \;\; \psi (0, \; y) \,, \qquad r = r_{ia} \,, \tag{19}$$

$$\left(\frac{\partial g}{\partial r} \right)_{r=0} = \frac{1}{2} \;\; \psi (0 \,, y) \,, \qquad r = r_{ij} \tag{20}$$

It is further possible to eliminate g since

$$\left(\frac{\partial \bar{\psi}}{\partial r} \right)_{r=0} = \left(\frac{\partial g}{\partial r} \right)_{r=0} \tag{21}$$

where $\bar{\psi}$ denotes the average of ψ over the sphere $r = constant$.

More detailed results have only been obtained in special cases. Hill showed that for the H atom in an analytic external potential [8a]

$$\psi \, (x) = f \, (x) + r \, g \, (x)$$

with analytic f and g. The problem of pointwise convergence of series expansions for the He ground state was treated by Morgan [8b] who also reviewed the previous literature . The wavefunction is not analytic in r_1, r_2 and r_{12}, but an expansion of the solution of the differential eq. (16) in powers of $R \ln R$ and $R^{1/2}$, $R = r_1^2 + r_2^2$, converges pointwise. However, it is not clear if this solution is square integrable for the exact E.

IV Asymptotic behaviour

A Preliminary considerations

Detailed results have been obtained for the structure of ψ, eq.(16), as one or more electrons have large distance from nuclei. For the sake of simplicity we only consider atomic systems since the treatment of molecules is analogous [4,5]. Let us recall that the textbook treatment of hydrogen atom states is based on the asymptotic behaviour of solutions of

$$\mathbf{H} \, f = - \, \varepsilon \, f \,, \qquad \varepsilon > 0 \tag{22}$$

for large r. This leads to

$$f(r) \approx r^{\gamma-1} \quad exp(-\eta r) \tag{23}$$

$$\eta = \sqrt{2\varepsilon} \tag{24}$$

$$\gamma = Z / \eta \tag{25}$$

This treatment can be carried over to n-electron states if one assumes separability of ψ for $r_1 = r >> r_i$, $i = 2, ..., n$

$$\psi(r, r_2 \ ..., r_n) \approx f(r)\,\Phi(r_2, ..., r_n) \tag{26}$$

$$\| \Phi \| = 1 \tag{27}$$

Integration of $\Phi(\mathbf{H} - E)\psi = 0$, with ψ from (16), (26), over $d\tau_2$ ..., $d\tau_n$ yields

$$\left(-\frac{\Delta}{2} - (Z - n + 1)\, r^{-1} + \varepsilon \right) f \approx 0 \tag{28}$$

where we have put

$$\sum_{i=2,n} r_{1i}^{-1} \approx (n-1)\, r^{-1} \tag{29}$$

$$\varepsilon = \langle \Phi \mid \mathbf{H}^{(n-1)} \mid \Phi \rangle - E \tag{30}$$

In eq. (30), and below, we indicate the m-electron Hamiltonian by $\mathbf{H}^{(m)}$ if confusion is possible. Eq. (28) is identical to (22) and leads to

$$f(r) \approx r^{\gamma-1} \exp(-\eta r) \tag{31}$$

$$\gamma = (Z - n + 1) / \eta \tag{32}$$

and η as in (24). The slowest possible decay results if ε is minimal, i.e. if Φ is the ground state of $\mathbf{H}^{(n-1)}$

$$\varepsilon = E_o^{(n-1)} - E \tag{33}$$

ε is simply the lowest ionization potential of the state ψ, E. The occurence of n-1 in (32) accounts for the screening of the nuclear charge Z by the inner n-1 electrons.

Unfortunately, it has been difficult to prove (26), (31)-(33) in a rigorous way. In subsection B we will derive explicit - and simple - upper bounds to expectation values of ψ which provide detailed information on the long range behaviour of ψ or the corresponding reduced K-particle density

$$\rho_k(r_1, \ldots, r_k) = \int |\psi(r_1, \ldots, r_n)|^2 \, d\tau_{k+1}, \ldots, d\tau_n \quad (34)$$

Improved results will be obtained in subsection C where we apply the technique of differential inequalities to derive upper bound for ρ_k. In subsection D we finally report - for the special case of the ground state of He-type ions - lower bounds for ρ_1 and ψ. The comparison of upper and lower bounds allows to assess their quality.

In subsequent derivations we will use frequently the rather obvious inequalities.

$$\begin{aligned} \mathbf{H} = \mathbf{H}^{(n)} &\geq \mathbf{H}^{(k)}(1,\ldots K) + \mathbf{H}^{(n-k)}(K+1,\ldots,n) \\ &\geq \mathbf{H}^{(n-k)} - Z \sum_{i=1,k} \mathbf{r}_i^{-1} \end{aligned} \quad (35)$$

The successive ionization potentials of the state ψ, E under consideration will be denoted by ε_i such that

$$\sum_{i=1,k} \varepsilon_i = E_o^{(n-k)} - E \quad (36)$$

We will only consider bound states

$$\varepsilon_i > 0 \quad (37)$$

B Bounds for $\| \mathbf{r}_1^{\mu} \mathbf{r}_2^{\nu} \ldots \psi \|$

Let $f(r_1, \ldots, r_k)$ denote a sufficiently smooth function. From the Schrödinger equation (16) one then derives by standard analysis [4]

$$\langle f\psi | \mathbf{H} - E | f\psi \rangle = \frac{1}{2} \sum_{i=1,k} \langle \psi | \, |\nabla_i f|^2 \, | \psi \rangle \quad (38)$$

The combination of (35), (36) and (38) yields

$$\begin{aligned} &\frac{1}{2} \sum_{i=1,k} \langle \psi | \, |\nabla_i f|^2 | \psi \rangle \geq \\ &\geq \langle f\psi | \sum_{i=1,k} (\varepsilon_i - Z \mathbf{r}_i^{-1}) | f\psi \rangle \end{aligned} \quad (39)$$

which is the basic relationship for this subsection.

For a start we put $f = r^{\mu}$, $r = r_1$, $\mu = 1,2,\ldots$ and get [4,6]

$$\varepsilon_1 \ \| \mathbf{r}^{\mu} \ \psi \|^2 \leq Z \ \| \mathbf{r}^{\mu} \psi \| \ \| \mathbf{r}^{\mu-1} \ \psi \| + + \frac{1}{2} \mu^2 \ \| \mathbf{r}^{\mu-1} \psi \|^2 \tag{40}$$

This inequality yields

$$\| \mathbf{r}^{\mu} \ \psi \| \leq \eta^{-\mu} \mathrm{F} \ (\mu, \gamma) \tag{41}$$

$$\mathrm{F} \ (\mu, \gamma) = \prod_{\lambda=1,\mu} \ [\gamma + (\gamma^2 + \lambda^2)^{1/2}] \tag{42}$$

where η and γ are as in (24), (25) for $\varepsilon = \varepsilon_1$. A slightly poorer but more convenient bound is

$$\| \mathbf{r}^{\mu} \ \psi \| \leq \mathrm{C} \ (\gamma) \ \eta^{-\mu} \ \Gamma \ (\mu + \gamma + 1) \tag{43}$$

$$\mathrm{C} \ (\gamma) = \prod_{\lambda=1,\infty} \ \left[1 + \frac{\gamma^2}{2 \ (\lambda + \gamma/2)^2} \right] / \ \Gamma \ (\gamma + 1) \tag{44}$$

The bounds (41)-(44) are simple and explicit.

A comparison with the exact $\| \mathbf{r}^{\mu} \ \psi \|$ for hydrogen - like ions shows that (43) has the correct dependence on η (or ε_1) and μ up to a factor $(\mu + \gamma)^{1/4}$. However, no account is taken of screening effects, compare (25) and (32), for $n > 1$. [4,6]

The bounds (43) are actually sufficient to prove the experimental decay of ψ or ρ_1 in the mean, since it follows that [6]

$$\| \mathrm{h} \ (\mathrm{r}) \ \psi \|^2 = \int \mathrm{h}^2 (\mathrm{r}) \ \rho_1 \ (\mathrm{x}) \ d\tau < \infty \tag{45}$$

for a regular function $h (r)$ which for large r behaves as

$$h \ (r) \approx r^{-\beta} \ \ exp \ (\eta \ r) \tag{46}$$

if

$$\beta > \gamma + 1 \tag{47}$$

Relationship (45) implies that ρ_1 vanishes like $r^{2\gamma-1}$ exp $(-2\eta \ r)$ in the mean. The exact behaviour for hydrogen - like ions just differs in the power of r which is $2 \ \gamma - 2$. (Screening is again not accounted for if $n > 1$.)

The considerations just presented can be extended to $f = r_1^{\mu_1} \ ... \ r_k^{\mu_k}$ and yield

$$\| \mathbf{r}_1^{\mu_1} \cdots \mathbf{r}_k^{\mu_k} \psi \| \leq \prod_{i=1,k} F(\mu_i, \gamma)$$

$$\Big[\sum_P \prod_{i=1,k} \eta_i^{-2P(\mu_i)} \Big]^{1/2} \tag{48}$$

$$\gamma = Z / \min_i \eta_i \tag{49}$$

η_i as in (24) with $\varepsilon = \varepsilon_i$, P runs over all permutation of S_k, and F is as in (42).

The bound (48) neglects screening again, but otherwise appears to give the right dependence on ε_i (or η_i) and μ_i as is indicated by comparison with exact results for the model case of atoms without interelectronic interaction [4] (again deviations by factors $(\mu_i + \gamma)^{1/4}$).

In other words, the bound (48) is typical for the following form of ρ_k

$$\rho_k(x_1, \ldots, x_k) \propto \mathbf{S} \prod_{i=1,k} \mathit{exp}(-2\eta_i r_i) \tag{50}$$

where **S** acts as a symmetrizer. However, the behaviour (50) fo ρ_k (in the mean) can only be inferred from (48) for regions in coordinate space where

$$\eta_i r_i \geq \eta_j r_j \quad \text{for } 1 \leq i \leq j \leq k \tag{51}$$

(we assume $\varepsilon_i < \varepsilon_j$ for $i < j$ without loss of generality) and similarly for permutations of labels. More satisfactory results are reported in the next subsection.

C Upper bounds for ρ_k

The most detailed results for the asymptotic behaviour of ρ_k - and hence ψ - have been obtained with the aid of differential inequalities, a really fascinating object. The basic idea is due to Maria and Thomas Hoffmann-Ostenhof [9] who proved that ρ_k of exact ψ, eq.(16), obey the differential inequality (in the generalized or distributional sense) [9,10]

$$\Big[\mathbf{H}^{(k)} + \sum_{i=1,k} \varepsilon_i \Big] \rho_k^{1/2} \leq 0 \tag{52}$$

Note: For n = k this reads $(\mathbf{H} - E) \, |\psi| \leq 0$, an obvious consequence of (16).

For a proof of (52) we start from (16) in the form $\psi^* \mathbf{H} \psi = E \, |\psi|^2$, integrate over particles k + 1 to n and use (35) and (36):

$$\int \psi^* \mathbf{H}^{(k)} \psi \, d\tau_{k+1} \ldots d\tau_n + \sum_{i=1,k} \varepsilon_i \, \rho_k \leq 0 \tag{53}$$

The kinetic energy term $\mathbf{T}^{(k)}$ in $\mathbf{H}^{(k)}$ is bounded by

$$\int \psi^* \mathbf{T}^{(k)} \psi \, d\tau_{k+1} \ldots d\tau_n \geq \rho_k^{1/2} \, \mathbf{T}^{(k)} \, \rho_k^{1/2} \tag{54}$$

which together with (53) proves (52).

The advantage of (52) is that one can now apply results of the theory of (sub or super) harmonic functions, e.g., the powerful comparison theorems which allow to bound solutions of (52). As an example we mention the following comparison theorem which follows straightforwardly from the maximum principle of harmonic functions, see e.g. refs. 7,10 and 11.

Let $\boldsymbol{B} \subset \boldsymbol{R}^m$ be an open subset, f and g continuous in $\bar{\boldsymbol{B}}$, $f \geq 0$ in $\boldsymbol{B}$, $g \leq f$ in $\partial \boldsymbol{B}$, $W_1 \geq W_2 \geq 0$ in $\boldsymbol{B}$, and

$$(-\Delta + W_1) \, g \leq 0$$

$$(-\Delta + W_2) \, f \geq 0$$

for sufficiently well behaved potentials W_1 and W_2, then

$$f \geq g \quad \text{in } \boldsymbol{B}$$

The application of this comparison theorem and (52) yields the following bound to ρ_1 [7]

$$[\rho_1 (x)]^{1/2} \leq C \, r^{\gamma-1} \exp (-\eta \, r) \tag{55}$$

if $r > a$, for sufficiently large constants C and a, where η and γ are as in (24), (25), for $\varepsilon = \varepsilon_1$.

The bound (55) demonstartes already the power of the techniques applied, since this was the first time that a pointwise bound which includes the equality sign in (24) was established. However, screening effects are still not accounted for, i.e., the preexponential factor in (55) appears not to be optimal.

Inclusion of screening effects has in fact been achieved but the corresponding deviations are technically complicated and require improvements of (52). Therefore we just report the final results.

<u>Theorem</u> [10] : Let $\varepsilon_i > 0$ denote successive ionization potentials, ψ a solution of (16) and ρ_k as defined in (34). Let us further assume $\varepsilon_1 \leq \varepsilon_2 \leq \ldots \leq \varepsilon_k$ (which is found "experimentally" but not proved in general so far). Then

$$[\rho_1(x)]^{1/2} \le C\,(1+r)^{\gamma-1}\,exp(-\eta\, r),\quad x \in \boldsymbol{R}^3 \tag{56}$$

$$[\rho_k(x)]^{1/2} \le C\,\mathbf{S} \prod_{i=1,k} (1+r_i)^{\gamma_i - 1}\,exp(-\eta_i\, r),\quad x \in \boldsymbol{R}^{3k} \tag{57}$$

$$\gamma_i = (Z - n + k)/\eta_i \tag{58}$$

η_i is as in (24) for $\varepsilon = \varepsilon_i$, γ as in (32), and **S** denotes the K-particle symmetrizer.

The bound (56) is optimal, see below, whereas (57), (58) include screening only partially (for n = k there is no screening in (57)). In the derivation of (57) the terms r_{ij}^{-1} in $\mathbf{H}^{(k)}$, see (52), have in fact been neglected. An improved bound has been obtained for ρ_2 [10].

$$[\rho_2(x_1, x_2)]^{1/2} \le C\,\mathbf{S}\,(1+r_1)^{\alpha}(1+r_2)^{\beta}\,exp(-\eta_1 r_1 - \eta_2 r_2) \tag{59}$$

$$\alpha = (Z - n + 1)/\eta_1 - 1 \tag{60}$$

$$\beta > (Z - n + 5/2)/\eta_2 - 1 \tag{61}$$

α refers to the "outer" electron ($\eta_1 \le \eta_2$ is assumed, without loss in generality) and appears to be optimal, vide supra, whereas this is not certain for β which refers to the "inner" electron.

The bound (52) is still independent of r_{12}. By virtue of (52) and the cusp conditions (20) it is clear that upper bounds for $\rho_2^{1/2}$ behave as $(1 + 1/2\, r_{12})$ for small r_{12}. Attempts to derive r_{12} - depending bounds for ρ_2 have had little success, but it is certain that ρ_2 vanishes faster for $r_{12} = 0$, implying $r = r_1 = r_2$, by some power $r^{-\alpha}$ as compared to (59). [10]

The present treatment has been restricted to Hamiltonions of the form (6)-(8). An extension to a wider class of potentials is possible and leads to similar results; the most general results are due to Agmon [12]. A review has been given by Simon [13].

D Lower bounds

The above mentioned comparison theorems allow for the derivation of lower bounds for ρ_1 and $|\psi| = \rho_2^{1/2}$ in the case of ground states of He - like ions (since $\psi > 0$ in this case) [10]. We skip again the interesting details of derivations and just mention and discuss the results.

Firstly it was shown that ρ_1 is bounded by

$$[\, \rho_1 (x) \,]^{1/2} \geq C \, (1 + r)^{\gamma - 1} \, exp \, (-\eta \, r) \tag{62}$$

which is identical to the upper bound (56), for a different C , of course. There is, hence, no possible improvement as long as the simple functional form is kept.

For the case of the He atom ground state it has been shown - a remarkable result - that separability in the sense of eq. (26) holds in fact asymptotically, where Φ is the ground state wavefunction for He^+ [14].

As for a lower bound to $|\psi|$, we define

$$f\,(u) = \sum_{j=0,\infty} u^j \,/\, (j! \, (j + 1) \,! \,) \quad , \quad u \in \boldsymbol{R} \tag{63}$$

which is a solution of

$$f'' + \frac{2}{u} f' = \frac{1}{u} f \;, \tag{64}$$

and behaves as $exp\, (2 \sqrt{u} \,)$ for large u.

One then gets for $r_1 \geq r_2$ (no loss in generality)

$$\begin{aligned} &\psi\, (x_1 , x_2) = \rho_2^{\,1/2} \, (x_1 , x_2) \geq \\ &C\,(1 + r_1)^{\alpha} \; exp\, (-\eta_1 \, r_1 - Z \, r_2) * f\,(r_{12}) \,/\, f\,(r_1 + r_2) \end{aligned} \tag{65}$$

(Note that $Z = \eta_2$, eq. (24), since $\varepsilon_2 = Z^2 / 2$). α is as in the upper bound for ρ_2, (59), (60). The upper and the lower bounds (59) and (65) differ in the term $(1 + r_2)^{\beta}$, see eq. (61), and in the term $f\,(r_{12}) \,/\, f\,(r_1 + r_2)$. If the electrons are on opposite sides of the nucleus, $x_1 / r_1 \approx - x_2 / r_2$, we have $r_{12} \approx r_1 + r_2$ and

$$f\,(r_{12}) \,/\, f\,(r_1 + r_2) \approx 1 \tag{66}$$

The bounds (59) and (65) are then vitually identical. For $r_{12} \approx 0$ one has

$$f\,(r_{12}) \,/\, f\,(r_1 + r_2) \approx exp\, (- \sqrt{r_1 + r_2} \,) \tag{67}$$

It is not clear if this result is meaningfull, i.e., if ψ really vanishes faster for $r_{12} \approx 0$ as indicated in (67). Our best upper bound indicates a faster decay by only some power of $r_1 \approx r_2$, as mentioned above.

The situation is less fortunate in general. It has been shown under rather mild conditions [15] that

$$\| \, exp \, (\alpha \, |x|) \, \psi \, \| \geq \infty \; , \text{ if } \alpha > \sqrt{(-E)}$$

This implies that ψ cannot vanish faster than exponentially but this result is rather crude in comparison to the upper bounds discussed above. However, examples have been given for which ψ vanishes faster than the upper bounds [15].

V Summary

Considering the technical difficulties of Hilbert space analysis it is fair to say that a good understanding of the coordinate space representation of wavefunctions has been achieved. Since the upper bounds for ρ_k and $|\psi|$, eqs (56)-(58), probably cannot be improved they strongly suggest the following structure of ψ

$$\psi \approx \Phi_o = \mathbf{P} \prod_{i=1,n} \varphi_i \, (x_i) \tag{68}$$

where $\mathbf{P}$ denotes an approximate projection operator, with η_i from (24) for $\varepsilon = \varepsilon_i$,

$$\varphi_i \, (x_i) \approx r_i^{\gamma_i} \; exp \, (-\eta_i \, r_i) \tag{69}$$

The structure indicated by (68), (69) quite certainly holds asymptotically for large r_i . Φ_o fulfills the cusp conditions at nuclei if this is the case for φ_i. The constants γ_i are still uncertain to some extent but γ_1, which refers to the outermost electron, appears to be fixed by eq. (32) since upper and lower bound for ρ_1 agree, eqs (59) and (65).

The ansatz (68) for ψ is of MCSCF type and the surprising accuracy of this technique indicates that (68) is not only valid asymptotically.

Φ_o still does not fulfill the interelectronic cusp condition, which would be met by

$$\psi \approx \Phi = \Phi_o \prod_{i<j} g \, (r_{ij}) \tag{70}$$

$$g \, (r_{ij}) = 1 + \frac{1}{2} r_{ij} + O \, (r_{ij}^{\,2}) \tag{71}$$

The long range behaviour of g - and its (probably) explicit dependence on the actual case - is still an open problem, see the discussion of eq. (67). The trial function (70) yields a bounded $\mathbf{H} \, \Phi$, which is still not continuous as it should be.

The ansatz (68) has a shell structure and supports the basic ideas of orbital approximation, which suffer mainly from the wrong behaviour at $r_{ij} = 0$.

The results reported in this article are in disagreement with a recent paper [16] which claims

$$\psi \sim \exp (- | 2 E \sum_i r_i^2 |^{1/2}) \qquad (72)$$

The asymptotic behaviour indicated in (72) is obviously not correct, the exactly known wavefunctions of atomic systems without interlectronic repulsion provide a counterexample.

We have not considered the problem of Hamiltonions used in relativistic treatments [1]. An interesting thorough discussion of "cusp conditions " has recently been carried out by Kutzelnigg [17].

References

[1] The following textbooks give an introduction to the field
(a) M. Reed and B. Simon, *Methods of modern Mathematical Physics* (Academic, N.Y., 1978), especially Vol. IV *Analysis of Operators.*
(b) T. Kato, *Perturbation Theory for Linear Operators* , (Springer, Berlin, 1966)
(c) W. Thirring, *Lehrbuch der Mathematischen Physik 3: Quantenmechanik von Atomen und Molekülen* (Springer, Wien, 1979)
References a and c give a nice account of the results published until around 1978.

[2] T. Kato, Trans. Am. Math. Soc. **70**, 195 (1951)

[3] T. Kato, Commun. Pure Appl. Math. **10**, 151 (1957)

[4] R. Ahlrichs, M. Hoffmann-Ostenhof, and T. Hoffmann-Ostenhof, J. Chem. Phys. **68**, 1402 (1978)

[5] R. Ahlrichs, Chem. Phys. Lett. **18**, 521 (1973)

[6] R. Ahlrichs, J. Math. Phys. **14**, 1860 (1973)

[7] T. Hoffmann-Ostenhof, M. Hoffmann-Ostenhof, and R. Ahlrichs, Phys. Rev. A**18**, 328 (1978)

[8a] R.N. Hill, J. Math. Phys. **25**, 1577 (1984)

[8b] J.D. Morgan, Theoret. Chim. Acta **69**, 181 (1986)

[9] M. Hoffmann-Ostenhof and T. Hoffmann-Ostenhof, Phys. Rev. A**16**, 1782 (1977)

[10] R. Ahlrichs, M. Hoffmann-Ostenhof, T. Hoffmann-Ostenhof, and J.D. Morgan III, Phys. Rev. A**23**, 2106 (1981)

[11] P. Deift, W. Hunziker, B. Simon, and E. Vock, Commun. Math. Phys. **64**, 1 (1978)

[12] S. Agmon, Proc. A. Pleijel Conf., Uppsala 1979

[13] B. Simon, Bull. Math. Soc. **7**, 447 (1982)

[14] J.M. Combes, M. Hoffmann-Ostenhof, and T. Hoffmann-Ostenhof, J. Math. Phys. **22**, 1299 (1981)

[15] R. Froese, I. Herbst, M. Hoffmann-Ostenhof, and T. Hoffmann-Ostenhof, Commun. Math. Phys. **87**, 265 (1982)
[16] K.A. Dawson, J. Chem. Phys. **83**, 2615 (1985)
[17] W. Kutzelnigg, to be published

BASIC MATHEMATICAL PROPERTIES OF ELECTRONIC WAVE FUNCTIONS IN MOMENTUM SPACE

Edwin N. Lassettre
Department of Chemistry
Carnegie Mellon University
4400 Fifth Avenue
Pittsburgh, Pennsylvania 15213

ABSTRACT. The analytic behavior of momentum eigenfunctions in the complex momentum plane is examined. One-electron models are considered including those based on local potentials as well as the Hartree-Fock equations. Methods for the location and characterization of singular points are described. Conformal mapping is employed to obtain eigenfunction expansions which are convergent for both real and complex values of the independent variable. Except by accident the singular points of momentum eigenfunctions are branch points except for atomic hydrogen where poles are encountered. The existence of branch points can eliminate certain series which might be employed to represent momentum eigenfunctions. Examples are discussed. An example is given in which the order of summation of basis sets affects convergence in the complex momentum plane. The connection between first order collision amplitudes and the Schrödinger equation in momentum space is discussed and singular points for the amplitudes are located. An exact local potential is described and the associated amplitude is examined. Other examples illustrating the importance of complex function theory in momentum space are described.

M. Defranceschi and J. Delhalle (eds.), Numerical Determination of the Electronic Structure of Atoms, Diatomic and Polyatomic Molecules, 17–48.

Introduction

In contrast to classical dynamics, where both coordinates and momenta of a system of particles are needed to specify a state of motion, quantum mechanics can be formulated in two alternative ways where the state of a system is specified either by a function depending on the positions of particles in space (r-space) or by a function dependent on momenta (p-space). Transformation theory provides the relation between the two formulations, wave functions in p-space are obtainable from those in r-space through a Fourier transform. The overwhelming majority of quantum chemical investigations have been carried out in r-space. However, momentum eigenfunctions provide a fertile field for the application of classical mathematical analysis, especially complex variable theory. Location of singular points is important since it is known that a function of a complex variable with no singularities is a constant. Considerable progress has been made in locating and characterizing singularities as described below.

A strong connection between first order collision amplitudes (the Born approximation) and momentum eigenfunctions becomes evident when the Schrödinger equation is transformed to momentum space. In momentum space the Schrödinger equation, for the motion of one electron in a local potential field, becomes an integral equation with a kernel function which is identical in functional form to the first order collision amplitude for direct scattering. For the Hartree-Fock equations the kernel function is the sum of first order collision amplitudes for direct and exchange scattering. Since the Hartree-Fock equations have played such an important role in the theory of chemical bonding it is of considerable interest that, for a bound state, they describe the steady state of an internal collision problem. This provides a very different viewpoint, one in which the wave nature of electron streams is brought to the foreground. An analytic study of first order collision amplitudes as well as momentum eigenfunctions is called for and a modest beginning on this program is described in the following sections.

The motion of an electron in a local potential field is normally considered to be a one-electron approximation. It is possible, however, to reformulate the problem so that the model is exact for a many electron system. Singular points can then be located in momentum space without approximation. The results are also described in the following sections.

1. Momentum Eigenfunctions. The Fourier Transform.

The momentum eigenfunction X can be obtained from the space function Ψ by means of the formula[1]

$$X = \frac{1}{h^{3n/2}} \int e^{-\frac{2\pi i}{h} \sum_{k=1}^{n} p_k \cdot r_k} \Psi d\tau \qquad (1)$$

where P_k, r_k are the momentum and position vectors of the kth particle and the integral extends over the positions of all particles. The number of particles is N.

Momentum eigenfunctions for atomic hydrogen were obtained in 1929 by Podolsky and Pauling[2] who used eq.(1), taking Ψ as the atomic hydrogen space eigenfunction. For the ground state momentum density they found

$$\frac{8}{\pi^2}\ \frac{1}{(1+P^2)^4}\ . \qquad (2)$$

Using advanced experimental techniques Lohman and Weigold[3] measured this distribution and accurately verified[4] formula (2).

Using the (e,2e) electron scattering technique momentum distributions for individual molecular orbitals have been determined experimentally for many atoms and molecules and the results compared with theoretical calculations using eq.(1).[4,5] These researches not only provide chemical information of fundamental interest but they decisively confirm the correctness of eq.(1). We now consider some analytic properties of the function X obtained from eq.(1).

2. SINGULAR POINTS OF MOMENTUM EIGENFUNCTIONS. ONE ELECTRON MODELS.

2.1 LOCATING SINGULARITIES

Consider eq.(1) for a single electron. In atomic units, which will be used hereafter,

$$X = \frac{1}{(2\pi)^{3/2}} \int e^{-iP.r}\ \psi(r)d\tau. \qquad (3)$$

Divide the integral into two parts X_i and X_e obtained by integrating respectively over the interior and exterior of a large sphere at radius c. Then

$$X_i = \frac{1}{(2\pi)^{3/2}} \int_{r<c} e^{-iP.r}\ \Psi d\tau. \qquad (4)$$

The eigenfunction Ψ is necessarily bounded. Let the upper bound be M. Also let $P.r = P\ r\cos\theta$ where θ is the angle between P and r. Expanding the exponential in a power series and integrating term by term gives a power series in P. The absolute values of the coefficients can be easily obtained and hence an upperbound series for (X_i) i.e.

$$|X_i| \leq \left(\frac{2}{\pi}\right)^{1/2} \sum_{n=0}^{\infty} |P|^n \frac{M\ c^n}{(n+3)n!}. \qquad (5)$$

But this series converges for all finite values of $|P|$ and hence the power series for X_i also converges and represents an analytic single valued function of P throughtout the finite complex p-plane. It follows that there are no singularities within X_i, all singular points are concentrated in X_e which is given by

$$X_e = \frac{1}{(2\pi)^{3/2}} \int_{r>c} e^{-iP.r} \psi d\tau. \tag{6}$$

We now consider how ψ is to be obtained for a special model.

2.2 EIGENFUNCTION ARISING FROM LOCAL POTENTIAL FUNCTIONS.

Consider the motion of one electron under the influence of a local potential V such that

$$V \longrightarrow -\frac{a}{r} \tag{7}$$

as r becomes large, i.e. $r>c$. In typical cases $V = -\frac{a}{r} - U$ where U decreases exponentially as r increases. The Schrödinger equation in r space is

$$-\frac{1}{2}\nabla^2 \psi + V\psi = E\psi \tag{8}$$

and when r is sufficiently large V is spherically symmetric (see egs. 6 and 7) and ψ takes the form

$$\psi = \frac{S}{r} Y_{\ell m} \tag{9}$$

where $Y_{\ell m}$ is a surface spherical harmonic normalized to unity and S satisfies the equation

$$\frac{d^2S}{dr^2} + \left(2E + \frac{2a}{r} - \frac{\ell(\ell+1)}{r^2}\right)S = 0. \tag{10}$$

Then X_e becomes

$$X_e = i\left(\frac{2}{\pi}\right)^{1/2} Y_{\ell m} \int_c^\infty S j_\ell (pr)\, r dr. \tag{11}$$

where j_ℓ is the spherical Bessel function.[7] That solution of (10) which, for a bound state, vanishes at $r = \infty$ is[8,9]

$$S = Cr^\mu e^{-Po} \left(1 + \frac{a_1}{r} + \frac{a_2}{r} + \ldots + \frac{a_m}{r^m} + \frac{R_0 \gamma_{M+1}}{r^{m+1}}\right) \tag{12}$$

with C a normalizing constant and

$$a_n = \frac{1}{n!(2Po)^n} \prod_{m=0}^{n-1} [(\ell+\tfrac{1}{2})^2 - (\mu-m-\tfrac{1}{2})^2],\ a_0 = 1 \tag{13}$$

$$Po = \sqrt{-2E} \tag{14}$$

$$\mu = a/Po \tag{15}$$

The function in parentheses in eq. (12) is shown as a finite sum plus a remainder. R_o is a constant and $\gamma_{n+1}(r)$ is a function of r which approaches zero as r increases.[8,9] The series cannot be extended to infinity since it is divergent. A finite number of terms of the series does, however, represent S with increasing accuracy as r increases. Hence eq.(12) can be employed in eq.(11) if c is large enough. Assume that c is so large that the first term of the series in eq.(12) is an accurate representation of S. Under these conditions j_ℓ is also represented by the first term of its asymptotic series

$$j_\ell \cong \frac{1}{Pr} \sin (Pr-\tfrac{1}{2}\ell\pi). \qquad (16)$$

Then

$$\int_c^\infty Sj_\ell(Pr)rdr \cong \frac{1}{2iP} [(-i)^\ell F^+ - (i)^\ell F^-] \qquad (17)$$

where

$$F^\pm = \int_c^\infty r^\mu e^{-(P_o+z)r}\, dr \qquad (18)$$

and z = iP.

F^+ satisfies the differential equation

$$\frac{d^2F^+}{dz^2} + \left(\frac{\mu+2}{z-Po} - C\right)\frac{dF^+}{dz} - \frac{C(\mu+1)}{z-Po}F^+ = 0. \qquad (19)$$

This linear, homogeneous, equation can have singular points only at singular points of the coefficients i.e. at z = Po which is a regular singular point. Expanding F^+ in powers of (z-Po), $F = (z-Po)^s \sum_{n=0}^{\infty} bn\,(z-Po)^n$ gives for the indical equation $s(s-1) + S(\mu+2) = 0$. This has two roots $S = 0$, $S = -(\mu+1)$. Thus eq.(19) has two solutions one of which, F_2^+, is analytic for all values of z and the other, F_1^+ is singular at z = Po. In fact, $F_1^+ = (z-Po)^{-(\mu+1)}$. Hence

$$F^+ = \frac{C_1}{(z-Po)^{\mu+1}} + C_2F_2^+ \qquad (20)$$

where C_1 and C_2 are constants. Similarly

$$F^- = \frac{C_1'}{(z+Po)^{\mu+1}} + C_2'F_2^- \qquad (21)$$

where F_2^- is analytic. Since z =iP it follows from (20) that $P = -iP_o$ is a singular point of F^+ and hence from eqs.(17) and (11) a singular point of X_e and X. Similarly from (21) $P = iP_o$ is also a singular point of X. For atomic hydrogen μ is an integer and the singularities are poles. For other substances μ is not an integer, except by accident, and the singularities are branch points. The analysis described above was first used to locate singular points for first order collision amplitudes.[6] Later it was extended to momentum eigenfunctions.[11,12] It might appear from (17) that P = 0 is a singular point. Note, however, that j_ℓ is represented by (16) only when Pr is large. If P = 0 then Pr also vanishes and (16) is not valid. The left side of eq.(17) is bounded at P = 0 and, in fact, is analytic there. P = 0 is not a singular point. The above analysis locates the singular point but it does not tell us how many terms of the series (12) are needed. That question is answered by the following analysis.

When the series (12) is substituted into (11) the following expression is obtained

$$\int_c^\infty Sj_\ell(Pr)rdr = C\int_c^\infty j_\ell(Pr)rdr\left[\sum_{n=0}^{N} A_n r^{\mu-n+1} + \sum_{n=N+1}^{m} A_n r^{\mu-n+1} + (Ro\gamma_{m+1})r^{\mu-m+1}\right]dr \tag{22}$$

In eq.(22) N is the largest value of n for which μ-N+1 is positive. In eq.(22) the powers of r in the first sum in brackets are all positive and hence as c increases all of these terms increase in magnitude. The powers of r in all of the remaining terms are negative and hence all of these terms decrease as c increases. In the limit of very large c only the first sum in brackets survives, the remaining terms vanish. This insertion of the series (12) into the integral of eq.(11) leads to an automatic termination of the series with the N+1 and higher terms eliminated. The singular points can now be located and characterized in the following way.

Consider a typical term in the residual series (22). Since μ-n+1>0 the integral from c to ∞ is expressible as the integral from 0 to ∞ minus that from 0 to c

$$\int_c^\infty Sj_\ell(pr)rdr = C\int_0^\infty j_\ell(pr)e^{-br}\left[\sum_{n=0}^{N} r^{\mu-n+1}\right]dr - C\int_0^c j_\ell(pr)e^{-br}\left[\sum_{n=0}^{N} r^{\mu-n+1}\right]dr. \tag{23}$$

By an argument similar to that employed in the treatment of Xi the second integral on the right of (23) can be shown to be an analytic function of p. The first integral on the right hand side of (23) contains the singular points. But that integral can be evaluated[10]

$$\int_0^\infty j_\ell(pr)e^{-Por}\left[\sum_{n=0}^{N} r^{\mu-n+1}\right]dr$$

$$= \left(\frac{\pi}{2}\right)^{1/2} \sum_{n=0}^{n} \frac{\Gamma(\ell+\mu-n+2)}{\Gamma(\ell+3/2)Po^{\mu-n+2}} \left(\frac{P}{2Po}\right)^\ell \left(\frac{Po^2}{P_o^2+P^2}\right)^{\mu-n+1}$$

$$\times\ F\left(\frac{\ell+n+\mu}{2},\frac{\ell+n-\mu+1}{2};\ell+3/2;-\frac{P^2}{P_o^2}\right). \tag{24}$$

As $p \longrightarrow \pm\, i\, P_o$ along the imaginary axis, the hypergeometric function approaches a finite limit (ref.9 page 282) and $(P/2\, P_O)^\ell \longrightarrow (\pm\frac{1}{2})^\ell$. Therefore, the position and nature of the singular point is determined by the factors

$$\left(\frac{Po^2}{P^2+Po^2}\right)^{\mu-n+1}. \qquad (25)$$

There are two singular points $P = +iP_0$ and $P = -iP_0$. For atomic hydrogen μ is an integer and the singular points are poles. In other cases μ is, except by accident, not an integer and hence the singularities are branch points.

The above analysis, which has been described previously,[11] has far reaching consequences which are enumerated below. (i)Since the series (24) is *finite* the singular points of the individual terms are singular points of X_e and hence of the momentum eigenfunction X. This would not necessarily be true of an *infinite* series. (ii)The eigenvalue P_O determines both the position and nature (pole or branch point) of the singularity. (iii)The position and nature of the singular point does not depend on ℓ or m, i.e. the *direction* of the vector **P** has no influence on the position or nature of the singular point. (iv)The integer N, which determines the number of terms needed to characterize the singular point (see eq.24), is determined by P_0 since N is the largest integer for which μ-N+1 is positive and $\mu = a/P_O$. (v)Since $V \to -a/r$ as r becomes large for both atoms and molecules the above argument applies to both. (vi)There are no singularities on the real axis. This follows because the integral on the left of eq.(24) is uniformly convergent when the imaginary part of p is less than Po and hence defines an analytic function of P. (vii)From eq.(24) or eqs.(20,(21) it is obvious that the exponential under the integral sign determines the position of the singularity while the power of r determines its nature (pole or branch point). Conversely, if we know the position and nature of a singular point we can immediately determine the exponential and the fractional part of the power of r which gave rise to it. This will be useful later in identifiying the exponentials in basis functions for atoms.

2.3 SECONDARY SINGULAR POINTS.

The points $P = \pm iP_O$ have been referred to as primary singular points[11] because they are closer to the origin in the P-plane than any others. There are, however, other singular points, also on the imaginary axis for bound states, further out. To establish the existence and location of these secondary singular points we proceed as follows.

Eq.(8) can be rearranged as follows

$$(-\nabla^2 + Po^2)\Psi = 2V\Psi. \qquad (26)$$

Multiply by $\frac{1}{(2\pi)^{3/2}}e^{-i\mathbf{P}\cdot\mathbf{r}}$ and integrate over r. Simultaneously replace Ψ on the left of eq.(26) by the inverse of the transform (2) i.e.

$$\Psi = \frac{1}{(2\pi)^{3/2}} \int e^{i\mathbf{P}\cdot\mathbf{r}}\, X(d\mathbf{P}). \qquad (27)$$

Then, using the Fourier integral theorem, we find

$$X = -\frac{2X'}{(2\pi)^{3/2}(P^2+Po^2)} \tag{28}$$

where

$$X' = \int e^{-iP.r}\, V\Psi d\tau \tag{29}$$

The same expression has been obtained previously[12] by a different method which employed a modified form of Green's theorem.

The term multiplying X' in eq.(28) is singular at $P = \pm iPo$, but this contributes nothing new since these points are already known to be singular. The new singular points are in X'. Divide the integral into two parts X'_i and X'_e over the interior and exterior of a large sphere of radius c. In considering X'_i there is one new factor. The product $V\Psi$ in eq.(29) approaches $-Z/r$ in the vicinity of a nucleus and hence $V\Psi$ is not necessarily bounded there and the method employed in section 2.1 cannot be used to show that X_i is an analytic function of P. Instead eq.(26) is used to replace $V\Psi$ in the integrand of X'_i by $\frac{1}{2}(Po^2-\nabla^2)\Psi$. Then

$$X'_e = \frac{Po^2}{2}\int_{r<c} e^{-iP.r}\Psi d\tau - \frac{1}{2}\int_{r<c} e^{-iP.r}\nabla^2\Psi d\tau. \tag{30}$$

The first term on the right of (30) can be shown to be an analytic function of P using same method as in section 2.1. The second term on the right can be transformed by means of Green's Theorem applied to a volume between the large sphere of radius c and small spheres of radius ε surrounding each nucleus. The result is a volume integral which differs from the first term on the right of eq.(30) only in that $P_0^2/2$ is replaced by $P^2/2$. That term is analytic in P. There remain the surface integrals. Those over the small spheres vanish as $\varepsilon \rightarrow 0$. Let c be so large that Ψ can be replaced by its asymptotic form on the large sphere i.e. a function of r multiplied by a spherical harmonic. Then the surface integral over the large sphere reduces to a linear combination of spherical Bessel functions. Since the Bessel functions are analytic functions of P it follows that X'_i is analytic for all finite values of P and all singular points are contained in X'_e. They can be located in the following manner.

Let $V = -a/r + U$ where U depends on the particular model chosen for the potential function. It might, for example, be that proposed by Slater[13] or the X_α potential.[14,15,16,17] In all of these cases the function U is expressible as a sum

$$U = U_1 + U_2 + \dots + U_M \tag{31}$$

such that the typical term approaches, at large r, to the following functional form

$$U_k \cong r^{\alpha_k} e^{-\beta_k r} Y_{\ell_k m_k} \tag{32}$$

where $Y_{\ell_k m_k}$ is a surface spherical harmonic and α_k and β_k are constants. The function X'_e then becomes

$$X_e' = \int_{r>c} -\frac{a}{r}\,\psi e^{-iP.r}\,d\tau + \sum_{k=1}^{m} \int_{r>c} r^{\alpha_k}\, e^{-\beta_k r - iP\cdot r} y_{\ell_k m_k} \psi d\tau. \quad (33)$$

With c very large ψ becomes equal to the function in eq.(9). Since the product $Y_{\ell kmk}\, Y_{\ell m}$ is expressible as a sum of spherical haromics, X_e' becomes a sum of terms each term being similar to the right side of eq.(11). From the analysis of section 2.2 it follows that the functions on the right of eq.(33) have singular points at

$$P = \pm i(P_m + \beta_k) \qquad k = 1,2 \ldots M \qquad (34)$$

where $P_n = \sqrt{-2E_n}$ for the $n^{\underline{th}}$ molecular orbital. The first term on the right of (33) gives the singular points at $P = \pm\, iP_n$ already obtained in section 2.2. The pair $P = \pm iP_n$ are the primary singular points, the remainder are secondary singularities.

As pointed out previously, the pair $P = \pm iP_n$ arose from the term $r^{\mu}e^{-P_n r}$ in the asymptotic expression for ψ. The pairs P $= \pm i(P_n+\beta_k)$ imply <u>additional</u> terms like $r^{a_k}e^{-(P_n + \beta_k)r}$ in the asymptotic expression for ψ. When these are included in ψ of eq.(33) additional singularities are found at

$$P = \pm\, i(P_n + \beta_k + \beta_j),\ j,\ k = 1,2,3\ldots + M. \qquad (35)$$

continuing this process singular points are found, in general, at the points

$$P = \pm i(P_n + \beta_i + \beta_j + \ldots + \beta_s),\ i,j,\ldots S = 1,2,3\ldots M. \qquad (36)$$

This can be continued to produce an infinite number of singular points which, for bound states, lie on the imaginary axis. These results were first obtained by Lassettre.[11]

The general technique for the location of singular points has now been developed using the local potential as an example. The technique can be applied in other cases described below.

2.4 HARTREE-FOCK EIGENFUNCTIONS.

The Hartree-Fock equations involve non-local potentials and this introduces a new factor into the limiting expression for the eigenfunctions at large distances, namely, at large distances the limiting expression involves not just one exponential, as is the case for local potentials, but a number of exponentials equal to the number of filled orbitals. This was first noticed by Handy, Marron, and Silverstone.[18] They also pointed out that for one case, Beryllium, the 1s and 2s orbitals do not mix. In other instances, however, all orbitals contribute to the asymptotic expression for <u>each</u> orbital. It is now possible to locate singular points for the Hartree-Fock orbitals.

2.4.1 Research of Huo and Lassettre. Huo and Lassettre[19] (1) located singular points for Hartree-Fock eigenfunctions and (2) showed that the exponentials corresponding to the singular points could also be used in a basis set of Slater type orbitals. Item (1) is discussed in the present section and item (2) in section 3.

The Hartree-Fock equations for a system with closed subshells can be expressed as follows

$$H\ (1)\ \Psi_i\ (1) + V_o\Psi i = E\Psi_i \tag{37}$$

where V_o is now an operator rather than a local potential function. The operator V_o is the sum of a local term, V, and a non-local term, -U, where

$$V = \sum_j \int \Psi_j \Psi_j^*(2) \frac{1}{r_{12}}\ d\tau_2 \tag{38}$$

and

$$U\Psi_i = \Sigma\Psi_j(1) \int \Psi_i(2)\Psi_j^*(2)\ \frac{1}{r_{12}} d\tau_2 \tag{39}$$

where U is an integral operator shown in (39) as operating on Ψ_i. Obviously, $V_o = V-U$. Equations (6) and (28) still hold but V in (28) is replaced by the operator V_o.

Handy, Marron, and Silverstone[18] showed that the leading term in the asymptotic expansion of the $i^{\underline{th}}$ orbital for a closed shell system is

$$\Psi_i \cong \sum_{j=1}^{N} f_{ij}(r) e^{-P_i r}\ Y_{\ell_i m_i} \tag{40}$$

where

$$P_i = \sqrt{-2E_i}. \tag{41}$$

The functions f_{ij} vary less rapidly than exponentials. In ref. (19) they were referred to as "polymials" in r although "power" would have been more appropriate. As long as only the <u>positions</u> of singular points are involved the distinction is unimportant. The spherical harmonic $Y_{\ell_i m_i}$ has been shown in previous sections to have no effect on the positions of singular points. Eq.(40) leads, using earlier analysis, to singularities at 2N points

$$P = \pm iP_1,\ \pm iP_2,\ \ldots,\ \pm iP_N. \tag{42}$$

Now using eq.(28), with V_o replacing V, additional singular points are found at

$$Pi = \pm\ i(P_i + P_j + P_k)\ \ i,j,k = 1,2,\ldots N \tag{43}$$

These singular points imply exponentials in the asymptotic expression for Ψ_i in addition to those of eq.(40) and hence new singularities arise. Repeating this process gives additional singularitie as follows

$$\left.\begin{aligned} P &= \pm\, i(P_i + P_j + P_k + P_\ell + P_m) \\ P &= \pm\, i(P_i + P_j + P_k + P_\ell + P_m + P_n + P_o) \end{aligned}\right\} \begin{aligned} &i,j,k,\ell,m \\ &= 1,2,3\ldots N \end{aligned} \qquad (44)$$

etc. In these equations i, j ... take all values from 1 to N. This continues indefinitely there being an infinite number of singular points.

The singular points arise from exponentials in the <u>asymptotic</u> expansion of Ψ_i. In section 3 it will be demonstrated that the same exponentials occur in a basis set of Slater type orbitals capable of representing the eigenfunction not just asymptotically but everywhere.

2.42 Asymptotic Expansion Of Handler, Smith and Silverstone. Handler, Smith and Silverstone[20] have obtained the coefficients f_{ij} of eq.(40) in explicit form for atoms with closed shells

$$f_{ij} = \begin{Bmatrix} r^{\alpha}{}_j & i = j \\ r^{\alpha}{}_j - \lambda_{ij} & i \neq j \end{Bmatrix} \qquad (45)$$

and

$$\alpha_j = \frac{a}{P_j} - 1 \qquad (46)$$

where a is the charge on the atom with one electron removed and $P_j = \sqrt{-2E_j}$. If all electrons are in S states f_{ii} remains the same and $f_{ij} = 0$. For atoms in ground electronic states this holds only for Beryllium. In eq.(45) λ_{ij} is a positive integer. Equations (17)-(21) can be applied to show that the singularities of eq.(42) are branch points if a/P_j is not an integer. Except for atomic hydrogen this ratio will not be zero except by accident and the singularity is a branch point.

2.43 Asymptotic Expansion Of Davis, Jensen, and Monkhorst. A much more complete asymptotic expansion of Hartree-Fock orbitals has been obtained by Davis, Jensen and Monkhorst.[21] Their expansion consists of linear combinations of a product of a spherical harmonics and terms like

$$r^{\mu} e^{-\alpha r} \qquad (47)$$

each multiplied by an asymptotic series in descending powers of r. This leads to singular points at $P = \pm\, i\alpha$. They found for α sums of P_i exactly like eqs. (42)-(44) and corresponding values of μ. For example, in one case singular points are found at

$$P = \pm i\alpha_{ijk} = \pm i(P_i + P_j + P_k) \quad i,j,k = 1,2,3...N. \qquad (48)$$

and to each α there corresponds a μ

$$\mu_{ijk} = (\mu_i + \mu_j + \mu_k) \quad i,j,k = 1,2,3...N \qquad (49)$$

with

$$\mu_i = 1/P_i. \qquad (50)$$

The remaining singularities are the same as those found by Huo and Lassettre[19] but the investigation by Davis, Jensen and Monkhorst goes further and determines the powers of r, r^{μ}. Their analysis shows that, except by accident, all singularities are branch points. They also carried out accurate numerical calculations which are discussed in the next section.

3. Series expansions. Mapping functions

Momentum eigenfunctions can be represented by series expansions. The most obvious expansion is a power series in P. Momentum eigenfunctions arising from local potential functions have singular points at $P = \pm iP_0$ and hence a power series in P converges only when $|P|<P_0$. This difficulty can be overcome by expanding in powers of an appropriately selected function of P. Appropriate functions can be obtained by conformal mapping. Some ex amples are given in the following sections. Atoms and molecules are considered separately.

3.1 ATOMS

For an atom the one-electron eigenfunction is given by

$$\Psi = \frac{S}{r} Y_{\ell m} \qquad (51)$$

where S is a function of r and $Y_{\ell m}$ is a spherical harmonic. Using eq.(2) the momentum eigenfunction is expressible as

$$PX = f(P) Y'_{\ell m} \qquad (52)$$

with

$$f(P) = \left(\frac{2}{\pi}\right)^{½} \int_0^{\infty} S \, j_\ell \, (Pr)Prdr \qquad (53)$$

and $Y'_{\ell m} = (-1)^\ell \, Y_{\ell m}$. $Y'_{\ell m}$ depends on the angles θ,Ψ defining the direction of the vector **P**. If the spherical Bessel function j_ℓ (Pr) is expanded in powers of Pr and integrated term by term it becomes immediately apparent that

$$f(P) = P^{\ell+1}\, g(P^2) \qquad (54)$$

where g is a function of P^2. Consider the momentum function arising from a local potential. There are primary singular points at $P = \pm iP_o$ and secondary singular points given by eqs.(34) to (36). Let

$$z = P^2. \qquad (55)$$

Then in the z plane all singularities are on the negative real axis, that at $z = -P_o^2$ being closest to the origin. For convenience let $P_o^2 = a$. Consider the transformation

$$t = \frac{P^2}{P^2+P_o^2} = \frac{z}{z+a} \qquad (56)$$

This can be considered a mapping function which maps the z-plane onto the t-plane. Let $z = -½a + iy$. Then as y is varied z traces out a straight line parallel to the imaginary axis in the z-plane. Eq.(56) maps this line onto a circle of unit radius in the complex t-plane. Eq.(56) also maps everything to the right of this line into the interior of the unit circle. The unit circle in the t-plane is shown in Fig. 1a and the area mapped into it in Fig. 1b. Since the area mapped into the unit circle contains no singular points it follows that the function g can be expanded in a power series in t and the series will converge when $|t|<1$. The physically attainable values of P are real and positive and are mapped into the unit t-circle by eq.(56). It follows that the power series in t converges and represents the function g for all physically attainable values of P. Since $t = 0$ when $P = 0$ it also follows that the series best represents g in the vicinity of $P = 0$. In eq.(56) the constant a was set equal to P_o^2 so that the transform t did not introduce any new singularity. This was also done in subsequent transforms.

Consider next the transformation

$$t' = \frac{a}{z+a} = \frac{P_o^2}{P^2+P_o^2}. \qquad (57)$$

Unlike the transformation (56) this transform is not suitable for expansion of g. The reason is apparent if (57) is considered as a mapping function Eq.(57) maps the area outside a circle of radius a, with center at $z = -a$, into the unit circle in the t'-plane. (See Fig. 1c). The primary singular point, $z=-a$, is a branch point of the function g. Encircling that point once does not return g to its initial value. However, encircling the point $z=-a$ once does return t', or any integral power of t' to its original value. Therefore a power series in t' can only represent single valued functions and this excludes g which is multiple valued.

There is a second reason why t' is an unsuitable variable for expansion of g. There are an infinite number of singular points along the negative z axis (see section 2) and some of these will be mapped into the unit t'-circle and a power series in t' cannot represent g throughtout the unit t'-circle. The variable t' is completely unsuited to the representation of momentum eigenfunctions of atoms.

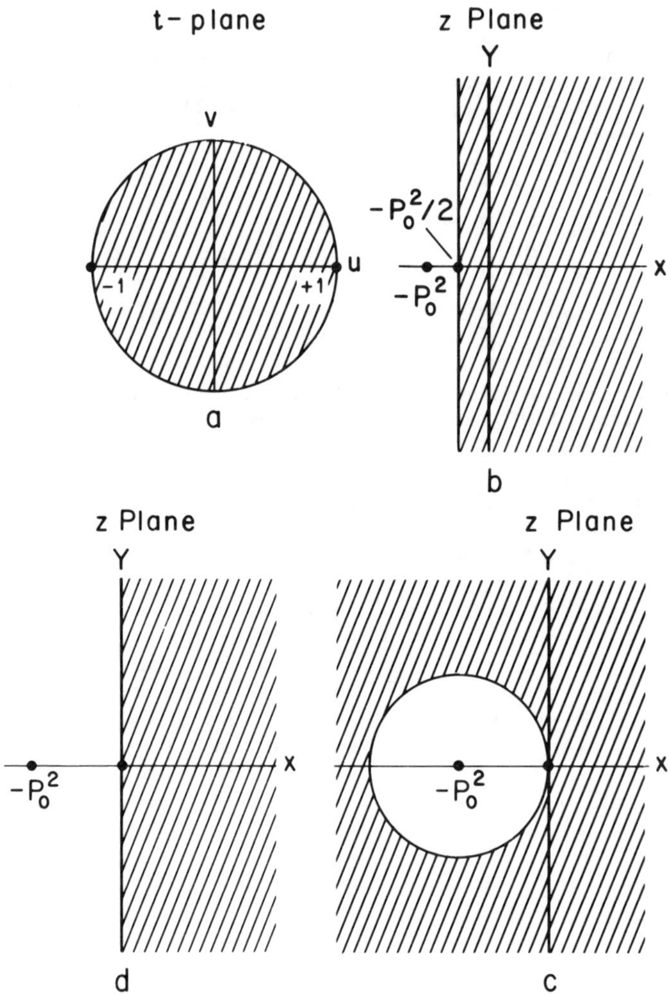

Fig. 1 $a = P_0^2$. (a) unit t-circle. (b) $t = z/(z+a)$ maps cross-hatched area into unit t-circle. (c) $t = a/(z+a)$ maps cross-hatched area into unit t-circle. (d) $t = (a-z)/(a+z)$ maps cross-hatched area into unit t-circle.

A transform suggested by Fock's treatment of atomic hydrogen[22] has also been studied. The transform is

$$t = \frac{a-z}{a+z} = \frac{P_o^2-P^2}{P_o^2+P^2} \qquad (58)$$

The entire half plane to the right of the imaginary z axis (cross hatched in Fig. 1d) is mapped into the unit t-circle. No singularities are mapped into the unit t-circle and hence the function g of eq.(54) can be represented by a convergent power series in t. When $z = a$(i.e. $P = P_o$), $t = 0$ so expansion in powers of t emphasizes the region near $P = P_o$, in contrast to the transform of eq.(56) which emphasizes the region around $P = 0$.

In fitting a function of a real variable with a polynomial it is known that the Taylor series is the most slowly convergent while the series of Chebyshev polynomials is the most rapidly convergent.[23,24] Other ultraspherical polynomials are intermediate in convergence rate. In the complex plane, however, the situation is reversed in the special case of momentum eigenfunctions for atoms. To see this clearly it is first necessary to single out the appropriate orthogonal polynomials. For this purpose the weight function is needed and can be obtained as follows. Following Fock[22] let

$$t = \frac{P_o^2-P^2}{P_o^2+P^2} = \cos\alpha. \qquad (59)$$

In Fock's treatment a four dimensional unit sphere is projected onto P-space by means of a stereographic projection and the element of solid angle in this 4-space is $d\Omega = \sin^2\alpha \sin\theta \, d\psi d\theta d\alpha$ where θ, ψ are the usual polar angles in P-space and α is a polar angle in 4-space. That portion involving P is $\sin^2\alpha d\alpha$ and, following Fock

$$\sin^2\alpha d\alpha = \frac{8P_o^2P^2dP}{(P^2+P_o^2)^3} = -\left(1-t^2\right)^{1/2}dt \qquad (60)$$

When $P = 0$, $t = 1$ and when $P = \infty$, $t = -1$. Integration from $P = 0$ to ∞ is equivalent to integration from $t = 1$ to $t = -1$. If this order is reversed dt in eq.(60) changes sign. Assuming integration from $t =-1$ to $t = 1$ the negative sign in eq.(60) must be omitted. This is assumed in future equations.

At large values of P the function f of eq.(54) decreases like $1/P^{\ell+4}$. This follows from a slight modification of the analysis of Rau and Fano.[25] This dependence can be accomplished by expressing f as follows

$$f = \frac{P^{\ell+1}}{(P^2+P_o^2)^{\ell+2}} G(P^2) \qquad (61)$$

where $G(P^2)$ is a function of P^2 which remains bounded as $P^2 \rightarrow \infty$. The term $P^2 + P_o^2$ in the denominator was selected because it approaches the proper limit without introducing any new singular points into f. It is known[12] that every one electron eigenfunction has $(P^2+P_o^2)^{-2}$ as a factor. This, together with the fact that Fock extracted such a factor in the treatment of atomic hydrogen, suggests the definition of a function F

$$f = (P^2+P_o^2)^{-2}\ F. \tag{62}$$

Then

$$F^2\ \mathrm{Sin}^2\alpha d\alpha = \frac{1}{2^{2\ell}P_o^{2\ell}}\ (1-t^2)^{\ell+\frac{1}{2}}\ G^2 dt. \tag{63}$$

The Gegenbauer (ultraspherical) polynomials $C_k^{\ell+1}$ constitute the set[26] orthogonal with respect to the weight function $(1-t^2)^{\ell+\frac{1}{2}}$ over the range -1 to $+1$. Hence an expansion

$$G = \sum_k a_k C_k^{\ell+1} \tag{64}$$

provides the best mean square fit to G if the a_k are given by the usual formula. This series is to be compared with the power series

$$G = \sum_{k=0}^{\infty} b_k t^k \tag{65}$$

where t is given by eq.(58). The power series (65), as noted earlier, converges for all physically attainable values of P. The series (64) is capable of representing G in the interval $-1 \leq t \leq 1$ subject to conditions similar to Fourier series. Less well known is the fact that (64) converges in the complex plane within the largest ellipse, with foci at ± 1, within which G(t) is analytic. Outside this ellipse the series diverges (see ref. 26, page 238, theorem a.1.1).

In section 2.3 singular points were located for momentum functions arising from both local and non-local (Hartree-Fock) potentials. The number is infinite and the distance from the origin extends to infinity. The pair $P = \pm\ iP_o$ are closest to the origin. Number the singular points 1,2,3.... according to the distance from the origin and designate by $\pm\ iC_nP_o$ the $n^{\underline{th}}$ pair where C_n is real and positive.

Then $t_n = \dfrac{1+C_n^2}{1-C_n^2}$ gives the position of the $n^{\underline{th}}$ singularity in the t plane; $C_1 = 1$ and $C_{n+1} > C_n$. Therefore t_n is negative and $|t_n| > 1$. Moreover, as $n \to \infty$ $C_n \to \infty$ and $t_n \to -1$. It follows that the convergence ellipse consists of the line between -1 and $+1$. The series (64) converges only for real values of t between -1 and $+1$. For all other values of t, real or complex, the series diverges. Momentum eigenfunctions are better represented in the complex plane by the power series (65) than by an orthogonal polynomial series (64).

3.2 MOLECULES.

In the case of molecules of low symmetry both even and odd powers of P may occur in the same series. In such cases mapping functions such as (56) and (58), which involve P^2, are not applicable. An appropriate transform, which has been previously used,[27] is

$$t = \frac{P}{(P^2+P_o^2)^{½}}. \qquad (66)$$

The transform can be made single valued by placing cuts from iP_o to $i\infty$ and from $-iP_o$ to $-i\infty$. As long as the cuts are not crossed the transform is single valued. The region mapped into the unit t -circle is cross hatched in Fig. 2. For a typical momentum eigenfunction the area mapped into the unit t -circle contains no singular points and therefore typical one-electon momentum eigenfunctions can be expanded in power series in t which converge when $t<1$.

A second transform,[11] which has not been applied to momentum eigenfunctions, is

$$t = \left[-P_o + (P+P_o^2)^{½} \right]/P \qquad (67)$$

with cuts from iP_o to $i\infty$ and from $-iP_o$ to $-i\infty$. The transform becomes single valued as long as the cuts are not crossed. The transform is singular at $P = \pm iP_o$. This, however, does not introduce any new singular points into a momentum eigenfunction since the function is already singular at those points. Every point in the P-plane is mapped within the unit t-circle except those points along the cuts.

3.3 BASIS SETS IN R-SPACE WHICH TRANSFORM INTO PROPERLY ANALYTIC MOMENTUM FUNCTIONS.

Research by Huo and Lassettre[19] on the location of singularities of Hartree-Fock momentum eigenfunctions has been mentioned previously. They also considered, and solved, the problem of obtaining appropriate basis functions for atoms in r-space. They showed that Slater type orbitals were a proper basis and, more important, the exponents are those quantities multiplying $\pm i$ in eqs.(43),(44),(45) etc. Therefore, exponents need not be optimized. They compared the theoretical exponents with those obtained by numerical optimization by Bagus, Gilbert and Roothaan for neon[28] and by Bagus for argon.[29] The theoretical and numerically calculated exponents agreed approximately in many instances. Huo and Lassettre[19] also did numerical calculations on neon and argon using the theoretical exponents and obtained good agreement with results obtained by others[27,28] with optimized exponents. Thus, by applying the theory of functions of a complex variable in P-space they showed that exponents obtained from asymptotic expansions in r-space were also appropriate for the representation of eigenfunctions not just asymptotically but throughtout r-space. This point was demonstrated even more emphatically by Davis, Jensen and Monkhorst[21] who did much more accurate and extensive calculations in addition to obtaining the complete asymptotic expansion previously referred to in section 2.43. Calculations were done for helium, beryllium and neon. The theoretical asymptotic exponents were multiplied by a common scale factor a and the energy calculated as a function of a. As the length of the basis set increases the value of a leading to minimum energy approaches a = 1 as it should if the theoretical exponents are also optimal. They also varied individual exponents with similar results.

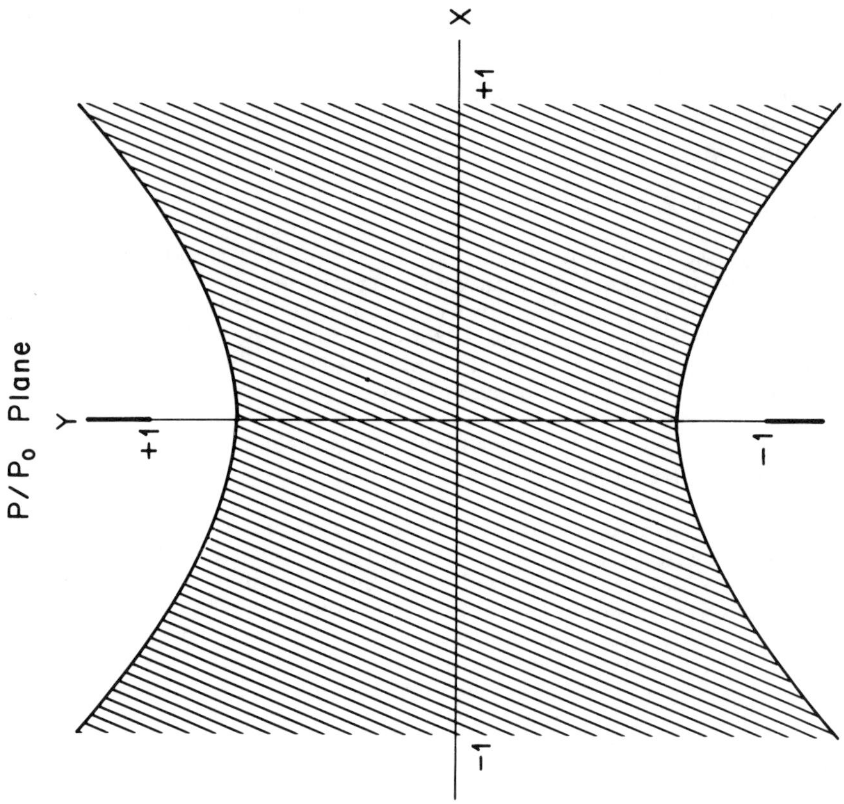

Fig. 2 $t = (P/P_o)/\sqrt{1+(P/P_o)^2}$ maps cross-hatched area into unit t-circle

The energy calculated for neon by Huo and Lassettre[19] was slightly higher than that obtained by Bagus, Gilbert, and Roothaan[28] and they assumed that more basis functions with theoretical exponents would be needed. Davis, Jensen, and Monkhorst[21] showed, however, that the discrepancy was due to a small numerical error (.0003 atomic energy units). The correct energy is actually smaller than that of Bagus et.al.

Davis and Monkhorst[30] have modified the basis set to contain functions like $r^n e^{-ar} M(a,b;-cr)$ where M is an appropriate confluent hypergeometric function so selected as to produce the correct asymptotic behavior. In previous calculations[21] with Slater type orbitals with theoretical exponents average values of r^k abruptly lost accuracy when k exceeded the number of orbitals in the basis set. With the new, asymptotically correct, basis set this deficiency was corrected.[30]

The mere fact that an eigenfunction in r-space is represented by a linear combination of Slater type orbitals with the correct theoretical exponents does not insure that transformation to p-space will lead to a series which converges in the complex p-plane.[31] The argument is as follows.

Let a typical Hartree-Fock orbital for an atom be expressed as a sum of terms each containing a pair of singularities from the sets of eqs.(42),(43),(44) etc. Designate a particular pair by $P = \pm i\zeta_a$ belonging to the term X_a of the sum

$$X = \sum_a X_a . \qquad (68)$$

For an atom each term in the sum is multiplied by the same spherical harmonic. Consider only the radial part of the product PX which is the function of eq.(52). Using eq.(61) and expanding G in powers of the variable t of eq.(56) gives

$$f_a = \frac{P^{\ell+1}}{(P^2+\zeta_a^2)^{\ell+2}} \sum_{n=o}^{\infty} a_n t^n = \frac{P^{\ell+1}}{(P^2+\zeta_a^2)^{\ell+2}} \sum_{n=o}^{\infty} a_n \left(\frac{P^2}{P^2+\zeta_a^2}\right)^n \qquad (69)$$

From section 3.1 the series (69) converges for all real positive P. Another series in powers of t' of eq.(56)

$$f_a = \frac{P^{\ell+1}}{(P^2+\zeta_a^2)^{\ell+2}} \sum_{n=o}^{\infty} b_n (t')^n = \frac{P^{\ell+1}}{(P^2+\zeta_a^2)^{\ell+}} \sum_{n=o}^{\infty} b_n \left(\frac{\zeta_a^2}{P^2+\zeta_a^2}\right)^n \qquad (70)$$

is known <u>not</u> to represent f_a within the unit t'-circle (see section 3.1) or, for that matter, in any circle within the unit t'-circle and concentric with it (see ref. 31 page 55).

Next consider the functions which transform into the nth term of each series (69) or (70). Define L_n by

$$L_n = \left(\frac{2}{\pi}\right)^{1/2} \frac{(2\zeta_a r)^{n+1} \left(\frac{\pi}{2\zeta_a r}\right)^{1/2} K_{n+1/2}(\zeta_a r)}{\zeta_a 2^{\ell+n+2} (n+\ell+1)!} \qquad (71)$$

where $K_{n+1/2}$ is the modified Bessel function of the third kind (ref.7, P.444). Also let M_n be defined by

$$M_n = \sum_{n=0}^{n} (-1)^k \binom{n}{k} L_n. \tag{72}$$

When transformed to p-space $r^{\ell+1}L_n$ gives the $n^{\underline{th}}$ term of the series in eq.(70) and $r^{\ell+1}\ M_n$ gives the $n^{\underline{th}}$ term of eq.(69). It follows that the two series

$$S_a' = \sum_{n=0}^{\infty} r^{\ell+1}\ L_n \tag{73}$$

and

$$S_a = \sum_{n=0}^{\infty} r^{\ell+1}\ M_n \tag{74}$$

when transformed to p-space, using eq.(53), give rise respectively to eqs.(70) and (69). The function L_n is a sum of Slater type orbitals all having the same eponential but different powers of r. The same is true of M_n. The difference between S_a' and S_a lies in the <u>order</u>, or grouping, in which the orbitals are summed. When transformed to p-space, however, S_a' leads to eq.(70) which cannot represent f_a while S_a leads to eq.(69) which is properly convergent. This leads to the following:

A. When an eigenfunction for an atom with closed subshells is expressed as a linear combination of Slater type orbitals in r-space then the convergence of the eigenfunction in complex p-space depends on the order in which the orbitals are summed.

This somewhat resembles conditional convergence of infinite series. Note that this arises only for complex variables. Both t^n and $(t')^n$ constitute a complete set due to Weirstrass' theorem that any continuous function can be represented by polynomials. Moreover, mean convergence is preserved under a Fourier transform and hence $r^{\ell+1}L_n$ and $r^{\ell+1}M_n$ are complete sets and can represent S_a <u>in the mean</u>. The same is true of f_a. However, representation of f_a as a <u>function of a complex variable</u> is a very different matter.

4. SCHRÖDINGER EQUATION IN MOMENTUM SPACE.

The Schrödinger equation in p-space is an integral equation which involves not only the momentum eigenfunction but also a kernel function which is closely related to the first order collision amplitudes which occur in electron scattering theory (the Born approximation). This relationship is most apparent for one-electron models and since such models have played an important role in chemical theory it seemed possible that new concepts would emerge from momentum space studies. This has been slow to develop partly because transferring all thinking to momenta is difficult but especially because in momentum space the wave nature of electron motion is brought strongly into the foreground. The present author has devoted more attention to analyzing the kernel function than to solving the Schrödinger equation. Since this is not the prime goal of the present conference it will be considered only briefly.

In the case of a local potential the one-electron Schrödinger equation in momentum space is

$$\tfrac{1}{2}(P^2+P_o^2)X + \int v(q)X(Q)(dQ) = 0 \tag{75}$$

where

$$v(q) = \frac{1}{(2\pi)^3} \int e^{i\,q\cdot r}V(r)(dr) \tag{76}$$

and

$$q = Q-P. \tag{77}$$

A useful relation for V(q) can be obtained by replacing $e^{iq\cdot r}$ by $-\nabla^2 e^{iq\cdot r}/q^2$ and applying Green's theorem. This gives

$$v(q) = \frac{1}{2\pi^2 q^2} \left(\int_{v'} e^{iq\cdot r}\eta d\tau - \sum_s Z_s\, e^{iq\cdot R_s} \right) \tag{78}$$

where

$$\eta(r) = -\frac{1}{4\pi} \nabla^2 V \tag{79}$$

the integral in (78) extends over the volume v′ between a very large sphere (infinite radius) and very small spheres around each nucleus. No matter how small the spheres around each nucleus the volume v′ never contains a nucleus. In (78) Z_s is the atomic number of the sth nucleus and R_s is a position vector to that nucleus.

The function η was introduced by Lassettre[32] who also described a local potential function suitable for exact calculation of ionization potentials and electron densities. Eq.(78) is identical in functional form to the first order collision amplitude for elastic electron scattering except that η replaces the charge density. For that reason η was referred to as the collision density. Pursuing the analogy, $\eta d\tau$ is the collision charge in the volume element $d\tau$. Integrating over the entire volume v′ and applying Green's theorem it follows that the collision charge in v′ is equal to N, the number of electrons in v′. The collision charge and the actual charge in v′ are the same. In general, this is not true in segments of v′. Although the above analysis was developed for application to the potential function discussed in ref. 32 it could be applied to any of the one-electron potential functions previously referred to.[13–17] For the purpose of locating singular points of v(q) eq.(78) is superior to eq.(76) for the following reason. In eq.(78) the second order pole at q = o is specifically in evidence. Moreover, the Fourier integral in eq.(78) does not contain that pole because at q = o the exponential in the integral is equal to 1 and it has just been pointed out that the integral of η over v′ is equal to the number of electrons and is therefore bounded. In the search for singular points attention can be concentrated on the Fourier transform in eq.(78). The location of singular points was first discussed for generalized oscillator strengths,[6] which involve the square of the first order collision amplitudes, and then for the collision amplitude.[27] Collision based expansions of electrostatic potentials were also obtained.[27] The same technique can be applied to eq.(78). Results will be described later. First, the Hartree-Fock equations in momentum space will be considered.

The Hartree-Fock equations are as follows

$$H(r)\Psi_i + 2\sum_j V_{ij}\Psi_i(1) - \sum_j V_{ij}\Psi_j(1) = E\Psi_i(1) \qquad (80)$$

where

$$V_{ij}(1) = \int \frac{\Psi_i(2)\Psi_j(2)}{r_{12}}\, d\tau_2 \qquad (81)$$

and H(1) is the sum of the kinetic energy operator for electron (1) and the potential energy of that electron in the field of the nuclei. The equations in momentum space are[33]

$$\tfrac{1}{2}(P^2+P_o^2)X_i + \int K(\mathbf{Q},\mathbf{P})X_i(\mathbf{Q})(d\mathbf{Q}) = 0 \qquad (82)$$

where

$$K(\mathbf{QP}) = 2F(\mathbf{QP}) - G(\mathbf{Q},\mathbf{P}) - f(\mathbf{Q},\mathbf{P}) \qquad (83)$$

and

$$F = \frac{1}{2\pi^2|\mathbf{Q}-\mathbf{P}|^2}\sum_{j=1}^{m} \int X_j^*(\mathbf{Q}-\boldsymbol{\eta})X_j(\mathbf{P}-\boldsymbol{\eta})d\eta \qquad (84)$$

$$G = \frac{1}{2\pi^2|\mathbf{Q}-\mathbf{P}|^2}\sum_s Z_s e^{i(\mathbf{Q}-\mathbf{P})\cdot\mathbf{R}_s} \qquad (85)$$

and

$$f(\mathbf{Q},\mathbf{P}) = \frac{1}{2\pi^2}\sum_j \int \frac{1}{\eta^2}\, X_j^*(\mathbf{Q}-\boldsymbol{\eta})X_j(\mathbf{P}-\boldsymbol{\eta})(d\boldsymbol{\eta}) \qquad (86)$$

M is the number of doubly occupied orbitals.

The rather formidable looking expression for K, eqs.(83)-(86), has a very simple physical significance; K <u>is the first order collision amplitude for the elastic scattering, with exchange, of an additional electron by the atom or molecule under consideration.</u>[33] At first sight this seems contradictory since it is known that the self potential is canceled by a term ($j = i$) in the exchange terms of eq.(80). This cancellation must also occur in eq. (52) but it does <u>not</u> occur in the kernel function K, i.e. no term in the exchange amplitude f cancels a term in the direct amplitude F. The cancellation occurs instead in eq. (82). The function X_i occurs twice in that equation, once multiplying K and again in the term $j = i$ of f. If that term of f with $j = i$ is singled out in eq.(82) and the order of integration over **Q** and $\boldsymbol{\eta}$ is reversed then, after two obvious changes of variable, the result will equal and be opposite in sign to a term $j = i$ of F in (82) and will cancel that term.

In one respect the function K(**Q**,**P**) differs from a first order collision amplitude. In a collision experiment the incident and scattered electrons are both in field free space and $|\mathbf{Q}| = |\mathbf{P}|$. But the kernel function K refers to a collision in the interior of an atom or molecule and **P** and **Q** are entirely independent. The same situation prevails for the amplitude of eq.(78).

5. Singularities of First Order Collision Amplitudes for the Hartree-Fock Equation.

The function F of eq.(84) is also given by the following formula[33]

$$F = \frac{1}{2\pi^2 q^2} \sum_j \varepsilon_{jj} \tag{87}$$

The summation extends over all filled orbitals, $q = |\mathbf{Q}-\mathbf{P}|$, and

$$\varepsilon_{jj} = \int e^{i\mathbf{q}\cdot\mathbf{r}} \psi_j \psi_j^*(d\mathbf{r}). \tag{88}$$

The principal singular point for a more general integral, in which the two functions in the integrand are different, has been considered previously.[6] By substituting the asymptotic expansion and using the results of section 2 the singular points can be located. Using the asymptotic expression of eq.(40) with coefficients from eq.(45) singular points can be located using the methods of Section 2. The singular points are

$$q = \pm i(P_i+P_j), \quad i,j = 1,2\ldots M \tag{89}$$

where $P_i = \sqrt{-2\varepsilon_i}$ and M is the number of doubly filled orbitals.

Except by accident all of these are branch points. Each filled orbital has singular points given by eq.(89). These apply, of course, to the Hartree-Fock orbitals. Eqs.(89) apply to filled orbitals for both atoms and molecules and are valid for the ground electronic states with closed shells with one exception, beryllium. For that substance the primary singular points are at

$$q = \pm 2iP_{1s}, \quad q = \pm 2iP_{2s} \tag{90}$$

for the 1s and 2s atomic orbitals respectively. If the complete asymptotic expression of Davis, Jensen and Monkhorst[21] is employed additional singularities are found. The positions are obvious and will not be explicitly enumerated.

Singular points for the exchange terms $f(\mathbf{Q},\mathbf{P})$ of eq.(86) have been studied by Huo.[34] The terms F, discussed above, depend only on $\mathbf{q} = \mathbf{Q}-\mathbf{P}$. The exchange terms depend on both **P** and **Q**. Huo found that the positions of singular points depended on the exponents of exponentials. Using the expression of eq.(40) the singular points are

$$Q = \pm iP_j, \quad P = \pm iP_j, \quad q = \pm 2iP_j, \quad j = 1,2,3\ldots M \tag{91}$$

where M is the number of doubly filled orbitals. All filled molecular orbitals have the same singular points. The results apply to both atoms and molecules. Huo assumed that the atomic and molecular orbitals were linear combinations of Slater type orbitals and hence the asymptotic form would be an integral power of r multiplied by an exponential. But we know that asymptotically the orbital approaches a fractional power of r times an exponential. In this case, as in cases previously treated, the positions of the singularities depend on the exponents and are believed to be reliable. The nature of the singularities will be affected.

Convergent series expansions of K(Q,P) can be devised but they are not very useful because the singularity nearest the origin is the same for all orbitals and, on physical grounds, this seems to insure that long expanisons are required.

Local potentials are treated in a later section.

6. A Local Potential and the Associated Momentum Wave Function.

An exact local potential function was described by Lassettre[32] which relied on the separation of the exact eigenfunction into the product of a function of the coordinates of one electron and a conditional probability amplitude which depends on the coordinates of all electrons in a many electron system. Previous research by Hunter[35] had introduced the idea of a conditional probability amplitude in separating nuclear and electronic motions in molecules. Later this was extended to electronic motions[36,37] and a potential similar to that of Lassettre was independently developed. In developing the potential function an important point, not previously stressed, is involved which bears directly on the subject matter of the present conference. Therefore, the steps involved in obtaining the potential are briefly outlined in the following section.

6.1 A LOCAL POTENTIAL.

Let $\Psi(1,N)$ be an eigenfunction in r-space for a system of N+1 electrons in a molecule with fixed nuclei. The coordinates of one electron are designated by 1 and the remaining electrons by N. The spin of electron 1, but not the position, is also included in the collection designated by N. Let $\psi(1)$ be defined by the equation

$$\psi^2(1) = \int |\Psi(1,N)|^2 \, d\tau_N \tag{92}$$

where $d\tau_N$ is the volume element for the N particles plus the spin variable for particle 1. Attention is restricted to the ground electronic state of a molecule with an even number of electrons in a singlet state. Since the integral in eq.(92) is real and positive extracting the square root will give $\pm\,\psi(1)$. Either sign could be selected. If ψ^2 vanishes along a nodal surface the sign is so selected that ψ and the gradient normal to the surface are continuous. This will be possible if ψ^2 is a minimum when it vanishes, a situation which prevails in all known cases. Generally the charge density does not vanish in the ground electronic state but other cases might arise in the future.

Next define a function $\Phi(1,N)$ by the equation

$$\Phi(1,N) = \Psi(1,N)/\psi(1). \tag{93}$$

If $\psi(1)$ vanishes along a nodal surface then Φ will be infinite unless Ψ also vanishes along the same surface. But this is exactly the situation. For, from (92) when $\psi(1)$ vanishes so does the integral and since the integrand is positive the integral can only vanish when Ψ vanishes. It will be shown later that ψ satisfies a linear second order differential equation and hence both ψ and its derivative normal to a surface cannot vanish simultaneously because otherwise the function ψ would vanish everywhere. Hence from L'Hospital's rule the right side of (93) approaches a non infinite limit along a nodal surface. Solving (93) for Ψ and integrating $|\Psi|^2$ over coordinates N gives, using (92) and the fact that Ψ is normalized to unity.

$$\int \Phi^2(1,N)d\tau_N = 1. \tag{94}$$

Since Ψ can always be selected as real, without loss of generality, the same holds for Φ. It is assumed to be real in what follows.

Let the Hamiltonian operator be expressed as the sum of two terms $H = h_o(1) + h(1,N)$ where h_o is the kinetic energy operator of electron 1 plus the potential of electron 1 in the field of the nuclei and h(1,N) contains the kinetic energy operators of the remaining N electrons and all remaining potential energy terms. Then substituting $\Psi = \psi(1)\Phi(1,N)$ in the equation $H\Psi = E\Psi$, multiplying by Φ and integrating over the coordinates of N electrons plus the spin of 1 we find

$$\left\{h_o(1) + \varepsilon(1) - \frac{h^2}{4\pi^2 m} R\cdot\nabla_1 - \frac{h^2}{8\pi^2 m} S\right\}\psi(1) = E\psi(1) \tag{95}$$

where

$$R = \int \Phi \; \nabla_1 \; \Phi \; d\tau_N \tag{96}$$

$$S = \int \Phi \; \nabla_1{}^2 \; \Phi \; d\tau_N \tag{97}$$

$$\varepsilon(1) = \int \Phi \; h(1,N)) \; \Phi \; d\tau_N \tag{98}$$

By applying the operator ∇_1 to eq.(94) it follows that $R = 0$. Applying $\nabla_1{}^2$ to the same equation leads to the relation

$$S = -\int_N (\nabla_1\Phi)\cdot(\nabla_1)\Phi \; d\tau_N = -k^2 \tag{99}$$

where eq.(101) defines k^2, a new symbol chosen to emphasize that S is negative. Using these facts eq.(95) becomes

$$\left(\frac{-h^2}{8\pi^2 M} \nabla_1{}^2 + U \right)\psi = E\psi \tag{100}$$

where

$$U = -\sum_s \frac{Z_s e^2}{r_s} + \varepsilon(1) + h^2k^2/8\pi^2 m. \tag{101}$$

When r_1 is large ε approaches a limit ε_∞. Substracting $\varepsilon_\infty \psi$ from sides of eq.(100) gives

$$\left(\frac{-h^2}{8\pi^2 m} \nabla_1{}^2 + V \right)\psi = W\psi \tag{102}$$

where

$$V = U - \varepsilon_\infty \tag{103}$$

and

$$W = -\varepsilon_\infty. \tag{104}$$

Eq. (102) is an equation of the Schrödinger type with a local potential function. There are no non local potentials like those in the Hartree-Fock equations which arise as the result of assigning each electron to a molecular orbital. Unlike the Hartree-Fock equations, which are approximations, eq.(102) is exact. Although the potential function would be difficult to obtain numerically it can be used in the investigation of analytic properties without excessive difficulty.

The eigenvalue W is equal to -I where I is the ionization potential. This can be established as follows. Lassettre[6] showed that when r_1 is very large the solution to the Schrödinger equation is separable i.e.

$$\Psi(1,N) \cong \Psi^+(N)\, r^{\mu_1 - 1} e^{-P_1 r}\, Y_{\ell m} \tag{105}$$

where $P_1 = \sqrt{2I}$, $\mu_1 = 1/P_1$, and $\Psi^+(N)$ is the eigenfunction of the resulting positive ion, and $Y_{\ell m}$ is a spherical harmonic. Katriel and Davidson[38] showed that $\Psi^+(N)$ is the eigenfunction for the ion in its lowest energy state. It follows from eq. (92)-(93) that when r_1 is very large

$$\Phi \cong \Psi^+(N). \tag{106}$$

This shows in turn that ε_∞ in eq.(103) is the energy of the positive ion and hence

$$W = -I \tag{107}$$

where I is the ionization potential, the ion being left in its lowest energy state. From the fact that Φ approaches Ψ^+ as r_1 becomes large it follows that V of Eq.(102) becomes,

$$V \cong -\frac{1}{r_1}. \tag{108}$$

Because of this limiting relation the analysis of Section 2 applies and, at large r_1

$$\psi \cong M\, r_1^{\mu-1} e^{-P_o r_1}\, Y_{\ell m} \tag{109}$$

where M is a constant, $P_o = \sqrt{2I}$, and $\mu = 1/P_o$. Ahlrichs et.al[39] have obtained an upper bound to the square root of the density as follows

$$\sqrt{\rho} \leq C\,(1+r)^{\mu-1}\, e^{-P_o r} \tag{110}$$

When r is large this has the same functional form as the radial portion of (109).

A crucial step in obtaining eq.(102) is the fact that the vector **R** of eq.(96) vanishes. This was established by applying the operator ∇_1 to eq.(94). The right hand side obviously vanishes and the left side is equal to 2R <u>if the order of differentiation and integration can be reversed</u>. There are, of course many examples in which this is invalid and the question arises as to whether this reversal is valid in the present application. In any region of space where Ψ is regular the reversal is certainly justified but if any small region of space is occupied by two or more particles in close proximity a singularity results and the reversal requires investigation. If in eq.(94), Φ is eliminated using (93) then the function $\psi^2(1)$ merely divides the integral and plays no role in the reversal. The important integral is

$$\int |\Psi(1,N)|^2\, d\tau_N. \tag{111}$$

Consider a small sphere of radius ε which contains two electrons. Pack and Byers-Brown[40] found that the eigenfunction Ψ behaved like $1 + \frac{1}{2} r_{12} + \mathbf{r}_{12}\cdot\mathbf{u}_{12}$ within such a region. The vector function $\mathbf{u}_{12}$ is regular in the neighborhood of $r_{12} = 0$ and therefore contributes nothing to the singularity. Using this result, and carrying out the integration over the vector $\mathbf{r}_2$ within the small sphere, no difference is found whether the operator ∇_1 is applied before or after integration and this result prevails as $\varepsilon \to 0$. The same result is found when the two particles are an electron and a nucleus.

A second type of singularity, in which two or more electrons are near a nucleus, is also important. Bartlett[41] was the first to notice that explicit account of the singularity was essential in order to obtain a convergent expansion for the helium eigenfunction. The singularity has been re-discovered, re-examined, and employed many times since 1937. In a recent numerical study Freund, Huxtable, and Morgan[42] have used basis functions, specifically tailored to fit this singular point, in highly accurate calculations on helium. They found for the leading term in Ψ near the "triple collision point" (a nucleus and two electrons in close proximity) the expression $1-Z(r_1+r_2)+\frac{1}{2}r_{12}+\ldots$ where r_1, r_2 are the respective distances of the two electrons from the nucleus and r_{12} is the distance between the two electrons. This is the expression which would prevail within a very small sphere, of radius ε, containing the nucleus and two electrons. Again, using this expression in eq.(111) to integrate over the small sphere the same result is obtained whether the operator ∇_1 is applied before or after integration over r_2 and this holds in the limit as $\varepsilon \to 0$. The conclusion is that **R** = 0 is not affected by the coalescence of two electrons, an electron and a nucleus, or two electrons and a nucleus. It is a fair inference that **R** = 0 in general for any atom or molecule.

The potential function V depends on the exact eigenfunction. Since this is not known the potential is not available for numerical calculation. Nevertheless, the function has value for an analytic study in p-space as outlined in the next section.

6.2 SINGULARITIES IN THE MOMENTUM EIGENFUNCTION.

The Schrödinger equation in p-space for a local potential has already been obtained (see eqs. 75 and 78). Since $V \to -1/r$ when r is large the analysis of section 2.2 is immediately applicable. The principal singular point of the momentum eigenfunction is located at

$$P = P_1 = \pm i\sqrt{2I_1} \tag{112}$$

where I_1 is the ionization potential. If $1/P_1$ is not an integer, the usual case, the singularity is a branch point. Since there are no approximations involved the theoretical ionization potential should equal that obtained by experiment. It is of considerable interest that these two primary singular points can be located and characterized by experiment. The complex momentum plane is not merely a mathematical artifice introduced to better understand a function (although it is that) it also has physical content attainable by experiment.

In order to locate secondary singular points the asymptotic expression for the local potential V of eq.(103) is needed when r, the distance of electron 1 from the positive ion, is large. The asymptotic expression consists of two parts one of which involves exponentials multiplied by powers of r and the other part involves inverse powers of r but no exponentials.[32] Of the inverse power terms contributed by $\varepsilon - \varepsilon_\infty$, which occurs in V, the first is $-1/r$, as already pointed out (eq.108), and the second is an inverse fourth power due to polarization of the ion by electron 1. The term k^2 (eq.99) varies inversely as the eighth power. Both the inverse fourth and eighth are too high order to interfere with the asymptotic expansion of eq.(12). These do not add any singular points. The exponential terms in the asymptotic series give rise to secondary singular points. One set is given by[32]

$$P = \pm i(P_1 + \alpha_{sk}) \tag{113}$$

with

$$\alpha_{sk} = P_s + P_k - 2P_1. \tag{114}$$

$P_s = \sqrt{2I_s}$ and $P_k = \sqrt{2I_k}$ where I_s and I_k are ionization potentials which leave the residual ion in excited electronic states. I_1 is the ionization potential which leaves the ion in its lowest energy state. There is one singularity for every pair of I's. Except by accident the singular points are branch points. There are additional singular points involving double ionization.[32] These are much further out on the imaginary axis. Since the model is exact the ionization potentials can be determined experimentally.

Although the inverse power terms have no significant effect on singularities of the momentum eigenfunction the same is not true of the kernel function $v(\mathbf{q})$ of eq.(78). These terms introduce odd powers of q in a power series expansion of $v(\mathbf{q})$. The Fourier transform of eq.(78) has singular points at[32]

$$q = \pm i\alpha_{sk} \tag{115}$$

where α_{sk} is given by eq.(114). Let $\pm i\alpha$ be the pair closest to the origin. Then the quantity in parentheses in eq.(78) can be expanded in a convergent series in powers of the variable

$$t = \frac{q}{(q^2+\alpha^2)^{1/2}}. \tag{116}$$

This follows because this transform maps a region like that of Fig. 2 (q being substituted for P) into the unit circle in the complex t-plane. Since the cross hatched region of the q-plane contains no singularities there are none in the unit t-circle and hence the power series in t necessarily converges. The first three terms of the series are, for an atom,

$$\begin{aligned} v(q) &= \frac{-1}{2\pi^2 q^2}\left(a_\circ + a_2 t^2 + a_3 t^3 + \ldots\right) \\ &= -\frac{1}{2\pi^2 q^2}\left(a_\circ + \frac{a_2 q^2}{q^2+\alpha^2} + \frac{a_3\, q^3}{(q^2+\alpha^2)^{3/2}} + \ldots\right) \end{aligned} \tag{117}$$

The functions $v(\mathbf{q})$ and $V(\mathbf{r})$ are related by the transform of eq.(76). From the inverse transform a formula for V is obtained as follows:

$$V(r) = -\frac{1}{r} - a_2\frac{e^{-\alpha r}}{r} - a_3\frac{L(\alpha r)}{r} \tag{118}$$

and

$$L(\alpha r) = \frac{2}{\pi}\int_0^\infty \frac{q^2 \sin(qr)}{(q^2+\alpha^2)^{3/2}}\, dq \tag{119}$$

or

$$L(\alpha r) = \sum_{m=0}^{\infty} \frac{(m+1)(-\alpha r)^m}{\left[\Gamma(\frac{m}{2}+1)\right]^2}, \tag{120}$$

see ref. 43 p.71 eq.28 for the integral (119). Since mean square error is preserved in a Fourier transform eq.(118) represents V in r-space with the same error that (117) represents $v(\mathbf{q})$ in q-space. However, by making use of the fact that $v(\mathbf{q})$ has the same functional form as a first order collision amplitude, it is possible to pinpoint the region of r-space where (118) is closest to $V(\mathbf{r})$. The argument is as follows.

The quantity q is the magnitude of the change in momentum **Q–P** on collision. Only a small force is required to change the momentum by a small amount and a small force occurs when the colliding electron is far away from the scatterer, i.e. when r is large. Since the series in parentheses of eq.(117) is most accurate when t (and q) are small it is to be expected that (118) will be most accurate when r is large. The term in t^3 exists because the polarization potential gives rise to odd powers of q in the expansion of the amplitude and odd powers begin with the cube term. The term involving L (αr) is due to the inverse fourth power polarization term. The polarization potential is, however, spread throughout the series rather than being concentrated in a single term.

7. The Importance of Selecting Appropriate Basis Functions.

In preceding sections considerable emphasis has been placed on the selection of expansions which are convergent in the complex momentum plane. Mapping functions have been employed to assist the selection. The series described have emphasized convergence in the sense of the theory of functions of a complex variables rather than convergence in the mean. In one simple example, involving the Yukawa potential, a method has been developed in which the Schrödinger equation in p-space is forced, by a power series method, to hold for complex as well as real variables.[44] Since Slater type orbitals convert into rational functions in p-space a general rational function was employed. When the power series method was employed the eigenvalue did not converge as more members were added to the basis set. But when the same basis was employed in a variation method, essentially an approximation in the mean, rapid convergence was obtained. The difficulty was traced back to the nature of the basis set. When forced to hold in the complex plane the basis set was simultaneously forcing the eigenfunction to be a meromorphic function. But a function possessing branch points cannot be meromorphic. Since this fact is embedded in the Schrödinger equation, convergence could not be obtained when the additional condition of convergence in the complex plane was imposed. A variation method, on the other hand employs real variables only, without reference to complex variables, and hence rapid convergence was obtained. In this example selection of the wrong basis set tended to conceal the actual analytic structure of the eigenfunction. It is believed that this is often the case in variational methods. It is to be hoped that guide lines can be developed which will eliminate this conflict. Study of the analytic structure of eigenfunctions should help in this endeavor but the final solution remains for future research.

Acknowledgement! The author appreciates the assistance in typing and arranging the manuscript by Mrs. Ruth Silverman.

References

1. R. C. Tolman, *The Principles of Statistical Mechanics,* Oxford University Press, New York (1938). See Eq. 54.2 p 194.
2. B. Podolsky and L. Pauling, *Phys. Rev.* **34,** 109 (1929).
3. B. Lohman and E. Weigold, *Phys. Lett.* **86** A, 139 (1981).
4. *Momentum Wave Functions-1982,* Editor E. Weigold, AIP Conference Proceedings, American Institute of Physics, New York (1982). See article by E. Weigold, Page 1.
5. *Momentum Wave Functions-1976,* Editor D. W. Devins, AIP conference Proceedings, American Institute of Physics, New York (1976).
6. E. N. Lassettre, *J. Chem. Phys.* **43,** 4479 (1965).
7. *Handbook of Mathematical Functions,* editors M. Abramowitz and I. Stegun National Bureau of Standards, Applied Math. series **55,** (1968); page 437.
8. E. L. Ince, *Ordinary Differential Equations,* Dover, New York, page 465.
9. E. T. Whittaker and G. N. Watson, *Modern Analysis,* Cambridge Univ. Press, Fourth Edition (1963) page 342.
10. G. N. Watson, *Bessel Functions,* Second Edition Cambridge University Press, New York (1944) Page 385.
11. E. N. Lassettre, *J. Chem. Phys.* **70,** 3468 (1979).
12. E. N. Lassettre, *J. Chem. Phys.* **64,** 4375 (1976).
13. J. C. Slater, *Phys. Rev.* **81,** 385 (1951); J. C. Slater *Phys. Rev.* **91,** 528 (1953).
14. J. C. Slater, *The Self Consistent Field for Molecules and Solids,* McGraw-Hill, New York (1974).
15. W. Kohn and L. J. Sham *Phys. Rev.* **140** A, 1135 (1965).
16. *Theory of the Inhomogeneous Electron Gas,* Editors S. Lundqvist and N. H. March; Plenum Press, New York, (1983).
17. *Solid State Physics* vol. 38, Ed. H. Ehrenreich, D. Turnbull and F. Seitz; Academic Press, New York (1984). See chapter on Density Functional Methods by J. Callaway and N. H. March, p. 136.
18. N. C. Handy, M. T. Marron and H. J. Silverstone *Phys. Rev.* **180,** 45 (1969).
19. W. M. Huo and E. N. Lassettre *J. Chem. Phys.* **72,** 2374 (1980).
20. G. S. Handler, D. W. Smith and H. J. Silverstone *J. Chem. Phys.* **73,** 3936 (1980).
21. C. L. Davis, H. Jensen and H. J. Monkhorst *J. Chem. Phys.* **80,** 840 (1984).
22. V. Fock, *Zeit. für Phys.* **98,** 145 (1935).
23. J. C. P. Miller, *Tables of Chebyshev Polynomials* $S_n(x)$ and $C_n(x)$, U. S. Government Printing office, Washington, D.C. (1952). see pages VII and VIII of the introduction for a comparison of power series and orthogonal polynomials convergence rate.
24. C. Lanczos, *Applied Analysis,* Prentice-Hall, Englewood Cliffs, N.J. (1956).
25. A. R. P. Rau and U. Fano, *Phys. Rev.* **162,** 68 (1967).

26. G. Szegö, *Orthogonal Polynomials* Am. Math. Soc. Colloquim Publications Vol. XXIII, Am. Math Soc., New York (1939). see page 80.
27. E. N. Lassettre, *J. Chem. Phys.* **57**, 4357 (1972).
28. P. S. Bagus, T. L. Gilbert and C. C. J. Roothaan *J. Chem. Phys.* **56**, 5195 (1972).
29. P. S. Bagus *Phys. Rev.* **A139**, 619 (1965).
30. C. L. Davis and H. J. Monkhorst, *J. Chem. Phys.* **81**, 5845 (1984).
31. E. N. Lassettre, *Comments At. and Mol. Phys.* **16**, 35 (1985).
32. E. N. Lassettre, *J. Chem. Phys.* **83**, 1709 (1985).
33. E. N. Lassettre, *J. Chem. Phys.* **58**, 1991 (1973).
34. W. M. Huo, *J. Chem. Phys.* **63**, 4013 (1975).
35. G. Hunter, *Int. J. Quant. Chem.* **9**, 237 (1975).
36. G. Hunter and Chin Chiu Tai, *Int. J. Quant. Chem.* **21**, 1041 (1982)
37. G. Hunter and Chin Chiu Tai, *Int. J. Quant. Chem.* Quantum Chemistry Symposium, **19**, 173 (1986).
38. J. Katriel and E. R. Davidson, *Proc. Nat'l Acad. Sci. U.S.A.* **77**, 4403 (1980).
39. R. Ahlrichs, M. Hoffman-Ostenhof, T. Ostenhof, and J. D. Morgan III, *Phys. Rev.* **A23**, 2106 (1981).
40. R. T. Pack and W. Byers Brown *J. Chem. Phys.* **45**, 556 (1966).
41. J. H. Bartlett Jr., *Phys. Rev.* **51**, 661 (1937).
42. D. E. Freund, B. D. Huxtable, and J. D. Morgan III, *Phys. Rev.* **A29**, 980 (1984).
43. *Table of Integral Transforms* Vol. I, Editor, A. Frdélyi McGraw-Hill, New York (1954) p. 67 eq. 35.
44. E. N. Lassettre, *J. Chem. Phys.* **82**, 827 (1985).

THE ANALYTIC STRUCTURE OF ATOMIC AND MOLECULAR WAVEFUNCTIONS AND ITS IMPACT ON THE RATE OF CONVERGENCE OF VARIATIONAL CALCULATIONS

John D. Morgan III
Department of Physics
University of Delaware
Newark, Delaware 19716
U.S.A.

INTRODUCTION

In these lectures I would like to present an overview of the analytic structure of atomic and molecular wavefunctions and how that structure determines the rate of convergence of a variational (Rayleigh-Ritz) calculation as the number of basis functions is increased. I shall present some illustrative examples of variational basis-set calculations on both simple model systems and more complicated systems of greater physical relevance, such as the helium atom and small multi-center molecules.

The fundamental works on these topics are those by Tosio Kato[1] and by Charles Schwartz[2] in the late 1950's and early 1960's. In the late 1970's Bruno Klahn (who with his advisor W. Bingel had recently published a pair of important articles on the convergence of the Rayleigh-Ritz method for basis sets used in quantum chemistry[2a]) and I were working on the connection between series expansions and basis set expansions for helium atom wavefunctions, and we rediscovered the fact first elaborated by Schwartz[2] that the rate of convergence of a basis set expansion for a particular function depends critically on how rapidly finite linear combination of the basis functions replicate the singularities of the function one is trying to approximate. Whereas Schwartz had explicitly worked through a number of specific cases which illustrate this point, Klahn and I found a general method for estimating the rate of convergence of the expansion of a function in a given orthonormal basis composed of the eigenfunctions of some operator.[3] Later my colleague Robert N. Hill found a generalization of the method used by Klahn and myself, and he applied his technique to analyze the rate of convergence of Legendre expansions for helium wavefunctions.[4] I shall briefly present these results and discuss their application to selecting bases for variational calculations on small atoms and molecules.

M. Defranceschi and J. Delhalle (eds.), Numerical Determination of the Electronic Structure of Atoms, Diatomic and Polyatomic Molecules, 49–84.

I. A SIMPLE EXAMPLE - FOURIER SERIES

Consider the approximation of a periodic function f(x) of period 2π

$$f(x+2\pi) = f(x), \tag{I.1}$$

and for simplicity suppose that f(x) is even so that only cosines appear in its Fourier series:

$$f(x) = \lim_{N\to\infty} f_N(x), \tag{I.2}$$

where the functions $f_N(x)$ are the partial sums

$$f_N(x) = \sum_{n=0}^{N} c_n \frac{\cos(nx)}{\sqrt{(1 + \delta_{n,0})\pi}} . \tag{I.3}$$

Since the expansion functions are orthonormal on the interval $[-\pi, \pi]$, the expansion coefficients c_n are simply

$$c_n = \int_{-\pi}^{\pi} dx\ f(x) \frac{\cos(nx)}{\sqrt{(1 + \delta_{n,0})\pi}} . \tag{I.4}$$

Assume that this Fourier approximation is convergent in the mean; i.e., that

$$\|f - f_N\|_2 = \left(\int_{-\pi}^{\pi} dx\ |f(x) - f_N(x)|^2\right)^{\frac{1}{2}} \to 0 \tag{I.5}$$

as $N \to \infty$. Then the error in the approximation of f by f_N is simply

$$\left(\sum_{n=N+1}^{\infty} |c_n|^2 \right)^{\frac{1}{2}}, \tag{I.6}$$

and the speed which with this quantity tends to 0 as $N \to \infty$ provides a measure of the rate of convergence of this expansion. Now suppose that the function f(x) has p continuous derivatives, and for simplicity assume that p is even. Then

$$f^{(p)}(x) = \left(\frac{d}{dx} \right)^p f(x) \tag{I.7}$$

will certainly be square integrable, and so the coefficients $c_n^{(p)}$ of its Fourier expansion

$$c_n^{(p)} = \int_{-\pi}^{\pi} dx \; f^{(p)}(x) \frac{\cos(nx)}{\sqrt{(1 + \delta_{n,0})\pi}} \tag{I.8}$$

must obey

$$\lim_{N\to\infty} \sum_{n=N+1}^{\infty} |c_n^{(p)}|^2 = 0. \tag{I.9}$$

If we now integrate by parts p times the formula for $c_n^{(p)}$ we see that

$$c_n^{(p)} = \int_{-\pi}^{\pi} dx \left(\frac{d}{dx} \right)^p f(x) \frac{\cos(nx)}{\sqrt{(1 + \delta_{n,0})\pi}}$$

$$= \int_{-\pi}^{\pi} dx \; f(x) \left(\frac{d}{dx} \right)^p \frac{\cos(nx)}{\sqrt{(1 + \delta_{n,0})\pi}}$$

$$= n^p \int_{-\pi}^{\pi} dx \; f(x) \frac{\cos(nx)}{\sqrt{(1 + \delta_{n,0})\pi}}$$

$$= n^p \; c_n. \tag{I.10}$$

(The boundary terms vanish since by hypothesis f(x) is periodic.) From condition (I.9), we see that

$$\lim_{N\to\infty} \sum_{n=N+1}^{\infty} n^{2p} |c_n|^2 = 0, \tag{I.11}$$

and this can be true only if $n^p c_n \to 0$ as $n \to \infty$, or equivalently, only if

$$c_n = o(n^{-p}) \tag{I.12}$$

as $n \to \infty$. Thus we see that the more derivatives of f(x) are continuous, the more rapidly its Fourier coefficients tend to 0, and the better it is approximated by a truncated Fourier series. In short, the analytic structure of f(x) determines the rate of convergence of its Fourier series.

II. A MORE COMPLICATED EXAMPLE: EXPANSIONS IN HERMITE ORTHONORMAL FUNCTIONS (Klahn and Morgan[3])

Consider the basis functions

$$\phi_n(r) = \frac{1}{2^{n+1}\sqrt{(2n+1)!}\ \pi^{3/4}} \frac{H_{2n+1}(r)}{r} e^{-r^2/2}, \tag{II.1}$$

which are orthonormal on the half-line $[0, \infty)$ with measure $r^2 dr$. They happen to be eigenfunctions of the harmonic oscillator Hamiltonian

$$\tilde{H} = -\frac{1}{2}\left(\frac{d^2}{dr^2} + \frac{2}{r}\frac{d}{dr}\right) + \tfrac{1}{2}r^2 \tag{II.2}$$

with eigenvalues $(2n + 3/2)$.

Now consider the approximation of some s-state eigenfunction $\psi(r)$ of some spherically symmetric Hamiltonian

$$H = T + V(r) \tag{II.3}$$

by finite linear combinations of the ϕ_n's:

$$\psi(r) \simeq \psi_N(r) = \sum_{n=0}^{N} c_n \phi_n(r). \tag{II.4}$$

Since the ϕ_n's are orthonormal and real, the coefficients c_n are given by

$$c_n = \int_0^\infty dr\, r^2\, \phi_n(r)\, \psi(r). \qquad \text{(II.5)}$$

One would like to try to find some analog of integration by parts to analyze this integral. The desired analog can be readily found by realizing that just as cos(nx) is an eigenfunction of $-d^2/dx^2$ with eigenvalue n^2, $\phi_n(r)$ is an eigenfunction of $\tilde{H}$ with eigenvalue $(2n + 3/2)$. Hence the analog of the integration by parts in (I.10) is the invocation of the Hermiticity of $\tilde{H}$:

$$c_n = \int_0^\infty dr\, r^2\, \phi_n(r)\, \psi(r)$$

$$= \int_0^\infty dr\, r^2 \left(\frac{1}{2n+3/2}\, \tilde{H} \right)^p \phi_n(r)\, \psi(r)$$

$$= \frac{1}{(2n+3/2)^p} \int_0^\infty dr\, r^2\, \phi_n(r)\, (\tilde{H})^p \psi(r). \qquad \text{(II.6)}$$

Provided that ψ is in the domain of $(\tilde{H})^p$; i.e., *provided that* $(\tilde{H})^p\psi$ is square-integrable, then the expansion coefficients of $(\tilde{H})^p\psi$ must tend to 0 as $n \to \infty$, and so dividing by $(2n + 3/2)^p$ shows that

$$c_n = o(n^{-p}) \qquad \text{(II.7)}$$

as $n \to \infty$.

In our 1984 article, Klahn and I[3] considered the case where ψ is the νth eigenfunction of the hydrogen atom with $\ell = 0$:

$$\psi^{(\nu)}(r) = \left(\frac{Z^3}{\pi\nu^5} \right)^{\frac{1}{2}} L_{\nu-1}^{(1)}(2Zr/\nu)\, \exp(-Zr/\nu). \qquad \text{(II.8)}$$

We demonstrated that the nth Hermite expansion coefficient $c_n^{(\nu)}$ for $\psi^{(\nu)}$ asymptotically obeys

$$c_n^{(\nu)} \sim \frac{(-1)^n}{\sqrt{2\pi}} \left(\frac{Z^5}{\nu^3} \right)^{\frac{1}{2}} \frac{1}{n^{7/4}} , \tag{II.9}$$

as $n \to \infty$, and hence that with

$$\psi_N^{(\nu)} = \sum_{n=0}^{N} c_n^{(\nu)} \phi_n(r), \tag{II.10}$$

the mean-square error

$$\|\psi^{(\nu)} - \psi_N^{(\nu)}\|_2^2 = \sum_{n=N+1}^{\infty} |c_n^{(\nu)}|^2$$

$$\simeq \frac{1}{2\pi} \frac{Z^5}{\nu^3} \sum_{n=N+1}^{\infty} \frac{1}{n^{7/2}}$$

$$\simeq \frac{1}{2\pi} \frac{Z^5}{\nu^3} \frac{2}{5} N^{-5/2} \tag{II.11}$$

as $N \to \infty$. We also showed that normalized expectation values of the kinetic energy operator and of the Hamiltonian, using $\psi_N^{(\nu)}$ as a trial function, have errors of $O(N^{-3/2})$. We also rigorously demonstrated and numerically illustrated the fact that if one uses $\psi_N^{(\nu)}$ to calculate expectation values of the operator r^k, then these converge like $N^{-(7-k)/2}$ for $2 \leq k < 7$, diverge like $N^{-(7-k)/2}$ for $k > 7$, and converge *to an incorrect value* for $k = 7$.

Besides the results described above, which were obtained using trial functions which for a given N minimise the r.m.s. error

$$\|\psi^{(\nu)} - \psi_N^{(\nu)}\|_2, \tag{II.12}$$

Klahn and I also studied rates of convergence for Rayleigh-Ritz approximations, in which one minimises the νth eigenvalue (= 0, 1, 2,...) of the truncated N X N Hamiltonian matrix. We found numerically that the rates of convergence were very similar, and later my colleague Robert N. Hill, in a particularly impressive article, placed on a rigorous footing the theory of rates of convergence for Rayleigh-Ritz calculations.[4]

III. A "REAL" EXAMPLE: EXPANSIONS OF ATOMIC AND MOLECULAR WAVEFUNCTIONS IN LEGENDRE SERIES (Hill[4])

Besides providing an explicit and rigorous analysis of the Rayleigh-Ritz problem studied by Klahn and myself, Hill applied his novel and powerful method to the analysis of the rate of convergence of the approximation of the ground state eigenvalue E and eigenfunction $\psi(r_1,r_2,r_{12})$ of the helium-atom Hamiltonian

$$H = -\tfrac{1}{2}\nabla_1^2 - \tfrac{1}{2}\nabla_2^2 - \frac{Z}{r_1} - \frac{Z}{r_2} + \frac{1}{r_{12}} \tag{III.1}$$

by a finite linear combination

$$\tilde{\psi}_L(r_1,r_2,r_{12}) = \sum_{\ell=0}^{L} \psi_\ell(r_1,r_2)\, P_\ell(\cos\Theta) \tag{III.2}$$

of Legendre polynomials in

$$\cos\Theta = \frac{\vec{r}_1\cdot\vec{r}_2}{r_1 r_2} = \frac{r_1^2 + r_2^2 - r_{12}^2}{2r_1 r_2} .$$

The slow convergence of the expansion (III.2) for a long time was explained qualitatively in terms of the rather slow power-law convergence of the expansion

$$r_{12} = r_> \sum_{\ell=0}^{\infty} \left(-\frac{(r_</r_>)^\ell}{2\ell-1} + \frac{(r_</r_>)^{\ell+2}}{2\ell+3} \right) P_\ell(\cos\Theta), \tag{III.3}$$

which shows explicitly that the cusp in the exact wavefunction at $r_{12}=0$ is slowly represented by an expansion in Legendre polynomials. A decade ago Carroll, Silverstone, and Metzger[5] had found numerically that the variational energy

$$\tilde{E}_L = \frac{\langle\tilde{\psi}_L|H|\tilde{\psi}_L\rangle}{\langle\tilde{\psi}_L|\tilde{\psi}_L\rangle} \tag{III.4}$$

converges to the "exact" E (as determined by highly accurate calculations by Frankowski and Pekeris[6]) with an error which obeys the incremental relation

$$\tilde{E}_L - \tilde{E}_{L-1} \simeq -0.0740(L + \tfrac{1}{2})^{-4} - 0.031(L + \tfrac{1}{2})^{-5}. \qquad \text{(III.5)}$$

Recently Hill proved that the error in the approximate $\tilde{E}_L$ obeys

$$\tilde{E}_L - E = C_3(L + 1)^{-3} + C_4(L + 1)^{-4} + O(L^{-5}), \qquad \text{(III.6)}$$

where the constants

$$C_3 = 2\pi^2 \int_0^\infty |\psi(r,r,0)|^2 \, r^5 \, dr \qquad \text{(III.7)}$$

and

$$C_4 = \frac{12\pi}{5} \int_0^\infty |\psi(r,r,0)|^2 \, r^6 \, dr \qquad \text{(III.8)}$$

are given by integrals over the exact wavefunction. A numerical evaluation of the constants using the extremely accurate helium wavefunction obtained by Freund, Huxtable, and me[7] yielded the values

$$C_3 \simeq 0.0247419 \qquad \text{(III.9)}$$

$$C_4 \simeq 0.00774727. \qquad \text{(III.10)}$$

To make the connection between Hill's formula (III.6) and the incremental formula (III.5), observe that

$$\begin{aligned}
\tilde{E}_L - \tilde{E}_{L-1} &= \tilde{E}_L - E - (\tilde{E}_{L-1} - E) \\
&= C_3\{(L + 1)^{-3} - L^{-3}\} + C_4\{(L + 1)^{-4} - L^{-4}\} + O(L^{-6}) \\
&= C_3\{((L + \tfrac{1}{2}) + \tfrac{1}{2})^{-3} - ((L + \tfrac{1}{2}) - \tfrac{1}{2})^{-3}\} \\
&\quad + C_4\{((L + \tfrac{1}{2}) + \tfrac{1}{2})^{-4} - ((L + \tfrac{1}{2}) - \tfrac{1}{2})^{-4}\} + O(L^{-6}) \\
&= C_3\{(-3)(L + \tfrac{1}{2})^{-4} + O((L + \tfrac{1}{2})^{-6})\}
\end{aligned}$$

$$+ C_4\{(-4)(L + \tfrac{1}{2})^{-5} + O((L + \tfrac{1}{2})^{-7})\} + O(L^{-6})$$

$$= -3C_3(L + \tfrac{1}{2})^{-4} - 4C_4(L + \tfrac{1}{2})^{-5} + O(L^{-6}). \qquad \text{(III.11)}$$

Since

$$-3C_3 \simeq -0.0742257 \qquad \text{(III.12)}$$

and

$$-4C_4 \simeq -0.0309891, \qquad \text{(III.13)}$$

we see that there is excellent agreement with Carroll, Silverstone, and Metzger's numerical values -0.0740 and -0.031, respectively, in eq. (III.5).

One can see the impact of cusp conditions on rates of convergence by going through Hill's derivation and realizing that the expression for C_3 actually comes from

$$C_3 = 8\pi^2 \frac{1}{2} \int_0^{\infty} \psi(r,r,0) \left[\frac{\partial\psi(r,r,r_{12})}{\partial r_{12}} \right]_{r_{12}=0} r^5 \, dr, \qquad \text{(III.14)}$$

where $8\pi^2$ comes from integrating out the Euler angles, and the Kato cusp condition

$$\left[\frac{\partial\psi(r,r,r_{12})}{\partial r_{12}} \right]_{r_{12}=0} = \tfrac{1}{2}\psi(r,r,0) \qquad \text{(III.15)}$$

was used to obtain the form (III.7), which is more suitable for numerical evaluation. The form (III.14) makes it clear that the leading order behavior of the rate of convergence of the variational energy is given explicitly by a weighted average of the discontinuity of $\vec{\nabla}\psi$ in Cartesian coordinates at the point $r_{12} = 0$, as illustrated in Fig. 1:

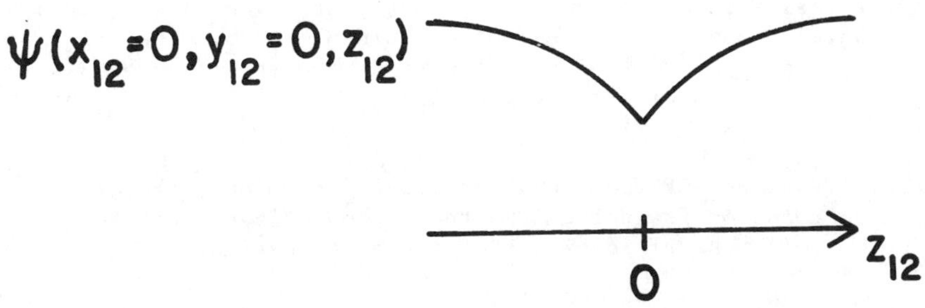

Fig. 1. Illustration of an electron-electron cusp.

It is enlightening to speculate how these results should generalize to the approximation of energies and eigenfunctions of many-electron atoms. If one is performing a variational calculation on a many-electron atom using a basis which involves products of one-electron spherical harmonics of degree up to L, then one might guess that the proper generalization of (III.14) would be

$$C_3 = \frac{1}{4} \int \left(\frac{\partial}{\partial r_{12}} \rho^{(2)}(\vec{r}_1, \vec{r}_2) \right) 4\pi\delta^{(3)}(\vec{r}_{12})\, r_1^3\, d^3\vec{r}_1\, d^3\vec{r}_2, \qquad \text{(III.16)}$$

where $\rho^{(2)}(\vec{r}_1, \vec{r}_2)$ is the two-electron density. Provided that the atom has at least one pair of electrons with antiparallel spins,

$$\left[\frac{\partial}{\partial r_{12}} \rho^{(2)}(\vec{r}_1, \vec{r}_2) \right]_{r_{12}=0} \qquad \text{(III.17)}$$

in practical situations will almost always be nonzero, and hence C_3 will be nonzero. This means that if one is using a basis of products of one-electron spherical harmonics to perform a variational calculation on a many-electron atom with at least one pair of antiparallel spins, the generic rate of convergence of the variational energy will be $(L + 1)^{-3}$. If one is trying to approximate a very diffuse atomic eigenfunction, such as that of a Rydberg state, it may happen that the constant C_4 exceeds C_3; in this case, the $(L+1)^{-3}$ convergence rate may not manifest itself until L is rather large.

For a molecule, one would also expect a generic $(L+1)^{-3}$ convergence rate to hold, but the formula analogous to (III.16) will be more complicated because of the need to take account of the spherical harmonic expansions about more than one center.

Now variational calculations on many-electron atoms and molecules are conventionally done according to the following procedure:

(i) Perform a self-consistent field (SCF) calculation, in which the variational trial function is a properly antisymmetrized linear combination of products of one-electron spin-orbitals. The spin-orbitals are products of spin functions and spatial orbitals, and in turn the spatial orbitals are products of radial functions and spherical harmonics.

(ii) After the SCF functions have been determined, the effect of electron correlation is included by the method of configuration interaction (CI). Matrix elements of the perturbation

$$W = \sum_{i>j} \frac{1}{r_{ij}} - \left\langle \sum_{i>j} \frac{1}{r_{ij}} \right\rangle \qquad \text{(III.18)}$$

are computed in a basis allowing for excitations of various numbers of electrons ("singles", "doubles", "triples", etc.). The eigenvalues of $H_o + W$ are calculated either directly, or by perturbation expansions.

It should be noted that in this approach (SCF followed by CI), all basis functions are expressible as a finite linear combination of products of one-electron spherical harmonics, and hence it should display the same relatively slow $(L + 1)^{-3}$ pattern of convergence, regardless of the atom or molecule under consideration (provided that it has at least two antiparallel spins). The same analysis will apply to multi-configuration self-consistent field (MCSCF) calculations. In short, attempting to describe electron correlation by finite linear combinations of products of spherical harmonics is generically a slowly-convergent $(L + 1)^{-3}$ procedure.

I should like to draw attention to a recent article by Werner Kutzelnigg,[7a] in which he explores the use of the ansatz

$$\psi = (1 + \tfrac{1}{2}r_{12})\Phi + \chi,$$

where Φ and χ are non-singular at $r_{12} = 0$, in variational calculations of the second-order Møller-Plesset energy and of the total energy for the ground state of the helium atom. He finds that this explicit inclusion of the electron-electron cusp in lowest order greatly accelerates the rate of convergence of the calculations, and he discusses how this ansatz can be generalized to treat larger systems while still avoiding the need to evaluate difficult integrals.

IV. IS THE ASYMPTOTIC BEHAVIOR OF EIGENFUNCTIONS RELEVANT TO RATES OF CONVERGENCE OF VARIATIONAL CALCULATIONS?

The answer to this question is a qualified "yes" - qualified, because although the asymptotic behavior of eigenfunctions is not completely irrelevant, it is much less important than a proper description of the small-distance cusp behavior. I should like to mention first a qualitative indication that this is so, and then proceed to a model system which explicitly demonstrates this fact.

In 1984 Freund, Huxtable, and I[7] published a study of the ground states of the helium isoelectronic sequence using scaled basis functions of the form

$$s^n t^\ell u^m (\ell n s)^j \, e^{-s/2}, \qquad \text{(IV.1)}$$

where s, t, and u are the Hylleraas coordinates defined by

$$s = r_1 + r_2$$

$$t = -r_1 + r_2$$

$$u = r_{12}. \tag{IV.2}$$

The basis functions (IV.1) do not have the correct asymptotic exponential behavior

$$\exp(-\zeta_1 r_1 - \zeta_2 r_2) + \exp(-\zeta_2 r_1 - \zeta_1 r_2), \tag{IV.3}$$

where ζ_1 and ζ_2 are related to the first and second binding energies E_1 and E_2 by $\zeta_i = \sqrt{-2E_i}$.[7b] However, the basis functions (IV.1) explicitly include the analytic structure of the wavefunction at all two-particle and three-particle coalescences. With 230 basis functions, we obtained for helium a variational upper bound of

$$-2.90372437703407 \text{ a.u.},$$

which seems to be accurate (see Section V below) to 4×10^{-14} a.u. At the time this was the most accurate calculation ever of the ground state energy of helium. I would challenge anyone to obtain with fewer variational parameters a lower energy using a basis which has the correct asymptotic behavior (IV.3) but which fails to have the correct analytic structure at two-particle and three-particle coalescences.

Let now us consider a simple one-dimensional model system, whose ground state wavefunction is rapidly approximated by an orthonormal basis whose elements have quite a different asymptotic behavior. Let

$$H = -\frac{1}{2}\frac{d^2}{dx^2} - \frac{V_o}{\cosh^2(\alpha x)} \tag{IV.4}$$

where $V_0 > 0$. The Schrödinger Equation for H is solvable in closed form, as is explicitly done in Landau and Lifshitz.[8] Since the details of the derivation can be found in a readily accessible textbook, I shall proceed directly to the result that the (unnormalized) ground state wavefunction is given by

$$\psi = \left(\frac{1}{\cosh(\alpha x)}\right)^{\sqrt{-2E}/\alpha}, \tag{IV.5}$$

where

$$E = -\frac{\alpha^2}{8}\left(-1 + \sqrt{1 + \frac{8V_o}{\alpha^2}}\right)^2 . \tag{IV.6}$$

Since as $x \to \pm\infty$, $1/\cosh(\alpha x) \to 2\exp(-\alpha|x|)$, one recognizes the $\exp(-\sqrt{-2E}|x|)$ fall-off characteristic of the exponential decay of an eigenfunction of a Hamiltonian for which $V(x)$ asymptotically vanishes.

Now consider the approximation of $\psi(x)$ by a finite linear combination of Hermite orthonormal functions:

$$\phi_n(x) = \pi^{-\frac{1}{4}} \frac{1}{\sqrt{(2n)!!}} H_n(x)\, e^{-x^2/2}, \tag{IV.7}$$

which are eigenfunctions with eigenvalues $(n + 1/2)$ of the harmonic oscillator Hamiltonian

$$\tilde{H} = -\tfrac{1}{2} \frac{d^2}{dx^2} + \tfrac{1}{2}x^2 . \tag{IV.8}$$

Since $\psi(x)$ and all its derivatives are bounded, continuous, and exponentially decreasing, it is straightforward to see that ψ is in the domain of any power of $\tilde{H}$. Hence for any non-negative integer p, the expansion coefficient

$$c_n = \int_{-\infty}^{\infty} dx\, \phi_n(x)\, \psi(x) \tag{IV.9}$$

obeys

$$c_n = \int_{-\infty}^{\infty} dx \left(\frac{1}{n+\frac{1}{2}} \tilde{H}\right)^p \phi_n(x)\, \psi(x)$$

$$= \frac{1}{(n+\frac{1}{2})^p} \int_{-\infty}^{\infty} dx\, \phi_n(x)\, (\tilde{H})^p \psi(x) . \tag{IV.10}$$

We therefore see that as $n \to \infty$ the expansion coefficients c_n decrease faster than any power of $1/n$, despite the mismatch between the asymptotic behavior of

$$\psi(x) \sim \exp(-\sqrt{-2E}\ |x|) \tag{IV.11}$$

and that of the nth Hermite orthonormal basis function

$$\phi_n(x) \sim x^n \exp(-x^2/2). \tag{IV.12}$$

Do the c_n's decrease exponentially, e.g., as exp(-bn) for some positive constant b? The answer is "no", and the proof is by contradiction. Suppose that for some positive constants C_0 and b the coefficients c_n obey the bound

$$|c_n| \le C_0 \exp(-bn) \tag{IV.13}$$

for all n. Then since

$$\psi = \sum_{n=0}^{\infty} c_n \phi_n, \tag{IV.14}$$

it is evident that

$$\begin{aligned} \exp(\beta\tilde{H})\psi &= \sum_{n=0}^{\infty} c_n \exp(\beta\tilde{H})\phi_n \\ &= \sum_{n=0}^{\infty} c_n \exp(\beta(n+\tfrac{1}{2}))\ \phi_n \end{aligned} \tag{IV.15}$$

would be an element of Hilbert space for $\beta < b$. However, by working directly with the kernel of $\exp(\beta\tilde{H})$, one can readily see that

$$\exp(\beta\tilde{H}) = U(\cosh^{\frac{1}{2}}\beta)\ \exp(\sinh\beta\ T)\ \exp(\sinh\beta\ V)\ U(\cosh^{\frac{1}{2}}\beta) \tag{IV.16}$$

where $T = \frac{1}{2}p^2$ and $V = \frac{1}{2}x^2$, and $U(\lambda)$ is the dilatation operator:

$$[U(\lambda)\psi](x) = \lambda^{\frac{1}{2}}\psi(\lambda x). \tag{IV.17}$$

Since $U(\cosh^{\frac{1}{2}}\beta)$ is unitary, $\exp(\beta\tilde{H})\psi$ is square-integrable if and only if

$$\exp(\sinh\beta\ T)\ \exp(\sinh\beta\ V)\ U(\cosh^{\frac{1}{2}}\beta)\psi \tag{IV.18}$$

is. Since $\exp(\sinh\beta\ k^2) \geq 1$,

$$f = \exp(\sinh\beta\ T)g \qquad (IV.19)$$

is square-integrable only if g is. But if $\psi(x) = (1/\cosh(\alpha x))^{\sqrt{-2E}/\alpha}$,

$$\exp(\sinh\beta\ x^2/2)\ U(\cosh^{\frac{1}{2}}\beta)\psi \qquad (IV.20)$$

is not square-integrable for any $\beta > 0$, because the quadratic exponential growth of $\exp(\sinh\beta\ x^2/2)$ overwhelms the linear exponential fall-off of $U(\cosh^{\frac{1}{2}}\beta)\psi$. Thus ψ is not in the domain of $\exp(\beta\tilde{H})$ for any $\beta > 0$, and hence the assumption (IV.13) is seen to be impossible for any $b > 0$.

If the Hermite expansion coefficients c_n of $\psi(x) = (1/\cosh(\alpha x))^{\sqrt{-2E}/\alpha}$ decrease faster than any power of $1/n$, but not as fast as $\exp(-bn)$ for any positive b, what is their asymptotic form as $n \to \infty$? I would conjecture that it is something like

$$\exp(-bn^\gamma), \qquad (IV.21)$$

where $0 < \gamma < 1$. In fact, an asymptotic analysis by my colleague Robert N. Hill of the overlap integral (IV.9) suggests that $\gamma = \frac{1}{2}$. This rate of convergence, although not as fast as the linear exponential $\exp(-bn)$, is still quite fast. We therefore see that

(i) if one's basis doesn't get the cusp behavior right, the rate of convergence tends to be a power law;

(ii) if one's basis gets the cusp behavior right but doesn't get the asymptotic behavior right (by having a different analytic form), the rate of convergence tends to be a modified exponential;

(iii) if one's basis gets both the cusp behavior and the asymptotic form right (by having the same analytic form, but perhaps with a different scale parameter), the rate of convergence tends to be a linear exponential.

Let us consider the implications for variational calculations, of the SCF followed by CI variety, on atoms and molecules. As discussed by Davis, Jensen, and Monkhorst,[9] who were following up earlier work by Huo and Lassettre,[9a] in doing an SCF calculation, it is advantageous to use basis functions with the same asymptotic exponents as the exact Hartree-Fock radial functions one is trying to approximate. (Of course, one also needs to make sure that these basis functions also have the right electron-nucleus cusp behavior.) In this case, one can expect to obtain an extremely rapid (linear exponential, or perhaps even faster) convergence of the SCF calculation. But what is the point of nailing down the SCF orbital energies to 10 or 12 or 15 digits if the subsequent CI displays an $(L + 1)^{-3}$ convergence pattern, and it requires an enormous amount of computer time to calculate a correlation energy accurate to 3 digits? In any serious (SCF-CI) calculation of the energy levels of an atom or a molecule, the error in the correlation energy due to a poor representation of the electron-electron cusps is likely to dominate the error due to an inexact representation of the exponentially decreasing tail of the wavefunction.

In this connection, a further point should be considered. The "hierarchy of exponents" derived by Huo and Lassettre[9a] for the atomic Hartree-Fock equations arises from the coupling of orbitals by the non-linear exchange operator. The multi-particle Schrödinger equation, however, is linear, and it is far from obvious that the exact solution of that equation will possess singularities in the complex momentum space at or near those arising from the "hierachy of exponents" in the solutions to the Hartree-Fock equations. It would at least be interesting to extend the method of Deift, Hunziker, Simon, and Vock[7a] to explore the complex momentum space for any additional singularities beyond those that determine the widths of the tubes within which the Fourier transform of the exact wavefunction is analytic.

V. BUILDING ANALYTIC STRUCTURE INTO BASIS FUNCTIONS - THE HELIUM ATOM (AND OTHER THREE-PARTICLE SYSTEMS)

The famous "Kato cusp condition"[1] for a non-relativistic many-particle system interacting by Coulomb forces states that in the limit as two particles of masses m_i and m_j and charges q_i and q_j approach each other

$$\left.\frac{\partial\hat{\psi}}{\partial r_{ij}}\right|_{r_{ij}=0} = \mu_{ij}\, q_i\, q_j\, \psi(r_{ij} = 0), \qquad \text{(V.1)}$$

where $\mu_{ij} = m_i m_j/(m_i + m_j)$ is the reduced mass of the two-particle system and $\hat{\psi}$ is the average of ψ over an infinitesimally small sphere centered at $r_{ij} = 0$. The proof of eq. (V.1) is long and complicated,[1] but one can understand whence (V.1) arises by considering the Schrödinger Equation for the multi-particle system and transforming the variables $\vec{r}_i$ and $\vec{r}_j$ into relative and center-of-mass coordinates

$$\vec{R}_{ij} = \frac{m_i\vec{r}_i + m_j\vec{r}_j}{m_i + m_j}$$

$$\vec{r}_{ij} = \vec{r}_i - \vec{r}_j \qquad \text{(V.2)}$$

Then as $r_{ij} \to 0$ and all other interparticle distances remain bounded away from 0, the only singularity in the potential energy term in the Schrödinger Equation is

$$\frac{q_i q_j}{r_{ij}} \psi, \qquad \text{(V.3)}$$

and the kinetic energy operator will contain a term

$$-\frac{1}{2\mu_{ij}} \left(\frac{\partial^2}{\partial r_{ij}^2} + \frac{2}{r_{ij}} \frac{\partial}{\partial r_{ij}} + \frac{\mathfrak{L}_{ij}^2}{r_{ij}^2} \right) \psi. \qquad \text{(V.4)}$$

If we now perform an average about an infinitesimally small sphere centered at $r_{ij} = 0$, this will eliminate from ψ all except the s-wave component, which is itself annihilated by $\mathfrak{L}_{ij}^2$. Assuming that $\partial^2\hat{\psi}/\partial r_{ij}^2$ is bounded, as it is for eigenfunctions of the hydrogen atom, then the coefficients of the $1/r_{ij}$ singularities in (V.3) and (V.4) should cancel, which leads to the Kato cusp condition (V.1).

Multi-particle coalescences are described by an expansion in hyperspherical coordinates developed by Fock[10] in 1954. It is simplest to describe in the case of S-states of the helium atom. If one defines a trio of hyperspherical variables by

$$R = r_1^2 + r_2^2 \qquad (\textit{not } R^2)$$

$$\alpha = 2 \arccos(r_2/R^{\frac{1}{2}})$$

$$\theta = \arccos(\vec{r}_1 \cdot \vec{r}_2 / r_1 r_2) \qquad \text{(V.5)}$$

and adds a fictitious variable ϕ on which physical eigenfunctions do not depend, the Schrödinger Equation for a helium atom of infinite nuclear mass becomes

$$2R \left(-\frac{\partial^2}{\partial R^2} - \frac{3}{R}\frac{\partial}{\partial R} - \frac{\square^*}{R^2} \right) \psi + R^{-\frac{1}{2}} U(\alpha,\theta)\psi = E\psi, \qquad \text{(V.6)}$$

where $-\square^*$ is the hyperspherical angular momentum operator and

$$U(\alpha,\theta) = -\left(\frac{Z}{\cos(\alpha/2)} + \frac{Z}{\sin(\alpha/2)} \right) + \frac{1}{\sqrt{1 - \sin\alpha \cos\theta}}$$

$$= R^{\frac{1}{2}} V(r_1, r_2, r_{12}). \tag{V.7}$$

If one tries to solve (V.6) using the ansatz of an expansion in powers of $R^{1/2}$

$$\psi = \sum_{j=0}^{\infty} R^{j/2} \psi_j(\alpha,\theta), \tag{V.8}$$

it is found that no well-behaved function solves the recursive differential equation for $\Psi_2(\alpha,\theta)$. Hence, in analogy with the Frobenius theory of regular singular points of ordinary differential equations, Fock proposed that powers of ℓnR be included in the expansion:

$$\psi = \sum_{j=0}^{\infty} \sum_{k=0}^{[j/2]} R^{j/2} (\ell nR)^k \psi_{j,k}(\alpha,\theta). \tag{V.9}$$

Fock showed that this expansion was at least consistent. In 1967 Macek[11] showed that this expansion converges in the mean for $R < 1/2$, and I[12] have proved that in fact it converges pointwise for all finite R. This question was also studied independently by Professor Jean Leray, who has sketched a proof of convergence in various conference proceedings,[13] but I believe that at the time of writing not all the details of his proof have appeared. A remaining question, whether for physically allowed energies Fock's expansion converges to the physical eigenfunctions and not to some unphysical solution, should be addressible by a multi-dimensional generalization of the Picard method developed by Robert N. Hill and me;[14] I would certainly bet that it does.

For singlet states, the first few coefficients in Fock's expansion (V.9) are

$$\psi_{0,0} = 1$$

$$\psi_{1,0} = -Z(\sin(\alpha/2) + \cos(\alpha/2)) + \tfrac{1}{2}(1 - \sin\alpha \cos\theta)^{\frac{1}{2}}$$

$$\psi_{2,1} = -Z \frac{\pi-2}{6\pi} \sin\alpha \cos\theta. \tag{V.10}$$

Observe that

$$\psi_{0,0} + R^{\frac{1}{2}}\psi_{1,0} = 1 - Z(r_1 + r_2) + \tfrac{1}{2}r_{12}, \tag{V.11}$$

as would be expected from the Kato cusp condition. The lowest order logarithmic term is

$$
\begin{aligned}
R\,\ell nR\,\psi_{2,1} &= -Z\,\frac{\pi-2}{6\pi}\,\sin\alpha\,\cos\theta\,R\,\ell nR \\
&= -Z\,\frac{\pi-2}{6\pi}\,(r_1^2 + r_2^2 - r_{12}^2)\,\ell n(r_1^2 + r_2^2) \\
&= -\,Z\,\frac{\pi-2}{6\pi}\,2\vec{r}_1\cdot\vec{r}_2\,\ell n(r_1^2 + r_2^2) \qquad \text{(V.12)}
\end{aligned}
$$

For over 25 years numerous efforts to solve for $\psi_{2,0}(\alpha,\theta)$ had been unsuccessful, until the quite recent extremely impressive work by John Gottschalk and Paul Abbott and their advisor E. N. Maslen,[15] who obtained a closed form expression for $\psi_{2,0}(\alpha,\theta)$ in terms of exponential integrals and Lobachevsky's functions. Work in two directions, obtaining higher-order $\psi_{j,k}(\alpha,\theta)$'s for helium, and extending the results to multi-electron atoms, should most certainly be pursued.

In doing variational calculations on the helium atom, it is desirable to include in the basis functions as much analytic structure as possible. Just as a Hylleraas basis converges much faster than an expansion in $P_\ell(\cos\Theta)$ because it explicitly includes the factor of r_{12} needed to describe the two-electron cusp, a basis which includes logarithmic terms and negative powers of the Hylleraas variable $s = r_1 + r_2$ converges faster because the wavefunction near the three-particle coalescence is described properly. This fact can be illustrated by comparing Pekeris' famous 1959 calculation[16] of the helium ground-state energy using 1078 Hylleraas basis functions, which yielded a variational energy of

$$-2.903724375 \quad \text{a.u.},$$

with his less well-known 1966 calculation with Frankowski[6] using 246 basis functions which included powers of ℓns and negative powers of s, which yielded a variational energy of

$$-2.9037243770326 \quad \text{a.u.}$$

By including in their basis terms suggested by Fock's expansion,[10] Frankowski and Pekeris were able to reduce their basis dimension by a factor of 4 and increase the accuracy of their variational energy by a factor of over 1000.

Pursuing this track further, Freund, Huxtable, and I[7] took a closer look at precisely what combinations of negative powers of s and powers of ℓns were predicted by Fock's expansion, and by including them and rejecting others excluded by Fock's expansion we were able, using only 230 basis functions of the form

$$s^n t^\ell u^m (\ell n s)^j e^{-s/2}, \qquad (V.13)$$

where s, t, and u are the Hylleraas coordinates defined by (IV.2), to obtain a variational energy of

-2.90372437703407 a.u.,

which was below even Frankowski and Pekeris' extrapolated energy. (It now takes less than five minutes to do this calculation in quadruple precision on the University of Delaware's IBM 3090.) One can do even better by including basis functions of the form

$$s^n t^\ell u^m (\ell n s)^j \left\{ \begin{matrix} \cosh(ct) \\ \sinh(ct) \end{matrix} \right\} e^{-s/2} \qquad (V.14)$$

introduced by Frankowski[17] to describe "in-out" electron correlation; using a composite basis of 476 functions, of the forms (V.13) and (V.14) Jon Baker and I[18] have obtained a variational energy of

-2.903724377034117 a.u.

This calculation took a couple of hours on the IBM 3081D at the University of Delaware; most of the CPU time was spent in evaluating matrix elements of the basis functions. By comparison, the estimates of Carroll, Silverstone, and Metzger[5] would suggest that to achieve comparable accuracy in a CI expansion would require the use of ca. 20,000 terms in a Legendre expansion and ca. 7×10^{12} configurations.

We have also performed calculations on the single discrete bound state of H^-, for which we[18] obtain, using a composite basis of 476 functions, a variational energy of

-0.52775101654437 a.u.

This quest for accuracy at the 10^{-15} a.u. level is of course of only academic interest, but it is worth noting that recent two-photon measurements[19] of wavelengths of transitions between excited states of helium with an accuracy of a few parts in 10^9 has stimulated comparably accurate variational calculations of non-relativistic energies and relativistic and QED corrections. In this work it is necessary to evaluate the relativistic corrections with an accuracy of ca. one part in 10^5 or 10^6, so the usual hand-waving argument about errors in the expectation value of an operator which doesn't commute with the Hamiltonian going like the square root of the error in the variational energy would suggest that one needs to be able to calculate non-relativistic energies with an accuracy of ca. one part in 10^{10} or 10^{12}. In such calculations it is very desirable to take full advantage of what is known about the analytic structure of helium eigenfunctions.

I should also mention the recent resurgence of interest in muon catalyzed fusion.[20] A crucial theoretical issue involves the precise calculation of the relative probability of a muon becoming stuck to an alpha particle in an alpha-muon bound state $\psi_{\alpha\mu}$:

$$\frac{\left| \int d^3\vec{r}\, \psi^*_{\alpha\mu}(\vec{r})\, \psi_{dt\mu}(r,r,r_{dt}=0)\, e^{i\vec{q}\cdot\vec{r}} \right|^2}{\int d^3\vec{r}\, |\psi_{dt\mu}(r,r,r_{dt}=0)|^2} ,$$

where $\psi_{dt\mu}(r_{d\mu},r_{t\mu},r_{dt})$ is the ground state of the deuteron-triton-muon molecular ion, and $q = 6.0616$ muon atomic units is the momentum of the recoiling α-particle. Since $q \simeq 6$, an important contribution to the sticking-amplitude integral comes from the region where $r \simeq 1/6$, so one needs an accurate representation of $\psi_{dt\mu}(r_{d\mu},r_{t\mu},r_{dt})$ where all three variables are small. Hence the use of basis functions which explicitly include the three-particle cusp behavior can be expected to lead to rapid convergence of the sticking probabilities.

VI. BUILDING ANALYTIC STRUCTURE INTO BASIS FUNCTIONS - THE LITHIUM ATOM (AND OTHER FOUR-PARTICLE SYSTEMS)

On going from a three-particle to a four-particle system interacting by Coulomb forces, there is no increase in the complexity of the analytic structure of the wavefunctions; the Kato cusp condition[1] continues to describe all two-particle coalescences, the Fock expansion[10] describes all three-particle coalescences, and, because there are no terms in the Hamiltonian arising from multi-body forces, a generalization of the Fock expansion to multi-particle systems describes the four-particle coalescence.[21] However, if one tries to include logarithmic terms in the basis, it leads to intractable integrals. In fact, it is not at all straightforward to use basis functions which include all the desired terms to describe explicitly just the two-body coalescences. It was only recently that my colleague Robert N. Hill and his student David Fromm[22] evaluated for lithium the generating integral for Hylleraas-type functions:

$$I(\alpha_1,\alpha_2,\alpha_3,\alpha_{12},\alpha_{23},\alpha_{13})$$

$$= \int d^3\vec{r}_1 d^3\vec{r}_2 d^3\vec{r}_3 \frac{\exp(-\alpha_1 r_1-\alpha_2 r_2-\alpha_3 r_3-\alpha_{12} r_{12}-\alpha_{23} r_{23}-\alpha_{13} r_{13})}{r_1\, r_2\, r_3\, r_{12}\, r_{23}\, r_{13}} , \tag{VI.1}$$

in terms of dilogarithms. Fromm and Hill's result allows one to use for lithium basis functions of the form

$$r_1^{n_1} r_2^{n_2} r_3^{n_3} r_{12}^{n_{12}} r_{23}^{n_{23}} r_{13}^{n_{13}} \times$$
$$\times \exp(-\beta_1 r_1 - \beta_2 r_2 - \beta_3 r_3 - \beta_{12} r_{12} - \beta_{23} r_{23} - \beta_{13} r_{13}) \qquad \text{(VI.2)}$$

Previously either not all the powers were allowed to differ simultaneously from 0, or the integrals were evaluated numerically. Hill is collaborating with Y. K. Ho on a study of the positronium molecule Ps_2 using these basis functions, and other four-particle systems such as Li, Be^+, and H_2 can be treated with these functions. Unfortunately, the fact that the powers of r are not allowed to be negative, and the inability to include logarithmic terms, means that it will not be possible to include an explicit description of the three-particle cusp behavior; however, calculations to spectroscopic accuracy on these systems should be feasible. To obtain for helium a non-relativistic energy accurate to 2×10^{-9} a.u. Pekeris needed to use 1078 basis functions; i.e., about 10 basis functions for each of the three degrees of freedom (after separation of the Euler angles); hence to obtain comparable accuracy for the lithium atom, which has six degrees of freedom (after separation of the Euler angles), one might need to use ca. 10^6 basis functions. These days this might not be an impossible calculation on a vector supercomputer, provided that the basis could be cast into a quasi-orthogonal form to reduce linear dependency problems and CPU time.

Actually, there is a better way to do such calculations, and it brings me to a discussion of approximations by means of a *flexible* basis, whose members change as the dimension of the basis increases.

VII. FLEXIBLE BASES VS. FIXED BASES

Our discussion so far has focussed on approximation of functions by linear combinations of basis functions. In these cases, as one augments the basis with additional basis functions, the ones already present are left unchanged (except perhaps for an overall scaling factor, which can be used to ensure that the variational energy satisfies the virial theorem, or perhaps, as in the calculations by Frankowski[17] and by us[18] on the helium atom, a pair of scaling factors). In such calculations, the number of linear variational parameters can be rather large (typically hundreds or a thousand in the helium calculations, or, in CI calculations, even millions!), while the number of *non-linear* variational parameters is quite small (typically one or two in the helium calculations, or, in the case of SCF calculations, perhaps one ("single zeta") or two ("double zeta") per radial function). However, it can be extremely advantageous to use multiple non-linear variational parameters.

VII. 1. Rapid duplication of a cusp by Gaussians (with *floating* exponents)

In 1965 Huzinaga[23] studied the representation of a single Slater orbital by a linear combination of Gaussians with floating exponential parameters (hereafter called exponents). Specifically, the Slater-type orbital (STO) $r^{n-1}\exp(r/n)$ is the ground-state eigenfunction of the radial operator

$$H_n = -\frac{1}{2}\left(\frac{d^2}{dr^2} + \frac{2}{r}\frac{d}{dr} + \frac{n(n-1)}{r^2}\right) - \frac{1}{r} \qquad \text{(VII.1.1)}$$

with eigenvalue $-1/2n^2$. Huzinaga studied the minimisation of

$$\frac{\langle\Psi_N|H_n|\Psi_N\rangle}{\langle\Psi_N|\Psi_N\rangle}, \qquad \text{(VII.1.2)}$$

where

$$\Psi_N = \sum_{j=1}^{N} c_j^{(N)}\, r^m \exp(-\zeta_j^{(N)}\, r^2) \qquad \text{(VII.1.3)}$$

is a linear combination of Gaussians with floating exponents, which can be optimised so that the expectation value of H_n is minimised. Huzinaga found that the variational energies E_N converged quite rapidly as N was increased. In a study undertaken by Susan Haywood, James Nicholson, and me,[24] we found that for $6 \le N \le 10$ Huzinaga's optimization of the exponents was incomplete, and we extended the optimization up to N = 12. Our results for n = 1 (where H_n is the ordinary hydrogen atom Hamiltonian) are listed below in Table 1:

Table 1
Convergence of Variational Energies with Gaussians

N	E_N	ΔE_N
1	-.424413181	.075586819
2	-.485812717	.014187283
3	-.496979253	.003020747
4	-.499278406	.000721594
5	-.499809832	.000190168
6	-.499945570	.000054430
7	-.499983298	.000016702
8	-.499994561	.000005439
9	-.499998136	.000001864
10	-.499999332	.000000668
11	-.499999751	.000000249
12	-.499999904	.000000096

Evidently the errors ΔE_N rapidly decrease, and in fact, we found empirically that they asymptotically obey the relation

$$\Delta E_N \simeq C' \, N^{\delta'} \exp(-\alpha' N^{\beta}), \qquad \text{(VII.1.4)}$$

where $C' \simeq 6.23$, $\delta' \simeq 0.386$, $\alpha' \simeq 4.07$, and β is very close to if not equal to the 'golden mean'

$$\frac{-1 + \sqrt{5}}{2} = 0.618034... \qquad \text{(VII.1.5)}$$

This rate of convergence of course is slower than a linear exponential, but it is still very fast. One attains millihartree accuracy with only 4 Gaussians and microhartree accuracy with 10 Gaussians.

One can understand why the linear combinations of Gaussians with floating exponents are so efficient by examining the dependence of the optimal exponents on N. This is best done graphically, as in Fig. 2, a log-log plot of $\zeta_j^{(N)}$ versus N. One sees that the largest $\zeta_j^{(N)}$ increase rapidly as a function of N. This yields more and more sharply peaked Gaussians, linear combinations of which can very efficiently approximate the cusp at $r = 0$, as is illustrated schematically in Fig. 3. I may add that Susan Haywood and I[24] discovered that the optimal $\zeta_j^{(N)}$ are asymptotically approximated very well by the relation

$$\zeta_j^{(N)} \simeq c \; j^{-\delta} \exp(\alpha(N^{\beta} - j^{\beta})), \qquad \text{(VII.1.6)}$$

where $c \simeq 0.35$, $\delta \simeq 0.71$, $\alpha \simeq 2.74$, and β is again the 'golden mean'. This sort of relation should prove useful in easily generating nearly optimal sets of Gaussian exponents for the approximation of Slater functions.

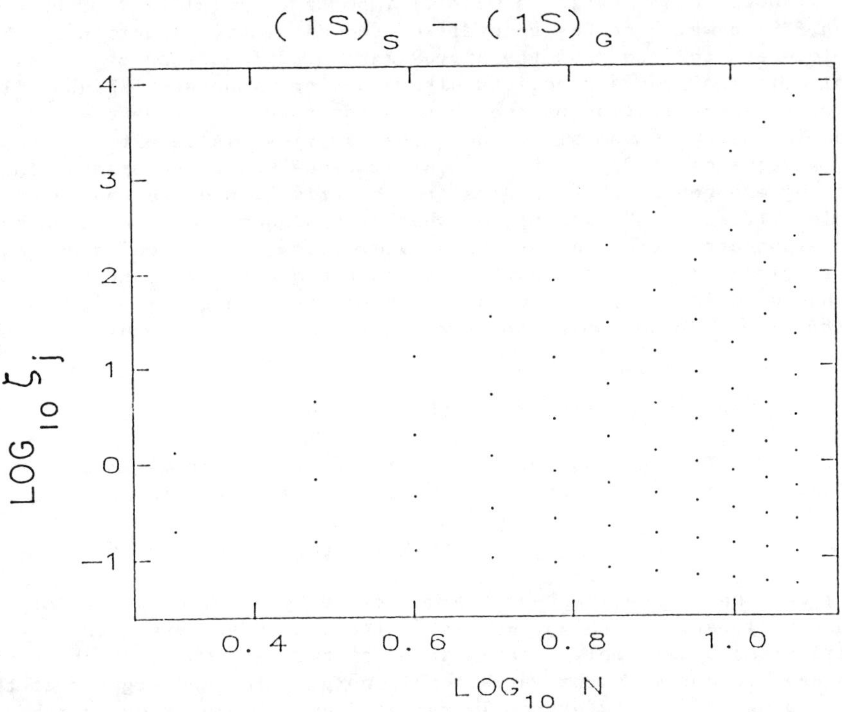

Fig. 2. Log-log plot of the optimal Gaussian exponents versus N.

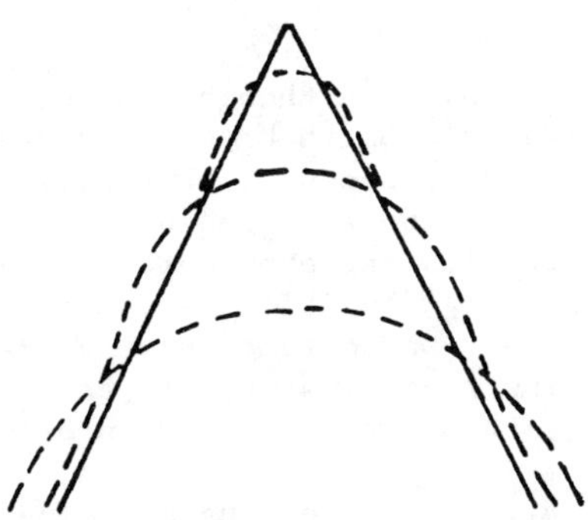

Fig. 3. The approximation of an electron-nucleus cusp by a superposition of sharply peaked Gaussians.

I should also mention a pair of important articles by Bruno Klahn,[24a] in which he has generalized the mathematical notion of completeness to encompass the approximation of functions by linear combinations of basis functions with floating exponents. Klahn first proves a generalization of the Müntz-Szász theorem on the completeness of basis sets of monomials, and he then applies his result to discuss the convergence of Rayleigh-Ritz calculations using Gaussians with floating exponents. His results are non-trivial because they provide a simple criterion for determining whether a sequence of truncated bases with exponents optimised for one problem (e.g., SCF calculations on atoms) yields convergent energies if this sequence of truncated bases is then used in a calculation on another system (e.g., a molecule), *without re-optimisation of the exponents,* as is usually the case in practice.

VII. 2.1 Correlated Slaters (Thakkar and Smith)

In 1977 Thakkar and Smith[25] published a study of various states of the helium atom and helium-like ions using basis functions of the form

$$\psi_k(r_1,r_2,r_{12}) = \exp(-\alpha_k r_1-\beta_k r_2-\gamma_k r_{12}) \qquad \text{(VII.2.1)}$$

where each basis function was characterized by a *different* set of exponential parameters (α_k, β_k, γ_k). These in turn were each parametrised by a simple rule in terms of two variables, which were then used to optimise the variational energy. The convergence of this method was found to be extremely rapid; e.g., with only 66 such basis functions (66 - 1 = 65 linear variational parameters and 2 x 3 = 6 non-linear variational parameters) Thakkar and Smith obtained for helium a variational energy of

$$-2.903724363 \text{ a.u.},$$

which is accurate to 1.4×10^{-8} a.u. Although none of the individual basis functions used by Thakkar and Smith have the correct analytic structure at the three-particle coalescence, they still obtained very rapid convergence for essentially the same reason that finite linear combinations of Gaussians with floating exponents rapidly duplicate the electron-nucleus cusp: by keeping flexible the length scale associated with some of the basis functions, those functions can rapidly become sharply peaked, and finite linear combinations of them can efficiently mimic a singularity in the exact function being approximated.

VII.2.2 Hylleras functions with multiple length scales (Drake)

At the 1987 SASHIA meeting, Gordon Drake[26] presented some impressive results on various states of helium using a composite basis of Hylleraas functions with multiple length scales, which for a given number of basis functions he optimises to minimise the variational energy. He obtains very fast convergence, as illustrated by his results on the ground state of helium quoted in Table 2:

Table 2

Convergence of Drake's Variational Energies

N	E_N
44	-2.9037241310211
68	-2.9037243515899
100	-2.9037243739261
140	-2.9037243766000
190	-2.9037243769652
250	-2.903724377020902
322	-2.903724377031537
406	-2.903724377033508
504	-2.903724377033960
616	-2.903724377034073
extrapolated	-2.903724377034105(28)
Baker & Morgan[18]	-2.903724377034117

Although the rate of convergence of Drake's calculation of course is not as fast as ours, because Drake does not include any of the negative powers of s or powers of ℓns to describe explicitly the analytic structure of the wavefunction at the three-particle coalescence, his use of multiple length scale factors nonetheless yields a quite rapid rate of convergence - certainly much faster than that of a Hylleraas-Pekeris basis with a single exponent.

If one ponders Drake's impressive result, it suggests that the right way to use Hylleraas-type functions for lithium, Be^+, Ps_2, etc., is not to use huge numbers of powers with a single set of fixed exponents, but instead to use multiple exponents, as has been done in a recent study of the weakly bound state of the $(td\mu)^+$ molecular ion.[26a] Since with 140 basis functions Drake achieves a helium energy accurate to 4×10^{-9} a.u., it would be reasonable to expect that comparable accuracy could be achieved for lithium by using roughly $(140)^2 \simeq 20{,}000$ basis functions, which is well within the capability of modern supercomputers.

It is possible that Hylleraas-type basis functions for many-electron systems could be handled by the evaluation of the corresponding generating integral (perhaps in terms of polylogarithms?). Certainly this possibility should be explored, but in the meantime, the only correlated basis functions for many-electron systems for which it currently is practical to evaluate all the integrals (besides the approach pioneered by Kutzelnigg[7a]) are explicitly correlated Gaussians.

VII.3. Explicitly Correlated Gaussians

In a famous article published in 1950, S. F. Boys[27] demonstrated that multi-center Coulomb integrals involving products of one-electron Gaussians are evaluable in terms of the error function. It is much less well known that a decade later in 1960 Boys[28] showed that multi-center Coulomb integrals involving *explicitly correlated Gaussians*

$$P(\vec{r}_1,\vec{r}_2,\ldots) \exp\left(- \sum_{i} \alpha_i r_{iA_i}^2 - \sum_{i \neq j} \gamma_{ij} r_{ij}^2 \right), \qquad \text{(VII.3.1)}$$

where P is a polynomial in Cartesian coordinates and

$$\vec{r}_{iA_i} = \vec{r}_i - \vec{A}_i \qquad \text{(VII.3.2)}$$

is the displacement of the ith electron coordinate from the center of the Gaussian at $\vec{A}_i$, also are evaluable in closed form in terms of the error function. The only catch is that one must perform multiple inversions of matrices whose off-diagonal elements are essentially the sums of the γ_{ij}'s for each pair of basis functions. For a system with a moderate number of electrons, this is not a major obstacle, since multiple inversions of small matrices can be performed very efficiently on a vector computer.

Just as ordinary Gaussians $\exp(-\zeta r^2)$ with floating exponents can rapidly duplicate the electron-nucleus cusp in e^{-r}, explicitly correlated Gaussians with floating sets of exponents α_i and γ_{ij} can duplicate rapidly all the electron-nucleus and electron-electron cusps in the exact multi-electron atomic or molecular wavefunctions at $r_i = 0$ and $r_{ij} = 0$. For simplicity, consider Regier and Thakkar's variational calculation on the ground state of helium using 16 explicitly correlated Gaussians.[29] The parameters in their 16 basis functions are listed in the Table 3. As one can see, some of the exponents α_1 and α_2 associated with the electron-nucleus distances r_1 and r_2 are quite large, thereby rapidly simulating the electron-nucleus cusps. Similarly, some of the exponents γ_{12} associated with the inter-electronic distance r_{12} are moderately large and positive and these basis functions are added in with *negative* linear coefficients C, thereby simulating the inverted electron-electron cusp by a *difference* of a smooth positive function and a sharply-peaked positive function, as illustrated in Fig. 4:

Fig. 4. Approximation of an inverted cusp by a difference of smooth Gaussians and sharply peaked Gaussians.

Table 3

Parameters in Regier and Thakkar's 16-term explicitly correlated Gaussian wavefunction for 1^1S helium:

α_1	α_2	γ_{12}	C
1.58295	1.58295	10.56617	-0.17518
656.89094	0.49496	-0.01973	0.19797
133.62311	1.73072	-0.00729	0.50764
29.89083	7.41791	-0.82519	0.63410
75.36497	0.28690	-0.00477	0.20076
1.67914	0.73883	1.46900	-0.20558
21.55071	0.80154	-0.01285	0.69011
6.60058	2.37524	-0.01447	0.99657
14.47331	0.16200	-0.00034	0.11175
5.64180	0.39838	-0.00073	0.54971
2.16889	0.91062	-0.03516	0.89688
0.67154	0.36221	0.33750	-0.09263
2.14749	0.16274	0.00081	0.16765
0.93226	0.40535	-0.01512	0.39929
0.64677	0.24682	-0.04497	-0.02949
0.51760	0.19326	-0.01849	0.07834

Because all cusps are simulated rapidly (albeit with analytic functions), the variational energy obtained with these 16 explicitly correlated Gaussian basis functions is quite accurate: it is -2.9034 a.u., which lies only 3×10^{-4} a.u. above the exact value -2.903724... a.u. By comparison, one would need to use about 40 configurations with the correct electron-nucleus cusp behavior to achieve comparable accuracy with a CI expansion for the helium wavefunction.[5]

The above comparison pitted explicitly correlated Gaussians against a CI expansion where the one-electron functions had the correct electron-nucleus cusp behavior, and the explicitly correlated Gaussians won handily despite having to overcome the handicap of no individual basis function having the correct electron-nucleus cusp behavior. If we now arrange a more even contest in which explicitly correlated Gaussians are matched against a CI expansion where the one-electron radial functions are Gaussians, the results are even more impressive.

In the late 1970's there was a flurry of interest in the H_3^+ molecular ion. A 1296-term CI expansion was found to give at the equilibrium geometry an energy of -1.342784 a.u.[30] The authors claimed that "the lowest energy previously reported for H_3^+ was 1.33764, by Schwartz and Schaad",[31] who in 1967 had used a 10-term CI expansion. However, the authors seem to have overlooked the remarkable 1973 article by Lydia Salmon and Ronald D. Poshusta,[32] who with only 15 explicitly correlated Gaussians (with 135 non-linear parameters, 3×15 exponents and 2×3×15 centers) obtained a variational upper bound of -1.34335 a.u., some 0.00057 a.u. *below* the result of ref. 30. Thus a 15-term expansion in explicitly correlated Gaussians with floating exponents does better for H_3^+ than a 1296-term CI expansion! And one can do better still. Recently R. Roeland, S. Alexander, K. Szalewicz, and H. J. Monkhorst[33] have obtained with 375 explicitly correlated Gaussians a variational upper bound of -1.343771 a.u., which may be compared with the estimate -1.34376(3) a.u. obtained using the Monte Carlo method by J. B. Anderson.[34] The more rapid convergence of the variational energy using explicitly correlated Gaussians is directly attributable to the fact that the floating correlation exponents γ_{ij} allow for a much more efficient representation of the electron-electron cusp in the exact wavefunction.

Besides being used in variational calculations on two-electron systems, such as the work on He and H_3^+ cited above and the work on H_2 in Ref. 35, explicitly correlated Gaussians have been used to include a description of correlation in many-body perturbation theory (MBPT) treatments of small atoms and molecules, such as Be, LiH, the helium dimer, and the neon atom.[36] Since the convergence of the 2nd and higher-order energy corrections is of course dependent on the quality of the approximation of the first-order correction to the wavefunction, which has cusps at electron-electron coalescences, it is not surprising that explicitly correlated Gaussians are quite effective in this situation.

VII.4. Tempering of Exponents

A major headache with using explicit correlated Gaussians with floating exponents, or, more generally, any basis expansion with multiple non-linear parameters, is trying to optimise them. As is well known, the minimisation of a non-quadratic function of several variables is fraught with difficulties, such as getting trapped in local minima, or wandering aimlessly on a nearly flat surface, or navigating through saddle-like regions. Hence it is desirable to find ways of reducing the number of independent non-linear parameters so that their optimisation is computationally feasible.

One approach, first set down in 1963 by Reeves and Harrison[37] and then rediscovered and expounded at length by Ruedenberg and his coworkers,[38] is that of even-tempered exponents, where for a given number N of one-electron Gaussians, the exponents $\zeta_j^{(N)}$, indexed in order of increasing size, are parametrised by

$$\zeta_j^{(N)} = \alpha_N(\beta_N)^j, \qquad \text{(VII.4.1)}$$

where α_N and β_N are N-dependent constants which obey

$$\lim_{N\to\infty} \alpha_N = 0.$$

$$\lim_{N\to\infty} \beta_N = 1. \qquad \text{(VII.4.2)}$$

It was found by Susan Haywood and me[24] that this scheme, in which for a given N the ratio of successive exponents is independent of j and dependent on N:

$$\zeta_{j+1}^{(N)}/\zeta_j^{(N)} = \beta_N, \qquad \text{(VII.4.3)}$$

is capable of improvement. We found, at least for the case of the approximation of Slater-type orbitals by linear combinations of Gaussians, that the ratio

$$\zeta_{j+1}^{(N)}/\zeta_j^{(N)} \qquad \text{(VII.4.4)}$$

of the optimal $\zeta_j^{(N)}$'s is actually independent of N and dependent on j, provided that the exponents are indexed in order of *decreasing* size. I had already noticed that the smallest exponents $\zeta_N^{(N)}$ obeyed $\zeta_N^{(N)} \simeq cN^{-\delta}$ with constants c and δ independent of N, and some inspired plotting of $\ell n\, \zeta_j^{(N)}$ versus powers of N by Susan Haywood led to our discovery of the asymptotic relation

$$\zeta_j^{(N)} \simeq cj^{-\delta} \exp(\alpha(N^\beta - j^\beta)) \qquad \text{(VII.4.5)}$$

with $\beta \simeq 0.618$ and the constants α, δ, and c all independent of N and j. Hence it is possible to optimise these 3 constants with a relatively small number of basis functions and then to generate much larger sets of Gaussian basis functions with nearly optimal sets of exponents.

It clearly would be desirable to extend these results to the approximation by Gaussians of more general radial functions, and if possible, to the approximation by explicitly correlated Gaussians of atomic and molecular wavefunctions. However, it probably will require much effort to reach these goals. In the meantime, other non-optimal but rapidly convergent tempering schemes are being found useful.

One such scheme is that of random tempered Gaussians.[39] In this approach, the sets of exponents of explicitly correlated Gaussians are distributed at random within a multi-dimensional box whose sides are varied so as to minimize the variational energy. This method, which had previously been applied to the generation of sets of exponents for correlated Slater-type basis functions,[25,40] has been found extremely useful.

SUMMARY

I have illustrated how the rate of convergence of a variational calculation is determined by the interplay between the analytic structure of the exact wavefunction and that of the basis functions being used to approximate it. I should like to conclude by offering the following summary on performing highly accurate variational calculations on atoms and molecules:

(i) First, one should ensure that one's basis has many elements which are concentrated in the region where the exact wavefunction is concentrated. E.g., if one wants to describe a highly excited Rydberg state of helium, the single configuration function

$$\frac{1}{\sqrt{2}}\,(1s\;n\ell \pm n\ell\;1s),$$

which does not have the correct cusp behavior at $r_{12} = 0$, will be much more accurate than a thousand term Pekeris expansion, which, although it has the correct cusp behavior at all two-particle coalescences, is insufficiently concentrated in those regions of configuration space where "in-out" correlation is occurring on a length scale of hundreds or thousands of Ångstroms.

(ii) Provided that (i) is satisfied, one should be concerned with efficiently describing the electron-nucleus cusp behavior, either by including it explicitly with, say, Slater-type functions, or by approximating it with a linear combination of several sharply peaked Gaussians.

(iii) Provided that (i) and (ii) are satisfied, one should then be concerned with describing the electron-electron cusp behavior, either by including it explicitly with Hylleraas-type basis functions, or by approximating it with a linear combination of several sharply peaked explicitly correlated Gaussians.

(iv) Provided that (i), (ii), and (iii) are satisfied, one's next concern is the analytic structure of the wavefunction at three-particle coalescences. At present, this can be explicitly put into basis functions only for systems with three particles. If one is studying a system with four or more particles, one should use basis functions concentrated at short length scales to mimic the analytic structure at the three-particle coalescences.

(v) Finally, provided that (i), (ii), (iii), and (iv) are satisfied, one should be concerned with describing the asymptotic behavior of the exact wavefunction.

In the case of an SCF calculation, one can of course skip over concerns (iii) and (iv) and proceed directly to (v). This may explain why the use of "asymptotic exponents" in Slater-type functions has been found particularly effective in SCF calculations on small atoms.[9,41] And of course, provided that the SCF equations involve only a single or at most a few variables, they can be solved numerically on a mesh with very high precision. But no matter how accurate an SCF calculation may be, attempts to obtain the physically and chemically relevant correlation energy via a CI-expansion will be plagued by the rather slow $(L + 1)^{-3}$ convergence due to the inefficient representation of the electron-electron cusps by truncated Legendre expansions.

Of course, for sufficiently small molecules, this may not preclude the attainment of chemical accuracy. Elsewhere in this volume Ludwik Adamowicz has presented some impressive calculations performed with R. J. Bartlett[42] using the McCullough method, which involves a mixed basis-set and numerical solution in elliptic coordinates,[43] on diatomic molecules such as LiH and HF. By using up to ϕ-functions, Adamowicz and Bartlett obtained correlation energies for LiH accurate to ca. 1 millihartree and for HF accurate to ca. 5 millihartrees. I should guess that perhaps half of these residual errors are due to the slow representation of the electron-electron cusps by the Legendre expansions.[44] Furthermore, as Reinhart Ahlrichs and Werner Kutzelnigg remarked to me, the McCullough method will converge rather slowly if the nuclear charges and/or the internuclear separation R become too large, essentially because functions like

$$e^{-Z_a r_a} = e^{-\frac{1}{2}Z_a R(\xi+\eta)}$$

and

$$e^{-Z_b r_b} = e^{-\frac{1}{2}Z_b R(\xi-\eta)},$$

which are sharply peaked near $\eta = \pm 1$ on an ever shorter scale as $Z_a R$ and $Z_b R$ increase, are ever more slowly represented by an expansion in $Y_\ell^m(\eta)$, which have no flexible scaling parameter. Hence for a given level of accuracy, it would require much longer partial wave expansions to use the McCullough method to study potential energy curves for large R (as in dissociation limits, or in long-range van der Waals interactions) or diatomic molecules with at least one nucleus with large Z.[45] Moreover, the McCullough method cannot readily be generalized in a useful way to treat polyatomic molecules.

Since a direct numerical solution of the Schrödinger Equation in a large number of unseparable variables is at present unfeasible and seems likely to remain so into the foreseeable future, it is my belief that for the calculation by generally applicable methods of highly accurate correlation energies of molecules, two approaches should be pursued: (i) the inclusion of explicit r_{ij} terms, as advocated by Kutzelnigg,[7a] and (ii) the use of explicitly correlated Gaussions, either in MBPT or directly in variational calculations.

ACKNOWLEDGEMENTS

I should first like to express my indebtedness to various collaborators over the course of the last decade: Bruno Klahn, Bob Hill, Bart Huxtable, Dave Freund, Jim Nicholson, Jon Baker, and Sue Haywood, all of whom have contributed to the results presented above. Financial support has been provided by a Chaim Weizmann Postdoctoral Fellowship and a National Science Foundation National Needs Postdoctoral Fellowship while I was at Princeton, and by grants from the University of Delaware Research Foundation, the Research Corporation, and the National Science Foundation (PHY-8317085 and 8608155) since my coming to Delaware. The University of Delaware has generously provided extensive amounts of computer time on a succession of mainframes. I should like to thank Bob Hill and Krzysztof Szalewicz for commenting on an earlier draft of this article, and I am particularly grateful to the participants in this NATO ARW for a number of helpful remarks.

REFERENCES

1. T. Kato, *Commun. Pure Appl. Math.* **10**, 151 (1957).
2. C. Schwartz, *Meth. Comp. Phys.* **2**, 241 (1963).
2a. B. Klahn and W. A. Bingel, *Theoret. Chim. Acta* **44**, 9, 27 (1977).
3. B. Klahn and J. D. Morgan III, *J. Chem. Phys.* **81**, 410 (1984).
4. R. N. Hill, *J. Chem. Phys.* **83**, 1173 (1985).
5. D. P. Carroll, H. J. Silverstone, and R. M. Metzger, *J. Chem. Phys.* **71**, 4142 (1979).
6. K. Frankowski and C. L. Pekeris, *Phys. Rev.* **146**, 46 (1966); **150**, 366 (1966) (E).
7. D. E. Freund, B. D. Huxtable, and J. D. Morgan III, *Phys. Rev. A* **29**, 980 (1984).
7a. W. Kutzelnigg, *Theoret. Chim. Acta* **68**, 445 (1985).
7b. J. D. Morgan III, *J. Phys. A: Math. Gen.* **10**, L91 (1977); P. Deift, W. Hunziker, B. Simon, and E. Vock, *Commun. Math. Phys.* **64**, 1 (1978); R. Ahlrichs, M. Hoffmann-Ostenhof, T. Hoffmann-Ostenhof, and J. D. Morgan III, *Phys. Rev. A* **23**, 2106 (1981); J. M. Combes, M. Hoffmann-Ostenhof,and T. Hoffmann-Ostenhof, *J. Math. Phys.* **22**, 1299 (1981).
8. L. D. Landau and E.M. Lifshitz, *Quantum Mechanics, Non-relativistic Theory* (Pergamon, Oxford 1965), pp. 72-73.
9. C. L. Davis, H.-J. A. Jensen, and H. J. Monkhorst, *J. Chem. Phys.* **80**, 840 (1984).

9a. W. M. Huo and E. N. Lassettre, *J. Chem. Phys.* **70**, 3468 (1980).
10. V. A. Fock, *Izv. Akad. Nauk SSSR, Ser. Fiz.* **18**, 161 (1954); English translation: *D. Kngl. Norske Videnskab. Selsk. Forh.* **31**, 138, 145 (1958).
11. J. H. Macek, *Phys. Rev.* **160**, 170 (1967).
12. J. D. Morgan III, *Theoret. Chim. Acta* **69**, 181 (1986).
13. J. Leray, *Actes du 6ème Congres du Groupement des Mathematiciens d'Expression Latine,* pp. 179-82, (Gauthier-Villars: Paris, 1982); *Methods of functional analysis and theory of elliptic operators* (Naples, 1982); *Lecture Notes in Physics* **195** (Springer, Berlin, 1985), pp. 235-247.
14. R. N. Hill and J. D. Morgan III, manuscript in preparation.
15. J. E. Gottschalk, P. C. Abbott, and E. N. Maslen, *J. Phys. A: Math. Gen.* **20**, 2077 (1987).
16. C. L. Pekeris, *Phys. Rev.* **115**, 1216 (1959).
17. K. Frankowski, *Phys. Rev.* **160**, 1 (1967).
18. J. D. Baker and J. D. Morgan III, manuscript in preparation.
19. L. Hlousek, S. A. Lee, and W. M. Fairbank, Jr., *Phys. Rev. Lett.* **50**, 328 (1983); P. Juncar, H. G. Berry, R. Damaschini, and H. T. Duong, *J. Phys. B: At. Mol. Phys.* **16**, 381 (1983).
20. For example, see the new journal *Muon Catalyzed Fusion* devoted to this topic.
21. Yu. N. Demkov and A. M. Ermolaev, *Zh. Eksper. Teor. Fiz.* **36**, 896 (1959); English translation: *Sov. Phys. JETP* **36**, 633 (1959).
22. D. M. Fromm and R. N. Hill, *Phys. Rev. A* **36**, 1013 (1987).
23. S. Huzinaga, *J. Chem. Phys.* **42**, 1293 (1963).
24. S. Haywood, J. D. Nicholson, and J. D. Morgan III, submitted to *J. Chem. Phys.*
24a. B. Klahn, *J. Chem. Phys.* **83**, 5749, 5754 (1985).
25. A. J. Thakkar and V. H. Smith, Jr., *Phys. Rev. A* **15**, 1, 16 (1977).
26. G. W. F. Drake, to appear in proceedings of the 1987 SASHIA conference.
26a. K. Szalewicz, H. J. Monkhorst, W. Kolos, and A. Scrinzi, *Phys. Rev. A* **36**, 5494 (1987).
27. S. F. Boys, *Proc. Royal Soc. (London) A* **200**, 542 (1950).
28. S. F. Boys, *Proc. Royal Soc. (London) A* **258**, 402 (1960).
29. P. E. Regier and A. J. Thakkar, *Phys. Rev. A* **30**, 30 (1984).
30. C. E. Dykstra, A. S. Gaylord, W. D. Gwinn, W. C. Swope, and H. F. Schaefer III, *J. Chem. Phys.* **68**, 3951 (1978); C. E. Dykstra and W. C. Swope, *J. Chem. Phys.* **70**, 1 (1979).
31. M. E. Schwartz and L. J. Schaad, *J. Chem. Phys.* **47**, 5325 (1967).
32. L. Salmon and R. D. Poshusta, *J. Chem. Phys.* **59**, 3497 (1973).
33. R. Roeland, S. Alexander, K. Szalewicz, and H. J. Monkhorst, presentation at March 1988 Sanibel Symposium.
34. J. B. Anderson, *J. Chem. Phys.* **86**, 2839 (1987).
35. B. Jeziorski and K. Szalewicz, *Phys. Rev. A* **19**, 2360 (1979); W. Kolos, H. J. Monkhorst, and K. Szalewicz, *J. Chem. Phys.* **77**, 1323, 1335 (1982).

36. L. Adamowicz and A. J. Sadlej, *J. Chem. Phys.* **67**, 4298 (1977); K. Szalewicz, B. Jeziorski, H. J. Monkhorst, and J. G. Zabolitzky, *J. Chem. Phys.* **78**, 1420 (1983); K. Szalewicz, B. Jeziorski, H. J. Monkhorst, and J. G. Zabolitzky, *J. Chem. Phys.* **79**, 5543 (1983); B. Jeziorski, H. J. Monkhorst, K. Szalewicz, and J. G. Zabolitzky, *J. Chem. Phys.* **81**, 368 (1984); K. Szalewicz, J. G. Zabolitzky, B. Jeziorski, and H. J. Monkhorst, *J. Chem. Phys.* **81**, 2723 (1984); K. B. Wenzel, J. G. Zabolitzky, K. Szalewicz, B. Jeziorski, and H. J. Monkhorst, *J. Chem. Phys.* **85**, 3964 (1986).

37. C. M. Reeves and M. C. Harrison, *J. Chem. Phys.* **39**, 11 (1963).

38. R. D. Bardo and K. Ruedenberg, *J. Chem. Phys.* **59**, 5956, 5966 (1973); R. C. Raffenetti and K. Ruedenberg, *ibid.*, **59**, 5978 (1973); R. D. Bardo and K. Ruedenberg, *ibid.*, **60**, 918, 932 (1974); D. F. Feller and K. Ruedenberg, *Theoret. Chim. Acta* **52**, 231 (1979); M. Schmidt and K. Ruedenberg, *J. Chem. Phys.* **71**, 3951 (1979).

39. S. A. Alexander, H. J. Monkhorst, and K. Szalewicz, *J. Chem. Phys.* **85**, 5821 (1986); **87**, 3976 (1987).

40. A. M. Frolov and V. D. Efros, *Pisma Zh. Eksp. Teor. Fiz.* **39**, 449 (1984); English translation: *JETP Letters* **39**, 544 (1984); *J. Phys. B: At. Mol. Phys.* **18**, L265 (1985); A. M. Frolov, *Yad. Fiz.* **44**, 589 (1986); English translation: *Sov. J. Nucl. Phys.* **44**, 380 (1986); S. A. Alexander and H. J. Monkhorst, *Phys. Rev. A*, in press (1988).

41. C. L. Davis and H. J. Monkhorst, *J. Chem. Phys.* **81**, 5845 (1984).

42. L. Adamowicz and R. J. Bartlett, in *Applied Quantum Chemistry*, edited by V. H. Smith, Jr., H. F. Schaefer III, and K. Morokuma (Reidel, Dordrecht, 1986), pp. 111-133.

43. E. A. McCullough, Jr., *Chem. Phys. Lett.* **24**, 55 (1974); *J. Chem. Phys.* **62**, 3991 (1975); *Computer Phys. Reports* **4**, 265 (1986).

44. E.g., for LiH, truncation at f-functions of the Legendre expansion for the Li^+ core, which is isoelectronic to He, should yield an error of ca. 0.4 millihartrees, and additional contributions to the error will come from valence-valence and valence-core electron-electron cusps.

45. There are recent calculations using up to 40 partial waves by E. A. McCullough, Jr., *Computer Phys. Reports* **4**, 307 (1986), on KO ($Z_a = 19$, $Z_b = 8$) at R = 4.4 atomic units, and by K. W. Richman and E. A. McCullough, Jr., *J. Chem. Phys.* **87**, 5050 (1987), on the chromium dimer (Z = 24) at R = 2.6, 2.8, 3.0, and 3.2 atomic units, i.e., with $\frac{1}{2}Z_{max} R_{max} \lesssim 40$. A system like CsI, with R ≃ 6.5 atomic units and $Z_{max} = 55$ (hence $\frac{1}{2}Z_{max} R \simeq 180$) would pose quite a challenge to this method!

STOCHASTIC METHODS IN QUANTUM MECHANICS

Michel Caffarel
Dynamique des Interactions Moléculaires
Université P. et M. Curie, Paris VI, Tour 22
4 Place Jussieu 75252 Paris Cedex 05, France

ABSTRACT. This paper discusses different ways of expressing quantum properties as stochastic averages. It is explained how such a formal mathematical connection between quantum mechanics and stochastic processes may be used to compute quantum observables (Monte Carlo simulation). The problems in applying stochastic methods to fermion systems are discussed and a few illustrative results for atoms and molecules are presented.

1. INTRODUCTION

In the present paper we are concerned with the treatment of the non-relativistic Schrödinger equation using stochastic (or Monte Carlo) methods. The possibility of solving the Schrödinger equation using such techniques may be easily understood when noticing the similarity between the time-dependent Schrödinger equation in imaginary time and a generalized diffusion equation. Most of the Monte Carlo methods developed so far rest on this remark and then make use of a diffusion-like stochastic process in configuration space. However, stochastic methods are more general and, as an example, stochastic processes such as Poisson processes may be used (see below). The main object of this paper is to emphasize on the possibility of expressing quantum properties as stochastic averages. A key point is that the underlying stochastic process from which averages are taken may be chosen in many different ways. To illustrate this idea we present a number of methods based on various stochastic processes. The methods presented here are the so-called variational, diffusion and pure diffusion Monte Carlo methods and a method based on the use of a generalized Poisson process. Other approaches have been developed (1-6) but there will not be discussed here for reasons of space. An important example is the so-called Green's function Monte Carlo method (1-4) which has led to some important results. Once the mathematical connection between quantum and stochastic mean values has been performed, the numerical evaluation of the stochastic averages may be achieved by using standard statistical tools. An important point associated with the practical implementation of a Monte Carlo method is that the efficiency of calculations (i.e. the accuracy obtained from a given

M. Defranceschi and J. Delhalle (eds.), Numerical Determination of the Electronic Structure of Atoms, Diatomic and Polyatomic Molecules, 85–105.

amount of computer time) is strongly dependent on the stochastic process used and on the property evaluated. To be able to exhibit a large number of different stochastic processes for evaluating a given quantum property is therefore important. In this paper the attention is mainly focused on the theoretical part of stochastic methods; the main features of the practical implementation are only outlined, for a detailed presentation the reader is referred to the original works. A few illustrative results for atoms and molecules are presented. The problems related to the extension of such results to more complex systems are not discussed here (for such a discussion see e.g. Refs. 7,8). Finally, in section 3 the problem of imposing the constraints resulting from the Pauli principle for fermion systems is discussed in the framework of Monte Carlo methods.

2. METHODS

Here we are concerned with quantum systems which can be described by a Hamiltonian written in the form

$$H = - \frac{1}{2} \vec{\nabla}^2 . + V(\vec{x}) \tag{1}$$

where $\vec{x}$ refers to the 3N particle coordinates. Atomic units are used throughout the paper.

2.1. Variational Monte Carlo method

The purpose of the method presented here is to evaluate the total energy associated with a given trial wave function ψ by using a stochastic procedure. This procedure represents an alternative approach to the standard method based on the expansion of ψ in terms of products of simple one-particle functions (gaussian in general) suitable for an easy evaluation of the many-electron integrals. Unfortunately, such an approach prevents using sophisticated forms for ψ . In contrast, in the variational Monte Carlo method no expansion is performed and a rather arbitrary form for ψ may be used without practical difficulties. In particular, it is possible to employ high-quality trial wave functions which explicitly contain the interelectron coordinates r_{ij} to properly describe the electron-electron interaction. The method presented here seems to have been first applied by McMillan (9) for boson systems and by Ceperley et al. (10) for fermion systems.

The variational energy is defined as

$$E(\psi) = \frac{\int \psi H \psi \, d\vec{x}}{\int \psi^2 \, d\vec{x}} \tag{2}$$

and may be rewritten as follows

$$E(\psi) = \int E_L(\vec{x})p(\vec{x})\, d\vec{x} \tag{3}$$

where the so-called local energy E_L (depending on $\vec{x}$) is written

$$E_L(\vec{x}) = \frac{H\psi}{\psi} \tag{4}$$

and $p(\vec{x})$ is a positive function given by

$$p(\vec{x}) = \frac{\psi^2}{\int \psi^2\, d\vec{x}} \tag{5}$$

It should be noted here that only real-valued trial wave functions are used in this paper since H (Eq. (1)) is real.

Now, the method consists in finding a stochastic process admitting $p(\vec{x})$ as stationary density. Such a process is by no means unique, two examples are then presented in the following. Once the process has been determined, the energy mean value may be rewritten as

$$E(\psi) = \langle E_L \rangle \tag{6}$$

where $\langle \ldots \rangle$ stands for the expectation with respect to the stationary density of the process.

This is a first (trivial) example of a quantum property (the energy mean value) expressed as a stochastic average. As concerns the stochastic process, a first possibility is to use the Markov chain in configuration space resulting from the famous algorithm of Metropolis, Rosenbluth, Rosenbluth, Teller, and Teller (11). An essential property of this algorithm based on an acceptation/rejection procedure is that the only quantity needed in the simulation is the ratio $p(\vec{x})/p(\vec{x}')$ (where $\vec{x}$ and $\vec{x}'$ are two arbitrary points in configuration space), a quantity which is independent of the unknown normalization term. For a presentation of this algorithm widely used in many fields of physics, the reader is referred to Refs. (11,12). A second possibility consists in defining a continuous diffusion process in configuration space. Diffusion processes are characterized by the set of their n-time probability densities which obey the following general Fokker-Planck equation: (13)

$$\frac{\partial W}{\partial t} = L\, W \tag{7}$$

where L, the Fokker-Planck operator, is given by

$$L = \sum_{i,j} \frac{\partial^2}{\partial x_i \partial x_j} (D_{ij}(\vec{x}).) - \sum_i \frac{\partial}{\partial x_i} (b_i(\vec{x}).) \tag{8}$$

b_i is the so-called drift vector and D_{ij} the so-called diffusion matrix. The stationary density $p(\vec{x})$ of the process is defined by

$$\frac{\partial p}{\partial t} = 0 \tag{9.a}$$

or, equivalently

$$Lp = 0 \tag{9.b}$$

Now, choosing

$$D_{ij} = \frac{1}{2} \delta_{ij} \tag{10}$$

the drift vector corresponding to the stationary density p is easily obtained, we find

$$b_i = \frac{1}{2} \frac{\partial}{\partial x_i} \operatorname{Log} p = \frac{1}{2} \frac{\partial p/\partial x_i}{p} \tag{11}$$

According to Eqs. (9-11), it is then always possible to construct a diffusion process admitting a given arbitrary positive function as stationary density. Now, the problem of evaluating the stochastic average given by Eq. (6) arises. For that, the energy mean value is rewritten as

$$E(\psi) \cong \frac{1}{N} \sum_{i=1}^{N} E_L(\vec{x}^{(i)}) \tag{12}$$

where N is large enough, the equality being valid only as N goes to infinity. $\{\vec{x}^{(i)}\}$ denotes a set of N points distributed according to $p(\vec{x})$. Such a set may be obtained by taking N points on any sample trajectory (random walk) of the diffusion process. Sample trajectories are generated by using the Langevin equation associated with the process. This equation is written (13)

$$d\vec{X}(t) = \vec{b}(\vec{X}(t))\, dt + d\vec{W}(t) \tag{13}$$

where $\vec{W}$ is the multidimensional Wiener process and $\vec{b}$ is the drift vector given by Eq. (11). A discretized form suitable for computer simulation is

$$\vec{X}(t+\Delta t) = \vec{X}(t) + \vec{b}(\vec{X}(t))\, \Delta t + \vec{\eta}\, \Delta t^{\frac{1}{2}} \tag{14}$$

where the $\vec{\eta}$' denote successive samples of a multidimensional Gaussian random vector whose components verify $\langle\eta_i\rangle = 0$ and $\langle\eta_i\eta_j\rangle = \delta_{ij}$. The integration of the Langevin equation (Eq. (13)) using the discretized form (Eq. (14)) leads to the so-called "short-time approximation" resulting from the use of a finite time step Δt instead of an infinitesimal one as required by the continuous form (Eq. (13)). Thus the estimator of the energy is only exact as the time step used in the simulation goes to zero. In contrast, no bias is introduced when using the Metropolis algorithm since it is based on a discrete Markov chain instead of a continuous diffusion process.

We present in Table I some applications of the variational Monte Carlo method for atoms and molecules. The results presented here are issued from a work of Moskowitz and Kalos. (12) The trial wave functions used in these calculations are of the form

$$\psi(\vec{x}) = \det D^{\alpha} \det D^{\beta} \exp\left(\sum_{i>j} U_{ij} \right) \tag{15}$$

where $\det D^{\alpha}$ and $\det D^{\beta}$ denote determinants of one-electron space orbitals for electrons of α and β spin, respectively, and the U_{ij} denote two-particle functions. The product of space-determinants is introduced in order to take account of the antisymmetry required by the Pauli principle. This particular form is presented in section 3.

Table I. Total ground-state energies obtained with the variational Monte Carlo method. (12) Energies are in a.u. Statistical uncertainties are indicated in parentheses. References for the quoted values may be found in Ref. 12.

	He	H_2	Li_2
E(Monte Carlo)	-2.9024(5)	-1.1568(5)	-14.932(3)
E(Hartree-Fock limit)	-2.86168	-1.13363	-14.8719
E(exact)	-2.90372	-1.1746	-14.9967

2.2 Diffusion Monte Carlo method

The so-called Diffusion Monte Carlo (DMC) method has been designed for removing the variational approximation inherent to the method presented above. This is accomplished by defining a generalized diffusion process admitting the following stationary density

$$p(\vec{x}) = \frac{\psi\phi_0}{\int \psi\phi_0 \, d\vec{x}} \tag{16}$$

where ϕ_0 stands for the exact unknown wave function, instead of $p(\vec{x}) = \psi^2 / \int \psi^2 \, d\vec{x}$ as used previously. This latter distribution was obtained as stationary density of the diffusion process corresponding to the following Fokker-Planck operator:

$$L = \frac{1}{2} \vec{\nabla}^2 - \vec{\nabla}(\vec{b}.) \tag{17.a}$$

with

$$\vec{b} = \frac{1}{2} \vec{\nabla} \mathrm{Log}\, \psi^2 \tag{17.b}$$

Now, a "generalized" Fokker-Planck operator is constructed by adding a multiplicative scalar operator to L. We write

$$L^* = L - (E_L(\vec{x}) - E_T) \tag{18}$$

where $E_L(\vec{x})$ is the local energy defined by Eq. (4) and E_T a trial energy. After simple algebra, it is easy to be convinced that the stationary density of this new process, defined by $L^*p = 0$, is $p(\vec{x})$ given by Eq. (16) if E_T is chosen to be the exact unknown energy E_0.

The new term added to the usual Fokker-Planck operator describes a branching process. Indeed, if this term were present alone in L^*, the resulting equation of evolution would be

$$\frac{\partial W}{\partial t} = - (E_L(\vec{x}) - E_T)\, W \tag{19}$$

that is, a rate equation. In practice, the branching process is thus simulated by creating a number of replicas of each point in configuration space according to the magnitude of the local energy. When E_T is adjusted to be the true ground-state energy, E_0, we have

$$< E_L - E_0 > = 0 \tag{20}$$

and thus the number of points is stabilized.

Finally, the numerical simulation consists in deplacing a finite set of points through configuration space using the Langevin equation(Eq.(14)) for the pure diffusion part of L^* and branching according to Eq. (19) for the scalar part. The stabilization of the total number of points leads to an estimator of the exact energy, E_0, according to Eq. (20). A detailed presentation of the method may be found in the basic paper of Reynolds et al. (14) (See also Anderson (15)). A few illustrative numerical results for atoms and molecules are displayed in Table II. The trial wave functions used in these calculations are of the form (15). The difference between the Monte Carlo results presented in Table II and the exact results is due to the so-called fixed-node approximation resulting from the use of trial wave functions with non-exact nodes. This approximation which is quite good is discussed in section 3 devoted to the many-fermion problem.

Table II. Total ground-state energies obtained with the diffusion Monte Carlo method. (14) Energies are in a.u. Statistical uncertainties are indicated in parentheses. References for the quoted values may be found in Ref. 14.

	H_2	LiH	Li_2	H_2O
E(Monte Carlo)	-1.174(1)	-8.067(2)	-14.990(2)	-76.377(7)
E(Hartree-Fock)	-1.1336	-7.987	-14.872	-76.0675
E(exact)	-1.17447	-8.0699	-14.9967	-76.4376

2.3. Pure diffusion Monte Carlo method

The method presented here is based on the pure diffusion process introduced above (Eqs. (17)). In contrast to the Diffusion Monte Carlo method, no branching is involved.

In the present approach it is convenient to rewrite the Hamiltonian of the system in the form

$$H = H^{(0)} + V_p \quad (21)$$

where $H^{(0)}$ and V_p are called "reference Hamiltonian" and "perturbing potential", respectively. The reference Hamiltonian is constructed from the trial wave function ψ as follows

$$H^{(0)} = -\frac{1}{2}\vec{\nabla}^2 + \frac{1}{2}\frac{\vec{\nabla}^2\psi}{\psi} \quad (22)$$

Stated differently, $H^{(0)}$ is the Hamiltonian of the form (1) verifying

$$H^{(0)}\psi = 0 \quad (23)$$

From Eqs. (21) and (22) we derive easily the expression of the perturbing potential in terms of the trial wave function. One obtains

$$V_p = V - \frac{1}{2}\frac{\vec{\nabla}^2\psi}{\psi} \quad (24)$$

or, equivalently

$$V_p = \frac{H\psi}{\psi} \quad (25)$$

Thus V_p coincides with the local energy introduced above (Eq. (4)).

The time-dependent Green's function (in imaginary time) is defined as

$$G(\vec{x},\vec{y};t) = \langle \vec{x}\,|e^{-tH}|\,\vec{y}\rangle \quad (26)$$

For our purpose, let us define a little more general quantity

$$I^{f,g}(t) = \langle f\psi\,|e^{-tH}|\,g\psi\rangle \quad (27)$$

where f and g are two arbitrary functions of $\vec{x}$. By choosing f and g respectively as $\delta_{\vec{x}}/\psi$ and $\delta_{\vec{y}}/\psi$ (where δ is the Dirac's δ-function) the usual Green's function (Eq.(26)) is recovered.

Now, it is shown how the generalized Green's function defined by Eq. (27) can be expressed as a stochastic average with respect to the reference diffusion process. We only give here the essential steps of the derivation. For the complete derivation the reader is referred to Ref. 16. A first basic ingredient consists in using the Trotter formula:

$$e^{A+B} = \lim_{n\to+\infty} (e^{A/n} e^{B/n})^n \tag{28}$$

where A and B denote two arbitrary operators (which do not commute in general). Thus $I^{f,g}(t)$ may be expressed as

$$I^{f,g}(t) = \lim_{n\to+\infty} \langle f\psi \mid (e^{-\Delta t H^{(0)}} e^{-\Delta t V_p})^n \mid g\psi \rangle \tag{29}$$

with $\Delta t = t/n$

Introducing the spectral resolution of operator $\vec{X}$, namely

$$\mathbb{1} = \int d\vec{x}\ |\vec{x}\rangle\langle\vec{x}| \tag{30}$$

between each operator involved in the matrix element of the right hand side of Eq. (29), we obtain after some manipulations

$$I^{f,g}(t) = \lim_{n\to+\infty} \int \prod_{i=0}^{n} d\vec{x}_i\ f(\vec{x}_0) g(\vec{x}_n)\ e^{-\Delta t \sum_{i=1}^{n} V_p(\vec{x}_i)} \times \{\psi(\vec{x}_0)\psi(\vec{x}_n) \prod_{i=0}^{n-1} \langle\vec{x}_i|e^{-\Delta t H^{(0)}}|\vec{x}_{i+1}\rangle\} \tag{31}$$

The key idea of the present approach is to interpret the quantity in parentheses as the (n+1)-time probability density of the diffusion process defined by Eqs. (17). For that, we first remark that the reference Hamiltonian $H^{(0)}$ is connected to the Fokker-Planck operator L by the following similarity transformation

$$L = -\psi H^{(0)} \frac{1}{\psi}\ . \tag{32}$$

From Eq. (32) it follows

$$\langle\vec{x}|e^{-\Delta t H^{(0)}}|\vec{y}\rangle = \frac{\psi(\vec{x})}{\psi(\vec{y})} \langle\vec{y}|e^{\Delta t L}|\vec{x}\rangle \tag{33}$$

In the latter formula $\langle\vec{y}|e^{\Delta tL}|\vec{x}\rangle$ is the time-dependent Green's function of the Fokker-Planck equation associated with L , i.e., the transition probability density of the diffusion process. We thus write

$$p(\vec{x}]\vec{y},\Delta t) = \langle\vec{y}|e^{\Delta t}|\vec{x}\rangle \tag{34}$$

Note that in this notation $\vec{x}$ is the conditioning variable. Using Eqs.(33) and (34) we then obtain

$$\psi(\vec{x}_0)\psi(\vec{x}_n)\prod_{i=0}^{n-1}\langle\vec{x}_i|e^{-\Delta tH^{(0)}}|\vec{x}_{i+1}\rangle = \psi^2(\vec{x}_0)\prod_{i=0}^{n-1}p(\vec{x}_i]\vec{x}_{i+1},\Delta t) \tag{35}$$

Now, by invoking the fundamental property for stationary Markovian processes , namely

$$p_n(\vec{x}_1,t_1;\ldots;\vec{x}_n,t_n) = p_1(\vec{x}_1)\prod_{i=1}^{n-1}p(\vec{x}_i]\vec{x}_{i+1},t_{i+1}-t_i) \tag{36}$$

where p_n is the n-time probability density of the diffusion process, Eq. (35) may be rewritten as

$$\psi(\vec{x}_0)\psi(\vec{x}_n)\prod_{i=0}^{n-1}\langle\vec{x}_i|e^{-\Delta tH^{(0)}}|\vec{x}_{i+1}\rangle = p_{n+1}(\vec{x}_0,0;\ldots;\vec{x}_n,n\Delta t=t) \tag{37}$$

Finally, expression (31) for $I^{f,g}(t)$ is written

$$I^{f,g}(t) = \lim_{n\to+\infty}\int f(\vec{x}_0)g(\vec{x}_n)\; e^{-\Delta t\sum_{i=1}^{n}V_p(\vec{x}_i)}$$
$$\times\; p_{n+1}(\vec{x}_0,0;\vec{x}_1,\Delta t;\ldots;\vec{x}_n,t)\prod_{i=0}^{n}d\vec{x}_i \tag{38}$$

The right hand side of Eq. (38) may be interpreted as the expectation value of the functional $f(\vec{x}(0))g(\vec{x}(t))\exp(-\int V_p(\vec{x}(s))ds)$ over the trajectories (or sample paths) of the reference diffusion process. It is of interest to note that, in contrast to other methods, not only the stationary density of the diffusion process is used, but also its dynamics through the transition probability density. In order to illustrate the basic idea of the paper the quantum matrix element is written in the form of the following stochastic average

$$I^{f,g}(t) = \langle\; f(\vec{X}(0))g(\vec{X}(t))\;\exp(-\int_0^t V_p(\vec{X}(s))ds)\;\rangle \tag{39}$$

where < ... > stands for the expectation value with respect to the diffusion process.

Another point of view is to consider Eq. (38) as defining a functional integral. (17,18) We thus write

$$I^{f,g}(t) = \int_{\Omega(0,t)} f(\vec{X}(0))g(\vec{X}(t)) \exp(-\int_0^t V_p(\vec{X}(s))ds)\, D^{\psi}X \quad (40)$$

where $D^{\psi}X$ denotes the diffusion functional measure associated with the diffusion process and $\Omega(0,t)$ the set of continuous trajectories defined in (0,t). In fact, the basic formula above (Eq. (40)) is a generalization of the well-known Feynman-Kac formula (17,18). This latter formula may be recovered by choosing the constant as function ψ and $\delta_{\vec{x}}/\psi$ and $\delta_{\vec{y}}/\psi$ as functions f and g, respectively. Now, it is possible to further extend the generalized Feynman-Kac (GFK) formula (Eq. (40)) into a so-called full generalized Feynman-Kac (FGFK) formula by inserting local operators $A_i(\vec{x})$ at different times into the evolution operator e^{-tH}. Essentially, the derivation is identical (see Ref. 16) and we obtain

$$I^{f,g}_{A_1 \ldots A_q}(t;t_1 \ldots t_q) = \langle f\psi | e^{-t_1 H} A_1 e^{-(t_2-t_1)H} A_2 \ldots A_q e^{-(t-t_q)H} | g\psi \rangle$$

$$= \int_{\Omega(0,t)} f(\vec{X}(0)) A_1(\vec{X}(t_1)) \ldots A_q(\vec{X}(t_q)) g(\vec{X}(t)) \exp(-\int_0^t V_p(\vec{X}(s))ds)\, D^{\psi}X \quad (41)$$

From the general quantum matrix element involved in the latter formula, the desired quantum observables may be extracted. (16) Let us present some examples. In the following formulas the subscript i_0 refers to the lowest eigenstate of H (denoted as ϕ_{i_0}) with non-zero overlap with functions $f\psi$ and $g\psi$.

Total energy:

$$E_{i_0} = \lim_{t\to+\infty} -\frac{1}{t} \,\mathrm{Log} < f(\vec{X}(0))g(\vec{X}(t))\exp(-\int_0^t V_p(\vec{X}(s))ds) > \quad (42)$$

First order mean values:

$$\frac{\langle \phi_{i_0} | A | \phi_{i_0} \rangle}{\langle \phi_{i_0} | \phi_{i_0} \rangle} = \lim_{t\to+\infty} \frac{< f(\vec{X}(0))A(\vec{X}(t_1))g(\vec{X}(t)) \exp(-\int_0^t V_p(\vec{X}(s))ds) >}{< f(\vec{X}(0))g(\vec{X}(t)) \exp(-\int_0^t V_p(\vec{X}(s))ds) >} \quad (43)$$

$$\forall t_1 \in \,]0,t[$$

Second order mean values:

$$C_{AB}(\tau) = \sum_k \langle\phi_{i_0}|A|\phi_k\rangle\langle\phi_k|B|\phi_{i_0}\rangle \, e^{-\tau(E_k-E_{i_0})}$$

$$= \lim_{t\to+\infty} \frac{\langle f(\vec{X}(0))A(\vec{X}(t_1))B(\vec{X}(t_1+\tau))g(\vec{X}(t))\exp(-\int_0^t V_p(\vec{X}(s))ds)\rangle}{\langle f(\vec{X}(0))g(\vec{X}(t))\exp(-\int_0^t V_p(\vec{X}(s))ds)\rangle} \quad (44)$$

$$\forall t_1 \in]0,t[$$

From $C_{AB}(\tau)$ it is possible to show that response properties such as static or dynamic polarizabilities may be derived. (16)

In principle, the numerical evaluation of the stochastic averages would require to sum over the unlimited number of stochastic trajectories in configuration space defined in (0,t). In practice, it is possible to escape from this difficulty by using the ergodic character of the reference diffusion process. This latter property holds only if the stationary density of the process is integrable. (16,19) In the present case it requires that the trial wave function is square-integrable, a condition which is easily fulfilled. Thus, by using the ergodic property the ensemble-averages (over the complete set of trajectories) may be expressed as time-averages along any single sample trajectory (or any finite number of sample trajectories) of the process

$$\langle f(\vec{X}(0))A_1(\vec{X}(t_1))\ldots A_q(\vec{X}(t_q))g(\vec{X}(t)) \exp(-\int_0^t V_p(\vec{X}(s))ds) \rangle$$

$$= \lim_{T\to+\infty} \frac{1}{T}\int_0^T f(\vec{X}^{(0)}(\tau)) \, A_1(\vec{X}^{(0)}(t_1+\tau))\ldots A_q(\vec{X}^{(0)}(t_q+\tau)) \, g(\vec{X}^{(0)}(t+\tau))$$

$$\times \exp(-\int_\tau^{t+\tau} V_p(\vec{X}^{(0)}(s))ds) \, d\tau \quad (45)$$

where $\vec{X}^{(0)}(s)$ denotes an arbitrary trajectory of the diffusion process. Such a trajectory may be obtained by using the Langevin equation (Eqs. (13) and (14)) as explained in subsection 2.1.

Finally, the main ideas of the present quantum Monte Carlo method can be summarized as follows

1. Express the desired quantum observables in terms of stochastic averages by using the FGFK formula.
2. Invoke the ergodic property to convert the ensemble-averages into time-averages along any sample path of the diffusion process.
3. Use the discretized Langevin equation for constructing step by step an arbitrary sample path.

We present in Tables III and IV some results (20) for small systems.

Table III. Total ground-state energies obtained with the pure diffusion Monte Carlo method. (20) Energies are in a.u. Statistical uncertainties are indicated in parentheses. References for the quoted values may be found in Ref. 20.

	He	Be^{++}	H_2
E(Monte Carlo)	-2.9038(10)	-13.655(11)	-1.175(1.7)
E(Hartree-Fock)	-2.8617	-13.6113	-1.1336
E(exact)	-2.90372	-13.6556	-1.1745

Table IV. Ground-state properties for H_2 molecule obtained with the pure diffusion Monte Carlo method. (20) All quantities are in a.u. and statistical uncertainties are indicated in parentheses. References for the quoted values may be found in Ref. 20.

	$R_{H-H} = 1.4$		$R_{H-H} = 4$	
	Monte Carlo	Exact	Monte Carlo	Exact
E_0	-1.175(1.7)	-1.1745	-1.0167(7)	-1.01637
$\langle r_{12}^{-1}\rangle$	0.587(2.6)	0.5874	0.263(1.5)	0.2630
$\langle r_{12}\rangle$	2.169(9)	2.1690	4.33(1.5)	4.327
$\langle r_{12}^2\rangle$	5.63(4.5)	5.632	20.6(1.3)	20.569
$\langle r_A\rangle$	1.550(6)	1.5499	2.86(1.3)	2.863
$\langle r_A^{-1}\rangle$	0.908(4)	0.9128	0.610(3.5)	0.6136
$\langle r_A^2\rangle$	3.04(2)	3.036	10.79(8.5)	10.815
$\langle r_A r_B\rangle$	2.71(2)	2.704	6.68(2.3)	6.663
$\langle r_{1A} r_{2A}\rangle$	2.33(1.5)	2.321	6.57(3)	6.551
$\langle r_{1A} r_{2B}\rangle$	2.39(1.5)	2.385	9.82(3)	9.806
$\langle z_1 z_2\rangle$	-0.156(5.5)	-0.1596	-3.39(6)	-3.392
$\langle x_1 x_2\rangle$	-0.055(3.5)	-0.0551	-0.039(4.5)	-0.0384
$\langle z^2\rangle$	1.02(2)	1.023	4.74(4)	4.708
$\langle x^2\rangle$	0.76(2)	0.762	1.05(1.5)	1.054
$\langle r^2\rangle$	2.55(4)	2.546	6.85(5.5)	6.815
Q(a)	0.9(1)	0.91	1.3(1.5)	1.38

(a) Q denotes the electric quadrupole moment of the molecule.

2.4. Use of generalized Poisson processes

Following Maslov and Chebotarev (21)(see also Ref. 22) quantum properties may be expressed as stochastic averages by using the time-dependent Schrödinger equation written in real time. This leads to introduce Poisson processes instead of diffusion processes. As an example, we shall derive here the expression of the Fourier transform of the time-dependent wave function in terms of an expectation value with respect to a suitably generalized Poisson process. Additional quantum properties (such as evolution of quantum observables) can be derived (23) but are not considered here. The numerical evaluation of stochastic averages based on Poisson processes seems to have never been considered (to our knowledge). In fact, this is not a straightforward task. The main reasons are i) the oscillatory character of the integrands and ii) the quite restrictive conditions imposed on the potential. Nevertheless, a precise analysis of the methodology motivated by computational purposes remains to be done.

Let us decompose the Hamiltonian H as follows

$$H = H_0 + V \tag{46}$$

where H_0 is the free Hamiltonian and V the potential energy of the system. The Fourier transform of the time-dependent wave function $\psi(\vec{x},t)$ is written

$$\hat{\psi}(\vec{p},t) = \int d\vec{x}\; e^{-i\vec{p}.\vec{x}}\; \psi(\vec{x},t) \tag{47}$$

the converse formula is written

$$\psi(\vec{x},t) = \frac{1}{(2\pi)^d} \int d\vec{p}\; e^{i\vec{p}.\vec{x}}\; \hat{\psi}(\vec{p},t) \tag{48}$$

where d is the dimension of the relevant space (3N for electronic problems). The time evolution of $\psi(\vec{x},t)$ is given by

$$\psi(\vec{x},t) = e^{-i(t-t_0)H}\; \psi(\vec{x},t_0) \tag{49}$$

Let us use the interaction representation. Defining $U(t,t_0)$ as follows

$$U(t,t_0) = e^{iH_0t}\; e^{-i(t-t_0)H}\; e^{-iH_0t_0} \tag{50}$$

we can rewrite Eq. (49) in the following form

$$\psi(\vec{x},t) = e^{-iH_0 t}\, U(t,t_0)\, e^{iH_0 t_0}\, \psi(\vec{x},t_0) \qquad (51)$$

Now, from definition (50) it is easy to derive the equation of evolution for $U(t,t_0)$

$$i\,\frac{\partial U(t,t_0)}{\partial t} = V(t)U(t,t_0) \qquad (52)$$

where $$V(t) = e^{iH_0 t}\, V(t)\, e^{-iH_0 t} \qquad (53)$$

Equation (52) may be rewritten in the equivalent form

$$U(t,t_0) = 1 - i\int_{t_0}^{t} V(\tau)U(\tau,t_0)\, d\tau \qquad (54)$$

By iterating Eq. (54) it is found that

$$U(t,t_0) = 1 + \sum_{n=1}^{\infty} (-i)^n \int_{t_0}^{t} dt_n \;\ldots\; \int_{t_0}^{t_2} dt_1\; V(t_n)\ldots V(t_1) \qquad (55)$$

Using Eq. (51) and expressions (55) and (53), the Fourier transform of $\psi(\vec{x},t)$ is written as

$$\hat{\psi}(\vec{p},t) = \int d\vec{x}\; e^{-i\vec{p}.\vec{x}}\; e^{-iH_0 t} \{ 1+\sum_{n=1}^{\infty}(-i)^n \int_{t_0}^{t} dt_n \ldots \int_{t_0}^{t_2} dt_1 \prod_{j=1}^{n} e^{iH_0 t_{n-j+1}}\, V$$

$$\times\; e^{-iH_0 t_{n-j+1}} \}\, e^{iH_0 t_0} (2\pi)^{-d} \int d\vec{p}\,' e^{i\vec{p}\,'.\vec{x}}\; \hat{\psi}(\vec{p}\,',t_0) \qquad (56)$$

Introducing the Fourier transform of V we finally obtain

$$\hat{\psi}(\vec{p},t) = e^{-ih_0(\vec{p})(t-t_0)}\; \hat{\psi}(\vec{p},t_0) + \sum_{n=1}^{\infty} (-i)^n \int_{t_0}^{t} dt_n \ldots \int_{t_0}^{t_2} dt_1 \int \frac{d\vec{p}_n}{(2\pi)^d}\; \hat{V}(\vec{p}_n) \ldots$$

$$\ldots \int \frac{d\vec{p}_1}{(2\pi)^d}\; \hat{V}(\vec{p}_1)\; e^{-ih_0(\vec{p})(t-t_n)} e^{-ih_0(\vec{p}-\vec{p}_1)(t_n-t_{n-1})} \ldots .$$

$$\ldots e^{-ih_0(\vec{p}-\vec{p}_1 \ldots -\vec{p}_n)(t_1-t_0)}\; \hat{\psi}(\vec{p}-\vec{p}_1 \ldots -\vec{p}_n, t_0) \qquad (57)$$

where $h_0(\vec{p}) = \vec{p}^2/2$ stands for the free Hamiltonian in momentum-coordinates.

Now, let us define a stochastic process which permits to interpret $\hat{\psi}(\vec{p},t)$ as an expectation value. In the pure diffusion Monte Carlo method (subsection 2.3), the expectation value was defined by

$$\langle F(\vec{X}(s))\rangle = \lim_{n\to+\infty} \langle F(\vec{x}_0,0;\vec{x}_1,\Delta t;\ldots;\vec{x}_n,t=n\Delta t)\rangle$$

$$= \lim_{n\to+\infty} \int F(\vec{x}_0,0;\vec{x}_1,\Delta t;\ldots;\vec{x}_n,t)\, p_{n+1}(\vec{x}_0,0;\vec{x}_1,\Delta t;\ldots;\vec{x}_n,t) \prod_{i=0}^{n} d\vec{x}_i \qquad (58)$$

where p_{n+1} denotes the (n+1)-time probability density of the diffusion process and F a rather arbitrary functional of the trajectory. Now, F is a function of $(n,t_1,\ldots,t_n,\vec{p}_1,\ldots,\vec{p}_n)$ where $n\in\mathbb{N}$, $t_i\in(0,t)$ and $\vec{p}_i\in\mathbb{R}^d$. The average is defined by

$$\langle F(n,t_i,\vec{p}_i)\rangle = e^{-t\mu}\sum_{n\geq 0}\int_{t_0}^{t} dt_n\ldots\int_{t_0}^{t_2} dt_1 \int\frac{d\vec{p}_n}{(2\pi)^d}\,\hat{V}(p_n)\ldots\int\frac{d\vec{p}_1}{(2\pi)^d}\,\hat{V}(\vec{p}_1)$$
$$\times\, F(n,t_1,\ldots,t_n,\vec{p}_1,\ldots,\vec{p}_n) \qquad (59)$$

This definition is valid only if $\frac{d\vec{p}}{(2\pi)^d}\,\hat{V}(\vec{p})$ is a positive (bounded) measure; μ stands for the "total mass", that is $\mu = \int\frac{d\vec{p}}{(2\pi)^d}\,\hat{V}(\vec{p})$

We verify that

$$\langle 1\rangle = e^{-t\mu}\sum_{n\geq 0}\frac{t^n}{n!}\mu^n = 1 \qquad (60)$$

The probability of having a given value n is obtained as follows

$$P(N=n) = e^{-t\mu}\int_{t_0}^{t} dt_n\ldots\int_{t_0}^{t_2} dt_1\int\frac{d\vec{p}_n}{(2\pi)^d}\,\hat{V}(\vec{p}_n)\ldots\int\frac{d\vec{p}_1}{(2\pi)^d}\,\hat{V}(\vec{p}_1)$$
$$= \frac{(t\mu)^n}{n!}\,e^{-t\mu} \qquad (61)$$

then N is a Poisson process.

The average defined above (Eq. (59)) is thus identified to the expectation value with respect to a generalized Poisson process whose trajectories are piecewise constant with a finite number n of jumps. The number of jumps is random and is given by a standard Poisson random variable. In addition, these jumps are uniformly distributed on the time interval (0,t) and their amplitudes are distributed according to the density $\hat{V}(\vec{p})/(2\pi)^d$. Finally, $\hat{\psi}(\vec{p},t)$ is written as the following expectation value

$$\hat{\psi}(\vec{p},t) = e^{t\mu} < e^{-i\int_{t_0}^{t} h(\vec{p}-\vec{P}(\tau))d\tau}(-i)^N \hat{\psi}(\vec{p}-\vec{P}(t_0),t_0) > \qquad (62)$$

where N stands for the usual Poisson process (Eq. (61)) and $\vec{P}(\tau)$ is given by

$$\vec{P}(\tau) = \sum_{i=1}^{n} \Theta(t_i-\tau)\, \vec{p}_i \qquad (63)$$

where Θ is the step function

$$\Theta(x)=\begin{cases} 0 & x<0 \\ 1 & x\geq 0 \end{cases}$$

3. THE MANY-FERMION PROBLEM

This section is devoted to the problem of imposing mathematical constraints resulting from the Pauli principle for fermion systems. As a general rule Monte Carlo methods do not involve spin-coordinates. The Pauli principle must be then expressed in a spin-free formalism. In usual space-spin formalism, it is known that the complete wave function (depending on space and spin coordinates) must be antisymmetric. Since only spin-free Hamiltonians are considered here, the wave function can be written as a product of a space function and of a spin function or as a linear combination of such products. On the other hand, the variables of the spin function can take only two values (spin-1/2 particles are treated here), and thus the spin function cannot be antisymmetrized with respect to more than two variables. It results from these constraints that symmetry properties of space functions (i.e. eigenfunctions of H) cannot be arbitrary. On this result is based the partition of eigenfunctions of H between so-called fermionic states (with physical symmetry) and bosonic states (with non-physical symmetry). An illustrative example is the lowest symmetric eigenfunction of H which becomes non-physical as the number of spin-1/2 particles is greater than two. Further explanations concerning the formulation of the Pauli principle in a spin-free formalism may be found in Refs. 24-26. Now, in order to obtain physical solutions to the Schrödinger equation it is necessary to construct space functions ϕ belonging to the different physical symmetry subspaces. The most current form used so far for such space functions is as follows (14,27)

$$\phi(\vec{r}_1,\ldots,\vec{r}_N) = \det\Big(\prod_{i=1}^{N/2+S} u_i(\vec{r}_i)\Big)\ \det\Big(\prod_{i=1}^{N/2-S} u_i(\vec{r}_{i+N/2+S})\Big) \tag{64}$$

where $\{u_i\}$ is a set of (N/2+S) different one-electron space orbitals, N the number of electrons, and S the total spin determining the desired type of symmetry.

Now, it is explained how to introduce such functions in the framework of Monte Carlo methods. For that, let us return to the very basic idea of this paper, namely, the possibility of expressing quantum properties as stochastic averages. From this idea follow two natural ways for introducing the symmetry requirements. First, they may be introduced into the underlying stochastic process. This approach is referred to as the fixed-node approach. Second, symmetry may be introduced into expectation values by using so-called projection functions. We then speak of projection methods. Such an alternative based on the basic idea of stochastic methods is common to all Monte Carlo approaches. However, for concreteness, we illustrate these two procedures in the particular case of the pure diffusion Monte Carlo method presented in subsection 2.3.

Let us consider the problem of evaluating the total energy. According to Eqs. (42) and (39) the quantity to evaluate is

$$I^{f,g}(t) = \langle f\psi|e^{-tH}|g\psi\rangle \tag{65}$$

Expanding the matrix element by using the resolution of the identity associated with the spectral resolution of H, we obtain

$$I^{f,g}(t) = \sum_i \langle f\psi|\phi_i\rangle\langle\phi_i|g\psi\rangle\ e^{-tE_i} \tag{66}$$

Now, symmetry is introduced by choosing space functions f,g and ψ so that $f\psi$ and $g\psi$ belong to the desired physical symmetry subspace. Indeed, once this has been done, scalar products $\langle f\psi|\phi_i\rangle$ and $\langle\phi_i|g\psi\rangle$ corresponding to non-physical levels vanish. The two approaches are

i. Fixed-node approach

This approach consists in choosing f=g=1 (no projection functions) and a trial wave function ψ of the form (64). In fact, more sophisticated choice for ψ may be used by multiplying (64) by a symmetric factor which does not change the type of symmetry (essentially for including cusp conditions as it is done in form(15)). According to its symmetry properties, such a function must have some nodal hypersurfaces. These nodal hypersurfaces, where the drift is infinite (see Eq. (17.b)), are seen as infinite barriers for any sample trajectory. The diffusion process is then decomposed into a juxtaposition of subprocesses in subdomains delimited by the nodes of ψ. These nodes are not changed in the Monte Carlo simulation. This is the reason why this method is referred to as a fixed-node approach. In addition, if the nodes of the trial wave function are not exact, which is usually the case since symmetry requirements do not determine entirely the nodes of the exact wave function, (28) a bias in the results will appear. This is the origin of the so-called fixed-node approximation. (14) An example of such an approximation may be found in Table II (see also Table V below).

ii. Projection method

In contrast to the fixed-node approach, no approximation occurs. The trial wave function is chosen to be nodeless (bosonic) and symmetry is introduced by choosing space functions f and g with adequate symmetry. A new aspect associated with this choice is that averaged quantities do not have a constant sign. In fact, it is possible to show that the positive and negative contributions nearly cancel. (6,16,29,30) From such a situation result serious numerical difficulties. However, by choosing properly the trial wave function and the projection functions it is possible to attempt to keep under control these difficulties. This is the basic idea of the so-called nodal relaxation method of Ceperley and Alder (30) and of the so-called release-node projection method of Caffarel and Claverie. (16,20) We present in Table V some very good results obtained by Ceperley and Alder in the framework of the Green's function Monte Carlo method using the nodal relaxation algorithm.

Table V. Total ground-state energies obtained with the Green's function Monte Carlo method using the nodal relaxation algorithm. (30) Energies are in a.u. Statistical uncertainties are indicated in parentheses. References for the quoted values may be found in Ref. 30.

	H_3	LiH	Li_2	H_2O
E(fixed-node)	-1.6606(2)	-8.067(1)	-14.990(2)	-76.39(1)
E(release-node)	-1.6617(2)	-8.071(1)	-14.994(2)	-76.43(2)
E(exact)	-1.66194	-8.0705	-14.9967	-76.4376

This work is devoted to the memory of Pierre Claverie who was responsible for its development but not for possible errors contained in the present report.

REFERENCES

1. M.H. Kalos, D. Levesque, and L. Verlet, Phys. Rev. A9, 2178 (1974).
2. D.M. Arnow, M.H. Kalos, M.A. Lee, and K.E. Schmidt, J.Chem.Phys. 77, 5562 (1982).
3. K.E. Schmidt and M.H. Kalos, "Monte Carlo Methods in Statistical Physics II", ed. by K. Binder (Springer, Berlin,1984), Chap. 4.
4. D.M. Ceperley, J.Comp.Phys. 51, 404 (1983).
5. I. Oksüz, J.Chem.Phys. 81, 5005 (1984).
6. V. Elser, Phys.Rev. A39, 2293 (1986).
7. J.W. Moskowitz and K.E. Schmidt, "Monte Carlo Methods in Quantum Problems", ed. by M.H. Kalos , NATO ASI Series C (Reidel, Dordrecht, 1982).
8. D.M. Ceperley, J.Stat.Phys. 43, 815 (1986).
9. W.L. McMillan, Phys.Rev. A138, 442 (1965).
10. D.M. Ceperley, G.V. Chester, and M.H. Kalos, Phys.Rev. B16, 3081 (1977); D.M. Ceperley, "Monte Carlo methods in Statistical Physics", ed. by K. Binder (Springer, Berlin, 1979), p.145.
11. N. Metropolis, A.W. Rosenbluth, M.N. Rosenbluth, A.M. Teller, and E. Teller, J.Chem.Phys. 21, 1087 (1953).
12. J.W. Moskowitz and M.H. Kalos, Int.J.Quantum Chem. 20, 1107 (1981).
13. H. Risken, "The Fokker-Planck Equation" (Springer-Verlag, Berlin, 1984).
14. P.J. Reynolds, D.M. Ceperley, B.J. Alder, and W.A. Lester,Jr. J.Chem.Phys. 77, 5593 (1982).
15. (a) J.B. Anderson, J.Chem.Phys. 63, 1499 (1975);(b) J.B. Anderson ibid. 73, 3897 (1980).
16. M. Caffarel and P. Claverie, J.Chem.Phys. 88, 1088 (1988).
17. I.M. Gelfand and A.M. Yaglom, J.Math.Phys. 1, 48 (1960).
18. S.G. Brush, Rev.Mod.Phys. 33, 79 (1961).
19. R.Z. Khas'minskii, Theory Prob. its Appl. (USSR) 5, 179 (1960).
20. M. Caffarel and P. Claverie, J.Chem.Phys. 88, 1100 (1988).
21. A.M. Chebotarev and V.P. Maslov. "Processus à sauts et leurs applications dans la mécanique quantique" in Feynman Path Integral Proceedings Marseille 1978. Lecture Notes in Physics 106 (Springer-Verlag, Berlin, 1979).
22. P. Combe, R. Rodriguez, R. Høegh-Krohn, M. Sirugue, and M. Sirugue-Collin Phys.Reports 77, 221 (1981).
23. J. Bertrand and G. Rideau in "Stochastic Processes in Quantum Theory and Statistical Physics", Lecture Notes in Physics 173 (Springer-Verlag, Berlin); J. Bertrand and G. Rideau, Letters in Math.Phys. 7, 327 (1983).
24. L.D. Landau and E. Lifshitz, "Quantum Mechanics" (Pergamon, Oxford, 1965), Chap. 9.
25. M. Hamermesh, "Group Theory" (Addison-Wesley, Reading, MA, 1962).

26. F.A. Matsen, Adv.Quantum Chem. 1, 60 (1964).
27. V.A. Fock, "Fundamentals of Quantum Mechanics" (Mir, Moscow, 1976), Part IV, p.261.
28. D.J. Klein and H.M. Pickett, J.Chem.Phys. 64, 4811 (1976).
29. M.H. Kalos, "Monte Carlo Methods in Quantum Problems", ed. by M.H. Kalos, NATO ASI Series C (Reidel, Dordrecht, 1982).
30. D.M. Ceperley and B.J. Alder, J.Chem.Phys. 81, 5833 (1984).

COMPUTATIONAL STRATEGIES AND NEW APPLICATIONS IN GREEN'S FUNCTION MONTE CARLO

B.H. WELLS
University of Oxford
Physical Chemistry Laboratory
South Parks Road
Oxford OX1 3QZ
England

ABSTRACT. The computational algorithm of the short-time fixed-node Green's function Monte Carlo method is discussed, and applications of the method to the study of small atomic and molecular systems, electron-positron systems and electron-hole pairs in semiconductor crystallites are reviewed.

1. Introduction

The short-time fixed-node Green's function Monte Carlo method is now a well-established technique for the study of the electronic structure of small molecules [1]. It is both computationally efficient and conceptually straight-forward, and is capable of yielding results of high accuracy. In this review the computational algorithm of the short-time fixed-node Green's function Monte Carlo method will be discussed in depth, emphasising the features that enhance the efficiency of the algorithm. Detailed consideration will be given to the functional form of the trial wavefunction and procedures by which the parameters of the trial wavefunction can be optimised. The application of the method to the study of electron-positron systems and electron-hole pairs in small semiconductor crystallites will then be presented, and finally the differential Green's function Monte Carlo method, which seeks to reduce the statistical error associated with a Monte Carlo calculation, will be discussed.

2. The Short-Time Fixed-Node Green's Function Monte Carlo Method

The short-time fixed-node Green's function Monte Carlo method has recently been reviewed in depth by Wells [1], and only a brief outline is given here. Because the majority of the calculations presented in this review employ wavefunctions whose spatial components are everywhere positive, the complications that arise when antisymmetric wavefunctions are employed will be considered only briefly at the end of this section.

M. Defranceschi and J. Delhalle (eds.), Numerical Determination of the Electronic Structure of Atoms, Diatomic and Polyatomic Molecules, 107–129.

It is useful to distinguish between two types of short-time Green's function Monte Carlo calculation: one type of calculation, called diffusion Monte Carlo [2], estimates an eigenvalue (usually the ground-state eigenvalue) of the hamiltonian $\hat{\mathcal{H}}$ of the system

$$E_{DMC} = \frac{< \psi_n|\hat{\mathcal{H}}|\psi_n >}{< \psi_n|\psi_n >} \tag{1}$$

where $\psi_n(\mathbf{r})$ is an eigenfunction of the hamiltonian $\hat{\mathcal{H}}$ (here we ignore the complications that arise if the spatial component of $\psi_n(\mathbf{r})$ has both positive and negative regions).

The other type of calculation, called variational Monte Carlo [3,4,5], estimates the variational energy of a given wavefunction $\phi(\mathbf{r})$

$$E_{VMC} = \frac{< \phi|\hat{\mathcal{H}}|\phi >}{< \phi|\phi >} \tag{2}$$

In diffusion Monte Carlo we consider the imaginary-time evolution of the wavefunction $\Psi(\mathbf{r},t)$ of the system of interest according to the following Schrödinger equation (written in atomic units)

$$-\frac{\partial \Psi(\mathbf{r},t)}{\partial t} = (\hat{\mathcal{H}} - E_T)\Psi(\mathbf{r},t) \tag{3}$$

where E_T is a shift in the zero of the energy scale which ensures non-negative eigenvalues. A distribution function $f(\mathbf{r},t)$, called the *importance function*, is then defined to be equal to the product of the time-dependent wavefunction $\Psi(\mathbf{r},t)$ and a known trial wavefunction $\phi_t(\mathbf{r})$ [3,6,7]

$$f(\mathbf{r},t) = \Psi(\mathbf{r},t)\phi_t(\mathbf{r}) \tag{4}$$

In the diffusion Monte Carlo calculation an ensemble of configurations is chosen to form a pointwise representation of the importance function $f(\mathbf{r},t)$. The term 'configuration' is defined as a set of 3N coordinates $(x_1, y_1, z_1, ...y_{3N}, z_{3N})$, where N is the number of diffusing particles. These diffusing particles are usually electrons but, in this review, may also be positively-charged species. A configuration represents one point $\mathbf{r}$ in the 3N-coordinate configuration space of the molecular system of interest and a set of such configurations (called an 'ensemble') therefore forms a pointwise representation of the importance function $f(\mathbf{r},t)$ at any time t. The ensemble is propagated in a fashion that is dependent on the Schrödinger equation (equation(3)), and we seek the long-time density distribution $f(\mathbf{r},t \rightarrow \infty)$, which, as will be shown below, allows the ground-state eigenvalue of the hamiltonian $\hat{\mathcal{H}}$ to be calculated. The imaginary-time evolution of this importance function may be represented by a Green's function $G(\mathbf{r} \rightarrow \mathbf{r},\tau)$ in the following fashion

$$\begin{aligned} f(\mathbf{r}',t+\tau) &= \int d\mathbf{r} G(\mathbf{r} \rightarrow \mathbf{r}',\tau) f(\mathbf{r},t) \\ &= \int d\mathbf{r} \left(\frac{\phi_t(\mathbf{r}')}{\phi_t(\mathbf{r})} K(\mathbf{r} \rightarrow \mathbf{r}',\tau) \right) f(\mathbf{r},t) \end{aligned} \tag{5}$$

where the Green's function $K(\mathbf{r} \rightarrow \mathbf{r}',\tau)$ obeys the same Schrödinger equation as $\Psi(\mathbf{r},t)$ [8]

$$-\frac{\partial K(\mathbf{r} \rightarrow \mathbf{r}',\tau)}{\partial \tau} = (\hat{\mathcal{H}}_{\mathbf{r}'} - E_T)K(\mathbf{r} \rightarrow \mathbf{r}',\tau) \tag{6}$$

but with the boundary condition

$$K(\mathbf{r} \rightarrow \mathbf{r}', 0) = \delta(\mathbf{r} - \mathbf{r}') \tag{7}$$

If the Green's function $K(\mathbf{r} \rightarrow \mathbf{r}', \tau)$ is expanded in the eigenfunctions $\psi_m(\mathbf{r})$ and eigenvalues E_m of the hamiltonian $\hat{\mathcal{H}}$, then it can be shown [1] that the long-time form of the importance function is

$$\lim_{t \rightarrow \infty} f(\mathbf{r}, t) = \phi_t(\mathbf{r}) \psi_n(\mathbf{r}) e^{-(E_n - E_T)\tau} \tag{8}$$

Where $\psi_n(\mathbf{r})$ is the lowest-energy eigenfunction that is not orthogonal to $\phi_t(\mathbf{r})$. If the trial energy E_T is adjusted to equal the eigenvalue E_n, then the asymptotic form of the importance function has the following steady-state distribution:

$$\begin{aligned} \lim_{t \rightarrow \infty} f(\mathbf{r}, t) &= \phi_t(\mathbf{r}) \psi_n(\mathbf{r}) \qquad\qquad (E_T = E_n) \\ &\equiv f_\infty(\mathbf{r}) \end{aligned} \tag{9}$$

In the discussions that follow, $\phi_t(\mathbf{r})$ will be chosen so that $\psi_n(\mathbf{r})$ is the ground-state eigenfunction $\psi_0(\mathbf{r})$, and the ground-state energy can then be estimated by sampling the 'local energy' $E_L(\mathbf{r}) = \hat{\mathcal{H}} \phi_t(\mathbf{r}) / \phi_t(\mathbf{r})$ from the time-asymptotic importance function $f_\infty(\mathbf{r})$ [1]

$$\begin{aligned} E_{DMC} &= \frac{\int d\mathbf{r} f_\infty(\mathbf{r}) [\hat{\mathcal{H}} \phi_t(\mathbf{r}) / \phi_t(\mathbf{r})]}{\int d\mathbf{r} f_\infty(\mathbf{r})} \\ &= \frac{\int d\mathbf{r} \psi_0(\mathbf{r}) \hat{\mathcal{H}} \phi_t(\mathbf{r})}{\int d\mathbf{r} \psi_0(\mathbf{r}) \phi_t(\mathbf{r})} \\ &= E_0 \end{aligned} \tag{10}$$

Therefore in a Monte Carlo calculation the configurations that form a pointwise representation of $f(\mathbf{r}, t)$ will propagate independently in a manner determined by the Green's function $G(\mathbf{r} \rightarrow \mathbf{r}', \tau)$, and at long times the ensemble will form a discrete representation of the time-asymptotic importance function $f_\infty(\mathbf{r})$. The local energy $\hat{\mathcal{H}} \phi_t / \phi_t$ can then be sampled from this distribution and the diffusion Monte Carlo energy E_{DMC} calculated.

In order to obtain quantitative expressions for the propagation of the configurations in the ensemble, we need to know the Green's function $G(\mathbf{r} \rightarrow \mathbf{r}', \tau)$. For non-trivial many-body systems the exact Green's function is not known, but there is a simple approximation to the exact Green's function, known as the *short-time approximation*, which has proved highly successful in calculations of molecular electronic structure. The short-time approximation for the Green's function $G(\mathbf{r} \rightarrow \mathbf{r}', \tau)$ may be derived by using the Trotter expression for the Green's function $K(\mathbf{r} \rightarrow \mathbf{r}', \tau)$ [1]

$$\begin{aligned} K(\mathbf{r} \rightarrow \mathbf{r}', \tau) &= < \mathbf{r}' | e^{(\hat{\mathcal{H}} - E_T)\tau} | \mathbf{r} > \\ &= < \mathbf{r}' | e^{-D\tau\nabla^2} e^{(V - E_T)\tau} e^{\hat{O}([-D\nabla^2, V - E_T]\tau^2)} | \mathbf{r} > \end{aligned} \tag{11}$$

where $D = 1/2M_p$, M_p being the mass of the diffusing particle. If the commutator terms of order τ^2 are omitted, we obtain

$$K(\mathbf{r} \rightarrow \mathbf{r}', \tau) \sim (4\pi D\tau)^{-3N/2} exp\left(\frac{-(\mathbf{r}' - \mathbf{r})^2}{4D\tau} \right) exp(-\tau(V(\mathbf{r}) - E_T)) \tag{12}$$

and if the value of the trial wavefunction at $\mathbf{r}'$ is expanded about the value at $\mathbf{r}$ in a Taylor series

$$\frac{\phi_t(\mathbf{r}')}{\phi_t(\mathbf{r})} = exp(Ln|\phi_t(\mathbf{r}')| - Ln|\phi_t(\mathbf{r})|)$$
$$= exp((\mathbf{r}' - \mathbf{r})\nabla Ln|\phi_t(\mathbf{r})| + ...) \tag{13}$$

then we obtain the following expression for the short-time Green's function $G(\mathbf{r} \to \mathbf{r}', \tau)$ employed in diffusion Monte Carlo [3,7,8]

$$G(\mathbf{r} \to \mathbf{r}', \tau) = (4\pi D\tau)^{-3N/2} exp\left(\frac{-(\mathbf{r}' - \mathbf{r} - 2D\tau\nabla\phi_t(\mathbf{r})/\phi_t(\mathbf{r}))^2}{4D\tau}\right) exp(-\tau(E_L(\mathbf{r}) - E_T)) \tag{14}$$

The Green's function determines the manner in which the configurations propagate through configuration space: the Gaussian part represents motion from $\mathbf{r}$ to $\mathbf{r}'$ in configuration space according to the following equation

$$\mathbf{r}' = \mathbf{r} + 2D\tau\nabla\phi_t(\mathbf{r})/\phi_t(\mathbf{r}) + \chi \tag{15}$$

where χ is a random number drawn from a Gaussian distribution with a mean of zero and a variance of $2D\tau$. After each move the exponential term $exp(-\tau(E_L(\mathbf{r}) - E_T))$ in the Green's function is evaluated and copies of the configuration are made according to the integer value of this term; these configurations then propagate independently of the parent configuration. If the integer term is zero, then the configuration is deleted after the move. After many moves the ensemble of configurations will form a discrete representation of the distribution $f_\infty(\mathbf{r})$, and the energy E_{DMC} can then be estimated by sampling the local energy $\hat{\mathcal{H}}\phi_t/\phi_t$ from this distribution.

In variational Monte Carlo the variational energy of a given wavefunction, which is usually taken to be the trial wavefunction $\phi_t(\mathbf{r})$ itself, is calculated by propagating the configurations that form a pointwise representation of the importance function according to the following Green's function [1,3]

$$G(\mathbf{r} \to \mathbf{r}', \tau) = (4\pi D\tau)^{-3N/2} exp\left(\frac{-(\mathbf{r}' - \mathbf{r} - 2D\tau\nabla\phi_t(\mathbf{r})/\phi_t(\mathbf{r}))^2}{4D\tau}\right) \tag{16}$$

i.e. the configurations are moved according to equation (15), but they are not copied or deleted after a move. The long-time form of this importance function is then

$$f_\infty(\mathbf{r}) = \phi_t(\mathbf{r})\phi_t(\mathbf{r}) \tag{17}$$

and sampling the local energy $\hat{\mathcal{H}}\phi_t/\phi_t$ from this distribution gives the variational energy E_{VMC} of the trial wavefunction $\phi_t(\mathbf{r})$

$$E_{VMC} = \frac{\int d\mathbf{r} f_\infty(\mathbf{r})[\hat{\mathcal{H}}\phi_t(\mathbf{r})/\phi_t(\mathbf{r})]}{\int d\mathbf{r} f_\infty(\mathbf{r})}$$
$$= \frac{< \phi_t|\hat{\mathcal{H}}|\phi_t >}{< \phi_t|\phi_t >} \tag{18}$$

When calculations are performed using wavefunctions whose spatial components are everywhere positive, the estimates of the energies E_{DMC} and E_{VMC} are subject to one source of inaccuracy and one source of imprecision. The inaccuracy is caused by the use of the short-time approximation in the derivations of the Green's functions (equations (14) and (16)), and can be eliminated either by calculating the energy at a number of timesteps τ and extrapolating to the limit $\tau \to 0$ [1,3,9,10], or by employing an algorithm that samples the exact Green's function [1,2,11,12,13]; the elimination of the short-time approximation is discussed at length in [1], and an illustration of the extrapolation method is given later in this review. The imprecision in the estimates of the energies is caused by the use of random number sequencies to sample the expectation values, and this gives rise to a statistical uncertainty in the energy estimate; this imprecision can be reduced either by increasing the length of the calculation (to increase the number of samples taken), or by employing the differential Green's function Monte Carlo method [14], which seeks to reduce the statistical error by means of correlated random walks.

We shall now consider the complications that arise in diffusion Monte Carlo when the spatial parts of the wavefunctions $\psi_0(\mathbf{r})$ and $\phi_t(\mathbf{r})$ have positive and negative regions that are separated by nodal boundaries. This situation may arise either because of the requirement of Fermi antisymmetry, or because of the presence of nodes in the spatial parts of the one-particle wavefunctions. If diffusion Monte Carlo calculations are performed on states whose spatial eigenfunctions have nodes, then the trial wavefunction $\phi_t(\mathbf{r})$ and the eigenfunction $\psi_0(\mathbf{r})$ must have the same nodal boundaries so that the importance function $f_\infty(\mathbf{r})$, which represents a density distribution, is everywhere positive

$$f_\infty(\mathbf{r}) = \psi_0(\mathbf{r})\phi_t(\mathbf{r}) \geq 0 \tag{19}$$

However, for non-trivial systems the positions of the nodes of an eigenfunction $\psi_0(\mathbf{r})$ are not in general known. A simple approximation that can be employed to overcome this uncertainty in nodal location constrains the nodes of the eigenfunction $\psi_0(\mathbf{r})$ to be the same as those of the trial wavefunction $\phi_t(\mathbf{r})$. This approximation is called the *fixed-node approximation* [1,3,7] and has proved successful in a number of studies of small molecules. The approximation is implemented computationally by deleting a configuration if a move from $\mathbf{r}$ to $\mathbf{r}'$ causes it to cross a node; that is, if the ratio $\phi_t(\mathbf{r}')/\phi_t(\mathbf{r})$ is negative. The energy E_{DMC} that is calculated by diffusion Monte Carlo within the fixed-node approximation is an upper bound to the true eigenvalue [3], and in the limit where the nodes of the trial wavefunction are the same as those of the eigenfunction $\psi_0(\mathbf{r})$, the fixed-node approximation is exact and E_{DMC} will not contain an error from this source.

3. The Computational Implementation of the Short-Time Fixed-Node Green's Function Monte Carlo Method

We shall now consider the computational algorithm of the short-time fixed-node Green's function Monte Carlo method, emphasising those features that enhance the efficiency of the algorithm. In an earlier review [1] the algorithm was presented in a form in which the diffusing particles in a configuration were moved one at a time; in the present review we shall consider a slightly different algorithm, in which they are moved simultaneously. This is more efficient when the number of diffusing particles is small.

Before starting a diffusion Monte Carlo calculation an initial distribution of $f(\mathbf{r}, t=0)$ of configurations must be chosen, and an initial estimate of the trial energy E_T is required (E_T is updated as the calculation proceeds and is ultimately adjusted to equal the estimate of the ground-state energy). Both the initial distribution $f(\mathbf{r}, 0)$ and the initial trial energy E_T can be obtained by means of a variational Monte Carlo calculation performed before the diffusion Monte Carlo calculation: a suitable initial distribution may be obtained by generating an ensemble of configurations with random positions and then propagating these by means of variational Monte Carlo. The multiplicity term $exp(-\tau(E_L(\mathbf{r})-E_T))$ is always unity in variational Monte Carlo and so no estimate of the trial energy need be made. The initial trial energy used in the diffusion Monte Carlo calculation can therefore be taken as the energy E_{VMC} obtained in a variational Monte Carlo calculation. The variational Monte Carlo algorithm is the same as the diffusion Monte Carlo algorithm described below, with the exception that configurations are not deleted if they cross a node of $\phi_t(\mathbf{r})$, and, as mentioned above, the multiplicity term is always unity and so there is no fluctuation in the ensemble size.

Figure 1 shows the algorithm employed for one 'block' of a diffusion Monte Carlo calculation in which each configuration is moved N_{move} times, subject to the constraints of deletion and copying. The initial number of configurations in the ensemble, N_{config}, is less than the maximum number of configurations dimensioned ($N_{ensemble}$) because the algorithm must allow for the possibility that the total size of the ensemble will increase if the number of configurations copied is greater than the number deleted. The details of the algorithm shown in Figure 1 are as follows:

(i) Configuration 1 is taken, with an initial position in configuration space $\mathbf{r} = (\mathbf{r}_1, \mathbf{r}_2, ...\mathbf{r}_n)$, $\mathbf{r}_i$ being the positions of the diffusing particles. The value of the trial wavefunction $\phi_t(\mathbf{r})$ at this initial position $\mathbf{r}$ is calculated, and the first and second derivatives $\nabla_i\phi_t(\mathbf{r})$ and $\nabla_i^2\phi_t(\mathbf{r})$ are calculated for each of the diffusing particles i. The local energy at this initial position

$$E_L(\mathbf{r}) = -\sum_i D_i \nabla_i^2 \phi_t(\mathbf{r})/\phi_t(\mathbf{r}) + V(\mathbf{r}) \tag{20}$$

is then calculated.

(ii) Each diffusing particle in the configuration is then moved according to the Gaussian part of the Green's function. Thus for particle 1 the move in the x direction is given by

$$\mathbf{x}_1' = \mathbf{x}_1 + 2D_1\tau \frac{1}{\phi_t(\mathbf{r})} \frac{\partial \phi_t(\mathbf{r})}{\partial \mathbf{x}_1} + \chi \tag{21}$$

and similarly for the y and z directions. The value of the timestep is generally of the order of 0.01 Hartree^{-1}. χ is a random number drawn from a Gaussian distribution with a mean of zero and a variance of $2D_1\tau$. For calculations on small systems the generation of random numbers in a Gaussian distribution is potentially one of the most time-consuming parts of the algorithm, and it is important to ensure that the generation of these random numbers is as efficient as possible. For the efficient generation of these random number sequences the following points should be borne in mind:

a)An efficient algorithm for generating random numbers in a Gaussian (or normal) distribution takes pairs of random numbers (u, v) which are uniform in the range $[0, 1]$ and from these generates pairs of normally-distributed random numbers (χ_1, χ_2) by means of the following equations

$$\chi_1 = [-4D\tau Ln(u)]^{1/2} sin(2\pi v) \tag{22}$$
$$\chi_2 = [-4D\tau Ln(u)]^{1/2} cos(2\pi v) \tag{23}$$

Other algorithms for generating normally-distributed random numbers have recently been reviewed by James [15].

Configuration = 1, $N_{ensemble}$

Calculate $\phi_t(\mathbf{r})$, $\{\nabla_i\phi_t(\mathbf{r})\}$, $\{\nabla_i^2\phi_t(\mathbf{r})\}$ and $E_L(\mathbf{r})$

Move = 1, N_{move}

$\mathbf{r}' = \mathbf{r} + 2D\tau\nabla\phi_t(\mathbf{r})/\phi_t(\mathbf{r}) + \chi$

Calculate $\phi_t(\mathbf{r}')$, $\{\nabla_i\phi_t(\mathbf{r}')\}$, $\{\nabla_i^2\phi_t(\mathbf{r}')\}$ and $E_L(\mathbf{r}')$

Delete if $\phi_t(\mathbf{r}')/\phi_t(\mathbf{r}) < 0$

Accept move with probability $Min[1, P(\mathbf{r} \rightarrow \mathbf{r}', \tau)]$

Calculate (i) $M = exp(-\tau(E_L(\mathbf{r}) - E_T))$

(ii) $M_{int} = INT(M + \mathcal{R})$

If M_{int} ≥ 2 make $(M-1)$ copies

≤ 0 delete

If move is accepted, update:

$\phi_t(\mathbf{r}) := \phi_t(\mathbf{r}')$; $\{\nabla_i\phi_t(\mathbf{r})\} := \{\nabla_i\phi_t(\mathbf{r}')\}$;

$\{\nabla_i^2\phi_t(\mathbf{r})\} := \{\nabla_i^2\phi_t(\mathbf{r}')\}$; $E_L(\mathbf{r}) := E_L(\mathbf{r}')$

Add $ME_L(\mathbf{r})$ and M to running totals

$E_{SA} =< ME_L > / < M >$

$E_T(new) = (E_T(old) + E_{SA})/2$

Print E_{SA} and $E_T(new)$

Figure 1. The computational algorithm for one 'block' of a diffusion Monte Carlo calculation.

b) If a vector-processing computer is used, then it is advantageous to generate an array of normally-distributed random numbers by means of a vectorised loop; this can reduce the time taken to generate the random numbers by more than an order of magnitude.

c) A typical diffusion Monte Carlo calculation may use of the order of 10^7 random numbers (u, v) during the course of a calculation, and it is important to ensure that the repeat cycle of the computer's random number generator is much greater than this; generally, the total of random numbers used in a calculation should not exceed the square root of the repeat cycle of the generator.

(iii) The value of the trial wavefunction at the new position $\mathbf{r}'$ is calculated, together with the first and second derivatives $\nabla_i \phi_t(\mathbf{r}')$ and $\nabla_i^2 \phi_t(\mathbf{r}')$ for each of the diffusing particles i, and the local energy $E_L(\mathbf{r}')$

$$E_L(\mathbf{r}') = -\sum_i D_i \nabla_i^2 \phi_t(\mathbf{r}')/\phi_t(\mathbf{r}') + V(\mathbf{r}') \tag{24}$$

(iv) If the move from $\mathbf{r}$ to $\mathbf{r}'$ has caused the configuration to cross a node of the trial wavefunction, then the ratio of the values of the trial wavefunction before and after the move will be negative

$$\frac{\phi_t(\mathbf{r}')}{\phi_t(\mathbf{r})} < 0 \tag{25}$$

If this is the case, then the configuration is deleted from the ensemble and the algorithm returns to stage (i) with the next configuration. Note that it is not necessary to know the exact locations of the nodes of $\phi_t(\mathbf{r})$ to determine whether node-crossing has occurred — we need to know only the ratio of the values of the trial wavefunction before and after the move.

(v) If node-crossing has not occurred, then the move from $\mathbf{r}$ to $\mathbf{r}'$ is accepted with a probability

$$Min[1, P(\mathbf{r} \rightarrow \mathbf{r}', \tau)] \tag{26}$$

where

$$P(\mathbf{r} \rightarrow \mathbf{r}', \tau) = \left(\frac{\phi_t(\mathbf{r}')}{\phi_t(\mathbf{r})}\right)^2 \frac{G(\mathbf{r}' \rightarrow \mathbf{r}, \tau)}{G(\mathbf{r} \rightarrow \mathbf{r}', \tau)} \tag{27}$$

This step ensures detailed balancing in the Monte Carlo procedure. Typically the acceptance rate is greater than 99%.

(vi) A multiplicity term M is calculated for the configuration according to the exponential part of the Green's function

$$M = exp(-\tau(E_L(\mathbf{r}) - E_T)) \tag{28}$$

This term is then rounded to an integer using the following equation

$$M_{int} = INT(M + \mathcal{R}) \tag{29}$$

where $\mathcal{R}$ is a random number uniform in the range [0,1]. If M_{int} is zero then the configuration is deleted from the ensemble; otherwise $(M_{int} - 1)$ copies of the configuration are made and these subsequently move independently of the parent configuration. A copy that is made after the Pth move of the parent configuration will be propagated for a further $(N_{move} - P)$ steps. Typically the rms value of $(M - 1)$ will be less than 0.01 if the trial wavefunction is well chosen.

(vii) If the move from $\mathbf{r}$ to $\mathbf{r}'$ has been accepted, the old values of the trial wavefunction, its first and second derivatives and the local energy at $\mathbf{r}$ are updated by the new values at $\mathbf{r}'$. This updating avoids the need to calculate these terms (as in step (i)) before the configuration is moved again.

(viii) The multiplicity M and weighted local energy $ME_L(\mathbf{r})$ are added to running totals.

(ix) The steps (ii)-(viii) are then repeated for the configuration in its new position, and the cycle continues until the configuration has made a total of N_{move} steps, or is deleted from the ensemble.

(x) Steps (i)-(ix) are then repeated for the next configuration in the ensemble.

(xi) When all the configurations, and copies of configurations, have made the requisite number of moves, the weighted average of the local energy for this 'block' (called the energy sub-average) is calculated

$$E_{SA} = \frac{< ME_L >}{< M >} \tag{30}$$

and the trial energy is updated

$$E_T(new) = (E_T(old) + E_{SA})/2 \tag{31}$$

(xii) The size of the ensemble is now re-set to its original value (N_{config}) by randomly deleting or copying configurations from the ensemble.

(xiii) The counters and running averages are then re-zeroed and steps (i)-(xii) are repeated for the next block of the calculation.

(xiv) Steps (i)-(xiii) are repeated until there is no systematic trend in successive values of E_{SA}.

(xv) When there is no systematic trend in successive energy sub-averages, the ensemble has reached its asymptotic steady state $f_\infty(\mathbf{r})$. The counters and running averages are then re-zeroed again and steps (i)-(xiv) are repeated until the Monte Carlo estimate of the energy, E_{MC}, (evaluated as the average of the energy sub-averages) has a statistical error that has been decreased to the desired tolerance

$$E_{MC} =< E_{SA} > \tag{32}$$

For typical calculations on small molecular systems the initial number of configurations in the ensemble would be $N_{config} \sim 400$ and the timestep would be $\tau \sim 0.01$ Hartree^{-1}. The number of moves made by a configuration in a block, N_{move}, is determined by a compromise between two criteria: on the one hand the number of moves should be small enough for there to be a sufficient number of 'blocks' to average over in the CPU time available; on the other hand N_{move} should be large enough for there to be little statistical correlation between the energy sub-averages obtained in successive blocks. Simple procedures have been developed for determining the statistical correlation between successive energy sub-averages in Monte Carlo calculations [16], and in the most extensive calculations M_{move} is generally set at $N_{move} \sim 2500$ [10].

4. The Optimisation of the Trial Wavefunction

The trial wavefunction $\phi_t(\mathbf{r})$ plays an important role in the short-time fixed-node Green's function Monte Carlo method: first, it forms the guiding function for the propagation of the configurations in the ensemble, and hence an accurate trial wavefunction will reduce the variance associated with the calculation of statistical averages. This can be seen by considering the form of the Green's function for diffusion Monte Carlo given in equation (14); the Gaussian part of the Green's function propagates the configurations by means of a random walk with superimposed drift (equation (15)) so as to sample preferentially the regions of configuration space where $|\phi_t(\mathbf{r})|$ is large. Furthermore, the multiplicity term $exp(-\tau(E_L(\mathbf{r}) - E_T))$ in the Green's function branches configurations according to the 'excess local energy' $(E_L(\mathbf{r}) - E_T)$, and if $\phi_t(\mathbf{r})$ is suitably chosen then $E_L(\mathbf{r})$ will be a smooth function of $\mathbf{r}$ and the multiplicity will not differ greatly from unity; the fluctuations in the ensemble size are thereby reduced and the convergence to the asymptotic steady state enhanced. In the limit where $\phi_t(\mathbf{r}) \rightarrow \psi_0(\mathbf{r})$, then $E_L(\mathbf{r}) \rightarrow E_0$ for all $\mathbf{r}$, and the fluctuations will vanish [1].

The second purpose of the trial wavefunction in the present formalism is to fix the nodes of the system, and this is associated with the accuracy that can be achieved in a calculation; the more accurately the nodes of $\phi_t(\mathbf{r})$ locate those of $\psi_0(\mathbf{r})$, the greater the accuracy of the calculation. In the limit where $\phi_t(\mathbf{r}) \rightarrow \psi_0(\mathbf{r})$, the fixed-node approximation becomes exact.

The choice of the trial wavefunction to be employed in a particular calculation is determined by a compromise between two opposing criteria: on the one hand we wish the trial wavefucntion to be as simple and compact as possible, so that the computation of the wavefunction and its first and second derivatives at points $\mathbf{r}$ in configuration space will not be time-consuming; on the other hand we wish the trial wavefunction to be as accurate as possible, since it will then lead to a smaller statistical variance for a given amount of sampling and will accurately locate the nodes of the system.

For molecular electronic systems a trial wavefunction that has been found [3,7] to be a suitable compromise between these criteria is one constructed as a product of a Hartree-Fock wavefunction, in the form of a Slater determinant, and a Jastrow factor [17], which allows explicitly for electron correlation

$$\phi_t(\mathbf{r}) = det|\phi_k(\mathbf{r})|exp(\sum_{i<j} U_{ij}) \tag{33}$$

where i and j label electrons, and

$$U_{ij} = \frac{ar_{ij}}{1 + br_{ij}} \tag{34}$$

The value of the constant a is determined by the requirement that the wavefunction has the correct cusp condition as two electrons approach each other: $a = 1/8D$ if the electrons have the same spin, and $a = 1/4D$ if they have opposite spin. The Slater determinant $det|\phi_k(\mathbf{r})|$ determines the nodes of the system; the $\phi_k(\mathbf{r})$ are molecular orbitals which are formed from a linear combination of atomic orbitals, the orbital exponents and expansion coefficients of this linear combination being optimised by *ab initio* Hartree-Fock calculations.

More sophisticated forms of trial wavefunction have also been considered, such as those which employ multi-configuration SCF wavefunctions [18], or which include electron-nucleus Jastrow factors [2,3]. These more sophisticated trial wavefunctions are considered in detail in reference [1]. The replacement of the Jastrow factor by Hylleraas-like terms has also been considered [41].

In this review we shall discuss the development of an efficient algorithm for the optimisation of the parameters of the trial wavefunction: if the trial wavefunction is constructed by augmenting a Hartree-Fock wavefunction by a Jastrow factor, then because the parameters

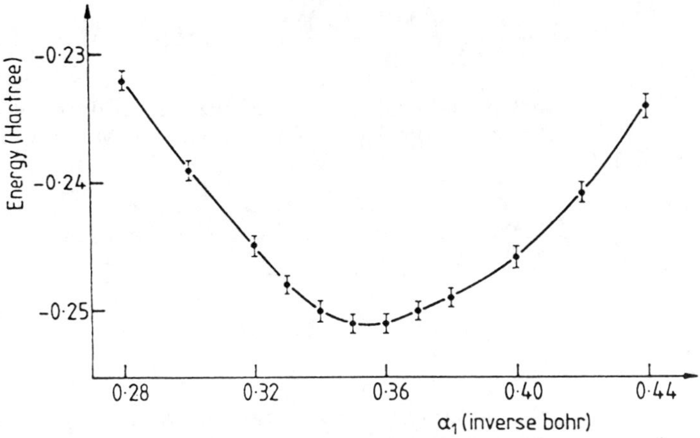

Figure 2. The optimisation of the trial wavefunction $\phi_t(\mathbf{r})$ (equation (35)) for $e_2^- e^+$: The variational Monte Carlo energy E_{VMC} plotted against the wavefunction parameter α_1

of the Hartree-Fock wavefunction were optimised in the absence of the Jastrow factor there is no guarantee that the resulting parameters of this product trial wavefunction will have their optimum values. It is therefore important to be able to optimise directly the parameters of the product trial wavefunction (that is, to find the set of values that minimises the variational Monte Carlo energy of $\phi_t(\mathbf{r})$): if the trial wavefunction contains a small number of parameters this may be done by means of a series of separate variational Monte Carlo calculations in which each parameter is varied in turn; i.e. holding all the parameters of the trial wavefunction except one fixed, the variational Monte Carlo energy is calculated at a range of values of the variable parameter in order to locate the value of that parameter which gives the lowest variational Monte Carlo energy. The procedure is then repeated for the other parameters in the trial wavefunction. This procedure is illustrated in Figure 2, in which a trial wavefunction of the form

$$\phi_t(\mathbf{r}) = e^{-\alpha_1 r_{13}} e^{-\alpha_1 r_{23}} e^{-\alpha_2 r_{12}/(1+\alpha_3 r_{12})} e^{-\alpha_4 r_{13}/(1+\alpha_5 r_{13})} e^{-\alpha_4 r_{23}/(1+\alpha_5 r_{23})} \tag{35}$$

is being optimised for calculations on the ion $e_2^- e^+$ [10], consisting of two electrons bound to a positron (1 and 2 label the electrons, 3 labels the positron); in the figure the variational energy is plotted as a function of α_1, with all the other parameters being held fixed.

As the number of parameters in the trial wavefunction increases it is clear that this procedure of optimisation will become very time-consuming, and it would therefore be advantageous to employ a more efficient optimisation procedure which optimised all the parameters of the wavefunction simultaneously and which used a knowledge of the gradients $\partial E_{VMC}/\partial \alpha_i$ to improve the rate of convergence to the optimised form.

It is possible to calculate the gradients $\partial E_{VMC}/\partial \alpha_i$ in the following way [19]: we consider a trial wavefunction $\phi_t(\mathbf{r}, \{\alpha_i\})$, which depends on a set of parameters $\{\alpha_i\}$ (both orbital coefficients and wavefunction exponents), and define the logarithmic partial derivative of the trial wavefunction with respect to an arbitrary parameter α_j as $P_j(\mathbf{r}, \{\alpha_i\})$, i.e.

$$\frac{1}{\phi_t(\mathbf{r},\{\alpha_i\})}\left(\frac{\partial\phi_t(\mathbf{r},\{\alpha_i\})}{\partial\alpha_j}\right)_{\alpha_{k\neq j}} = P_j(\mathbf{r},\{\alpha_i\}) \tag{36}$$

It is possible to show that if the hamiltonian $\hat{\mathcal{H}}$ of the system does not depend on the parameters $\{\alpha_i\}$ then the partial derivative of the variational Monte Carlo energy with respect to α_j is [19]

$$\left(\frac{\partial E_{VMC}}{\partial\alpha_j}\right)_{\alpha_{k\neq j}} = 2(< P_j E_L > - < P_j >< E_L >) \tag{37}$$

where

$$< X >= \frac{\int d\mathbf{r}\phi_t(\mathbf{r})^2 X}{\int d\mathbf{r}\phi_t(\mathbf{r})^2} \tag{38}$$

But $< E_L >$ is simply the variational Monte Carlo energy, and $< P_j >$ and $< P_j E_L >$ can be obtained as expectation values during the course of the variational Monte Carlo calculation. The first derivatives $\partial E_{VMC}/\partial\alpha_i$ are therefore easily calculated, and can then be used to improve the convergence of the trial wavefunction to its optimised form.

The computational algorithm employed for optimising the trial wavefunction by this procedure is as follows: A set of values $\{\alpha_i\}$ is chosen as a first estimate of the parameters of the trial wavefunction, and a calculation consisting of a single variational Monte Carlo 'block' is then performed using this trial wavefunction. In this block each configuration in the ensemble is moved N_{move} times (typically $N_{config} \sim 400$ and $N_{move} \sim 2500$), and after each move the local energy E_L and the logarithmic derivatives P_i are calculated, and E_L and each of the quantities P_i and $P_i E_L$ are added to running totals. At the end of the block the average $< E_L >$ and the sets of averages $< P_i >$ and $< P_i E_L >$ are evaluated, and from these the derivatives $\partial E_{VMC}/\partial\alpha_i$ are calculated. From a knowledge of the set of parameters $\{\alpha_i\}$ used in the variational Monte Carlo calculation and the derivatives $\{\partial E_{VMC}/\partial\alpha_i\}$, a new improved set of parameters can be calculated. These new parameters are then used in the next variational Monte Carlo block, and the procedure iterates until there is no discernable trend in successive sets of $\{\alpha_i\}$ values. The optimised wavefunction can then be used as the trial wavefunction in diffusion Monte Carlo calculations.

As an example of this procedure, we shall consider the optimisation of a trial wavefunction for atomic helium [19]:

$$\phi_t(\mathbf{r}) = \left(\sum_{i=1,2}(Ae^{-\alpha_1 r_i} + Be^{-\alpha_2 r_i} + Ce^{-\alpha_3 r_i})\right)e^{ar_{12}/(1+br_{12})} \tag{39}$$

This trial wavefunction consists of a triple-zeta Hartree-Fock wavefunction augmented by a Jastrow factor. The initial values of the Hartree-Fock parameters were taken from the work of Clementi and Roetti [40], and the Jastrow parameter a was given the correct value for the electron-electron cusp condition. The values of the starting parameters are given in Table I(a). The parameters of this trial wavefunction were optimised using the procedure described above, and the values of the optimised parameters are also given in Table I(a). It is notable that the Jastrow parameter a has departed from its cusp-condition value, and the increase in the exponent α_1 decreases the electron density at large electron-nuclear separations. Table I(b) compares the variational Monte Carlo energy for the wavefunction before optimisation with that after optimisation, and it is clear that optimisation has produced a marked lowering of the variational energy. Because the spacial part of the

TABLE I(a)

Parameter	Before Optimisation	After Optimisation
A	1.45276	1.43241
B	0.8496	0.8856
C	0.07485	0.06813
α_1	1.45286	1.56941
α_2	2.77954	2.7759
α_3	4.346	4.3434
a	0.5	0.66483
b	0.833	0.69832

Table I(a). The parameters of the triple-zeta wavefunction for atomic helium before and after variational Monte Carlo optimisation.

TABLE I(b)

	$E_{VMC}(H)$	$E_{DMC}(H)$
Before Optimisation	-2.8882(6)	-2.9033(6)
After Optimisation	-2.8992(2)	-2.9037(3)

Table I(b). The variational and diffusion Monte Carlo energies for helium obtained using the triple-zeta wavefunctions before and after optimisation. The figures in parentheses indicate the error in the last decimal place

ground-state wavefunction of helium contains no nodes, the two wavefunctions should give the same diffusion Monte Carlo energy to within statistical error. This is indeed the case, as shown in Table I(b), but it is clear that the optimised trial wavefunction gives a diffusion Monte Carlo energy which, for a given size of calculation, has a smaller statistical error.

5. The Ground-State Energy of the $e_2^- e^+$ System

We shall now consider the application of the short-time Green's function Monte Carlo method to the calculation of the ground-state energy of the three-body system $e_2^- e^+$ (or Ps^-) consisting of two electrons bound to a positron (or, equivalently, the system $e_2^+ e^-$ or Ps^+, consisting of two positrons bound to an electron). This system has been the subject of a number of recent studies using hyperspherical coordinates [22,24,25]. A study of this system using the short-time Green's function Monte Carlo method has recently been published [10], and this work is extended in the present review.

We consider two trial wavefunctions for the system $e_2^- e^+$:

$$\phi_t^a(\mathbf{r}) = e^{-\alpha_1 r_{13}} e^{-\alpha_1 r_{23}} e^{-\alpha_2 r_{12}/(1+\alpha_3 r_{12})} e^{-\alpha_4 r_{13}/(1+\alpha_5 r_{13})} e^{-\alpha_4 r_{23}/(1+\alpha_5 r_{23})} \tag{40}$$

$$\phi_t^b(\mathbf{r}) = (e^{-\alpha r_{13}} e^{-\beta r_{23}} + e^{-\alpha r_{23}} e^{-\beta r_{13}})(1 + A r_{12} e^{-\gamma r_{12}}) \tag{41}$$

TABLE II(a)

$E_{VMC}(\phi_t^a(\mathbf{r}))$	$E_{VMC}(\phi_t^b(\mathbf{r}))$
-0.2487(6)	-0.2607(1)

Table II(a). The variational Monte Carlo energies (in Hartree) for the $e_2^- e^+$ system obtained using the trial wavefunctions $\phi_t^a(\mathbf{r})$ and $\phi_t^b(\mathbf{r})$. The figures in parentheses indicate the error in the last decimal place.

TABLE II(b)

Timestep (H^{-1})	$E_{DMC}(\phi_t^a(\mathbf{r}))$	$E_{DMC}(\phi_t^b(\mathbf{r}))$
0.001	-0.26212(17)	
0.002	-0.26222(22)	
0.005	-0.26246(18)	-0.261996(59)
0.010	-0.26262(14)	-0.261999(39)
0.020	-0.26289(13)	-0.261995(25)

Table II(b). Diffusion Monte Carlo energies (in Hartree) for the $e_2^- e^+$ system as a function of timestep for the two trial wavefunctions $\phi_t^a(\mathbf{r})$ and $\phi_t^b(\mathbf{r})$. The figures in parentheses indicate the error in the last two decimal places.

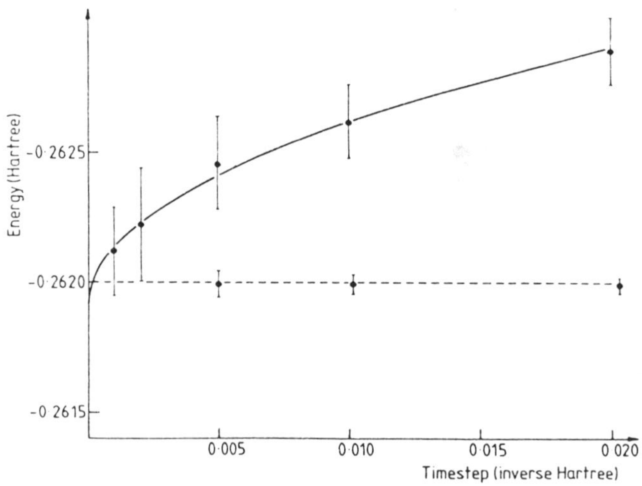

Figure 3. The diffusion Monte Carlo energy of the $e_2^- e^+$ system as a function of timestep τ: — trial wavefunction $\phi_t^a(\mathbf{r})$; - - - - trial wavefunction $\phi_t^b(\mathbf{r})$

TABLE III

	Energy (H)
$E_{DMC}^{\tau\to 0}(\phi_t^a(\mathbf{r}))$	-0.2619(2)
$E_{DMC}^{\tau\to 0}(\phi_t^b(\mathbf{r}))$	-0.26199(3)
Frey [22]	
Upper Bound	-0.2601
Lower Bound	-0.2654
Bhatia and Drachman [23]	-0.262005
Burden [24]	-0.2417
Mignaco and Roditti [25]	-0.248105
Kolos *et al* [26]	-0.2619956

Table III. The ground-state energies of $e_2^- e^+$ obtained by Green's function Monte Carlo using the two trial wavefunctions $\phi_t^a(\mathbf{r})$ and $\phi_t^b(\mathbf{r})$ (extrapolated to the limit $\tau \to 0$) compared with the results obtained in previous studies

where 1 and 2 label the electrons and 3 labels the positron. The trial wavefunction $\phi_t^a(\mathbf{r})$ is the product of two exponential wavefunctions $e^{-\alpha_1 r_{i3}}$ (i=1,2) and three Jastrow factors $e^{-\alpha_m r_{ij}/(1+\alpha_n r_{ij})}$. The exponential functions and the two Jastrow factors $e^{-\alpha_4 r_{i3}/(1+\alpha_5 r_{i3})}$ (i=1,2) describe the electron-positron correlation, and the Jastrow factor $e^{-\alpha_2 r_{12}/(1+\alpha_3 r_{12})}$ describes the electron-electron correlation. The optimised parameters of this wavefunction are given in reference [10]. The trial wavefunction $\phi_t^b(\mathbf{r})$ was taken from the work of Ferrante and Geracitano [20]. The diffusion Monte Carlo calculation of the ground-state energy of $e_2^- e^+$ gives a result which is independent of the trial wavefunction; however, as discussed above, the variance in the diffusion Monte Carlo energy can be reduced by using a well-optimised trial wavefunction.

For both trial wavefunctions the variational Monte Carlo energies E_{VMC} were calculated, and diffusion Monte Carlo calculations were then performed at a number of timesteps τ [10,21]. The results of the calculations are given in Tables II(a) and II(b). To estimate the error associated with the average energies, each calculation was divided into a number of blocks, and the error quoted is the standard error of the mean of the sub-average energies obtained at the end of each block. The number of moves in each block, N_{move} was increased as the timestep decreased to ensure that the standard error did not underestimate the true error of the mean because of the serial correlation between successive sub-averages.

In Figure 3 the energies calculated by diffusion Monte Carlo for the two trial wavefunctions are plotted as a function of timestep. For $\phi_t^b(\mathbf{r})$ there is no systematic timestep dependence that can be discerned to within the accuracy of the calculation. However, for $\phi_t^a(\mathbf{r})$ there is some systematic timestep dependence, and as the time corrections to the short-time Green's function may be expressed as a power series in $\tau^{1/2}$ the results using this wavefunction were fitted to the equation $E(\tau) = E(0) + A\tau^{1/2}$. The estimates of the extrapolated ground-state energies for the two trial wavefunctions are given in Table III, where they are compared with the results of other workers. The standard errors associated with these energies have been taken to be of the same order as those associated with the energies calculated at specific timesteps. It can be seen that the results of the diffusion Monte Carlo calculations are in good agreement with the best analytical results for this system.

6. The Electron-Hole Exciton in Small Semiconductor Crystallites

In recent years there has been considerable interest in the properties of semiconductor particles containing up to a few hundred atoms [27]. These crystallites display electronic properties that are strongly dependent on the size and shape of the crystallite; for instance, the ultra-violet absorption maxima of cadmium suphide suspensions in water are at much shorter wavelengths than those of the bulk material [28,29,30] because the kinetic energy of the electron-hole exciton produced by light absorption is increased by the confinement to a small crystallite, and this results in an increased band gap.

Our understanding of the optical properties of these small semiconductor crystallites has been aided by theoretical studies which seek to calculate the eigenvalues of the electron-hole exciton system. The simplest theoretical model of the confined electron-hole exciton describes a two-body system — a conduction-band electron of effective mass m_e and a valence-band hole of effective mass m_h — which is confined by an infinite potential to a spherically-symmetric continuum of radius R and dielectric coefficient ϵ [31,32,33]. The Schrödinger equation for this model is, therefore

$$\left(-\frac{1}{2m_e}\nabla_e^2 - \frac{1}{2m_h}\nabla_h^2 - \frac{1}{\epsilon r_{eh}}\right)\Psi(\mathbf{r_e},\mathbf{r_h}) = E\Psi(\mathbf{r_e},\mathbf{r_h}) \tag{42}$$

where the wavefunction $\Psi(\mathbf{r_e},\mathbf{r_h})$ satisfies the boundary condition

$$\Psi(\mathbf{r_e},\mathbf{r_h}) = 0 \qquad\qquad |\mathbf{r_e}|,|\mathbf{r_h}| \geq R \tag{43}$$

Approximate eigenvalues for this model have recently been calculated by Brus [31,32], and by Schmidt and Weller [33]. Brus performed variational calculations in which the wavefunction $\Phi(\mathbf{r_e},\mathbf{r_h})$ was expressed as a linear combination of the products $\phi_i(\mathbf{r}_e)\phi_j(\mathbf{r}_h)$ of particle-in-a-sphere eigenfunctions. Variational calculations have also been performed by Schmidt and Weller, who used a Hylleraas-type wavefunction to account explicitly for electron-hole correlation.

The objectives of the Green's function Monte Carlo study [34,35,36] reviewed here are two-fold: first, to calculate accurate eigenvalues of the Schrödinger equation for the confined electron-hole exciton (equation (42)); secondly, to investigate the effect of improvements to this simple model, in particular the effect of electron tunnelling through the surface of the crystallite. In this review the discussion will be restricted to a study of the ground state of the electron-hole exciton in CdS crystallites suspended in water; extensive results for other systems are to be the subject of a future publication [36].

For the *hard-wall* model, in which both the electron and the hole are confined to the crystallite, the trial wavefunction is taken to be

$$\phi_t(\mathbf{r}) = \phi_e(\mathbf{r}_e)\phi_h(\mathbf{r}_h)e^{-\alpha r_{eh}} \tag{44}$$

where $\phi_e(\mathbf{r}_e)$ and $\phi_h(\mathbf{r}_h)$ are particle-in-a-sphere eigenfunctions for the electron and hole respectively, and the term $e^{-\alpha r_{eh}}$ allows explicitly for electron-hole correlation. The particle-in-a-sphere eigenfunctions may be expressed as products of spherical Bessel functions and spherical harmonics [37], and for the ground state of the electron-hole exciton the trial wavefunction takes the form

$$\phi_t(\mathbf{r}) = \left(\frac{sin(\pi r_e/R)}{\pi r_e/R}\right)\left(\frac{sin(\pi r_h/R)}{\pi r_h/R}\right)e^{-\alpha r_{eh}} \tag{45}$$

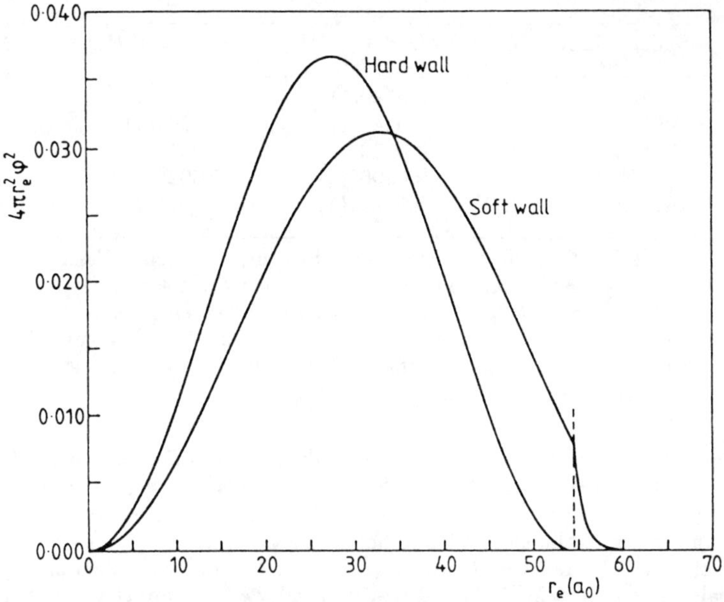

Figure 4. The radial distribution function for the electronic component of the trial wavefunction $\phi_e(\mathbf{r}_e)$ for the hard-wall model, in which the electron is confined to the crystallite, compared with that for the soft-wall model, in which it is allowed to tunnel. The radius of the crystallite was set at $R = 54.545$ a.u.

For the *soft-wall* model, in which the electron is allowed to tunnel through the surface of the crystallite, the trial wavefunction still has the form given in equation (44), but the electronic component $\phi_e(\mathbf{r}_e)$ is no longer a particle-in-a-sphere eigenfunction. The single-particle model that is used to obtain a suitable electronic component of the trial wavefunction assumes that when the electron is inside the crystallite it is in a spherically-symmetric well of potential $-V_0$ [37]. The Schrödinger equation obeyed by the electron is therefore

$$(-1/2m_e)\nabla_e^2\phi_e(\mathbf{r}_e) = (E - V_0)\phi_e(\mathbf{r}_e) \qquad r_e < R \tag{46}$$

$$(-1/2m_0)\nabla_e^2\phi_e(\mathbf{r}_e) = E\phi_e(\mathbf{r}_e) \qquad r_e > R \tag{47}$$

Where m_0 is the free-electron mass. The form of the electronic wavefunction is therefore [32,37]

$$\phi_e(\mathbf{r}_e) = A sink r_e/r_e \qquad r_e < R \tag{48}$$

$$\phi_e(\mathbf{r}_e) = Be^{-Kr_e}/r_e \qquad r_e > R \tag{49}$$

The details of the calculation of the coefficients (A,k,B,K) are given by Brus [32] and by Flügge [37]. The electronic wavefunction radial distribution functions $4\pi r^2\phi_e^2(\mathbf{r}_e)$ for the hard-wall and soft-wall models are compared in Figure 4. The discontinuity in the radial

TABLE IV

Sphere Radius (a.u.)	Hard Wall (H)	Soft Wall (H)
18.182	0.0789(1)	0.0326(1)
36.364	0.0146(1)	0.0067(1)
54.545	0.00400(4)	0.0012(6)
72.727	0.00062(5)	-0.002(1)
181.820	-0.00201(1)	-0.00217(1)

Table IV. Preliminary results for the diffusion Monte Carlo energies for the hard-wall model, in which the electron is confined to the crystallite, compared with the energies of the soft-wall model, in which the electron is allowed to tunnel through the crystallite surface. The figures in parentheses indicate the error in the last decimal place.

distribution function for the soft-wall model is due to the different values of the electronic mass inside and outside the crystallite; the gradient $1/m(\partial\phi_e(\mathbf{r}_e)/\partial\mathbf{r}_e)$ is continuous.

For the study of the ground-state energy of the electron-hole exciton in CdS crystallites, the effective masses of the electron and hole were taken as $m_e = 0.19$ a.u. and $m_h = 0.8$ a.u. respectively [33], and the dielectric coefficient of CdS was taken as $\epsilon_{CdS} = 5.6$ [33]. For the calculations in which the electron was allowed to tunnel through the crystallite surface, the CdS crystallites were modelled as being suspended in water, and V_0 was taken as the energy difference between the lower edge of the conduction band in bulk CdS and the empty conduction band of water (these calculations used $V_0 = 4$ ev), and the dielectric coefficient of water was taken to be $\epsilon_{H_2O} = 1.7$.

Preliminary calculations have been performed on the two models at a range of CdS crystallite radii. The results are shown in Table IV. The second column of Table IV shows the diffusion Monte Carlo results for the hard-wall model, in which the electron and hole are both confined to the crystallite; these results are in good agreement with those of Schmidt and Weller [33] (who used the same values of ϵ, m_e and m_h, but slightly different values of R). The third column of Table IV shows the diffusion Monte Carlo results for the soft-wall model, in which the electron can tunnel through the crystallite surface. It is clear that the effect of tunnelling is to lower the energy of the system, the energy lowering being greater at the smaller crystallite radii.

7. The Differential Green's Function Monte Carlo Method

Green's function Monte Carlo calculations on small molecules give results for the average energy which have associated with them a standard error which is generally of the order of 10^{-3} H to 10^{-4} H [1], and this limits the extent to which Green's function Monte Carlo methods can be applied to some aspects of the study of molecular electronic structure, such as the detailed investigation of the energy changes that occur on chemical bonding. Although the standard error could be reduced by increasing the length of the calculation (to increase the number of statistically-independent energy estimates) it is not always computationally tractable to do so; for example, if the standard error is to be reduced by a factor of three then the length of the calculation must be increased by a factor of nine. The differential Green's function Monte Carlo method provides an alternative means of reducing the error of a calculation; in this method the reduction is achieved by a modification of the

Green's function Monte Carlo algorithm, rather than by an increase in the length of the calculation.

The differential Green's function Monte Carlo method uses correlated sampling to calculate the difference between two energy expectation values. It has been used to estimate the dipole moment of LiH by performing a direct calculation of the energy change that occurs when the molecule is placed in a weak electric field [1,14]; it has also been applied to the more general problem of the calculation of total energies [34,38] by means of direct calculations of the difference between the known (variational) energy of an *ab initio* wavefunction and the diffusion Monte Carlo estimate of the exact energy.

The differential Green's function Monte Carlo algorithm [1,14] is concerned with the calculation of the energy difference ΔE between two expectation values, which will be defined as E^A and E^B

$$\Delta E = E^A - E^B \tag{50}$$

Each of the expectation values is evaluated by propagating an ensemble of configurations according to the appropriate Green's function and then, once the asymptotic distribution has been attained, sampling the local energy. It is possible to employ the techniques of correlated sampling when evaluating the two expectation values in order to reduce the statistical error with which ΔE is estimated. There are two criteria that must be satisfied for the sampling of the two expectation values to be correlated: first, the two ensembles of configurations used to sample the two expectation values must be propagated by the same Green's function; secondly, the two ensembles of configurations must have the same initial positions. The first criterion logically comprises three separate sub-criteria: first, the same trial wavefunction should appear in the Gaussian parts of the two Green's functions; secondly, the corresponding configurations in the two ensembles should be propagated by the same random number sequence; thirdly, the multiplicity terms should be the same for the two Green's functions. The first and second of these sub-criteria ensure that corresponding configurations in the two ensembles make identical moves, while the third ensures that the copying and deletion of configurations are the same in the two ensembles, and so the behaviour of the configurations in the two ensembles is exactly correlated.

The mathematical formulation of the differential Green's function Monte Carlo procedure for calculating the energy difference ΔE is as follows: The importance functions $f^A(\mathbf{r},t)$ and $f^B(\mathbf{r},t)$, which represent the ensembles of configurations used in the evaluation of the expectation values E^A and E^B respectively, have the same initial forms

$$f^A(\mathbf{r},0) = f^B(\mathbf{r},0) \tag{51}$$

$f^A(\mathbf{r},t)$ is then propagated according to the following equation

$$f^A(\mathbf{r}',t+\tau) = \int d\mathbf{r} G^A(\mathbf{r}\rightarrow\mathbf{r}',\tau) f^A(\mathbf{r},t) \tag{52}$$

where

$$G^A(\mathbf{r}\rightarrow\mathbf{r}',\tau) = (4\pi D\tau)^{-3N/2} exp\left(-\frac{(\mathbf{r}'-\mathbf{r}-2D\tau\nabla\phi_t(\mathbf{r})/\phi_t(\mathbf{r}))^2}{4D\tau}\right) exp(-\tau(E_L^A(\mathbf{r}) - E_T^A)) \tag{53}$$

and

$$E_L^A(\mathbf{r}) = \hat{\mathcal{H}}^A\phi_t(\mathbf{r})/\phi_t(\mathbf{r}) \tag{54}$$

$\hat{\mathcal{H}}^A$ being the hamiltonian of the system A. The weighted average of the local energy sampled from the asymptotic distribution $f^A_\infty(\mathbf{r})$ will give the expectation value E^A

$$E^A = \frac{< M^A E_L^A >}{< M^A >} \tag{55}$$

where $M^A = exp(-\tau(E_L^A(\mathbf{r}) - E_T))$. The ensemble of configurations representing the importance function $f^B(\mathbf{r},t)$ is propagated in a similar fashion

$$f^B(\mathbf{r}',t+\tau) = \int d\mathbf{r} G^B(\mathbf{r} \to \mathbf{r}',\tau) f^B(\mathbf{r},t) \tag{56}$$

where

$$G^B(\mathbf{r} \to \mathbf{r}',\tau) = (4\pi D\tau)^{-3N/2} exp\left(-\frac{(\mathbf{r}' - \mathbf{r} - 2D\tau\nabla\phi_t(\mathbf{r})/\phi_t(\mathbf{r}))^2}{4D\tau}\right) exp(-\tau(E_L^B(\mathbf{r}) - E_T^B)) \tag{57}$$

and

$$E_L^B(\mathbf{r}) = \hat{\mathcal{H}}^B \phi_t(\mathbf{r})/\phi_t(\mathbf{r}) \tag{58}$$

$\hat{\mathcal{H}}^B$ being the hamiltonian for system B.

However, if $G^B(\mathbf{r} \to \mathbf{r}',\tau)$ is used to propagate the ensemble $f^B(\mathbf{r},t)$ then the sampling of the two expectation values E^A and E^B will not be exactly correlated because the two Green's functions have different multiplicity terms; for the walks to be exactly correlated the importance function $f^B(\mathbf{r},t)$ must be propagated using $G^A(\mathbf{r} \to \mathbf{r}',\tau)$, and this may be achieved formally by introducing $G^A(\mathbf{r} \to \mathbf{r}',\tau)$ into equation (56).

$$f^B(\mathbf{r}',t+\tau) = \int d\mathbf{r}\left(\frac{G^B(\mathbf{r} \to \mathbf{r}',\tau)}{G^A(\mathbf{r} \to \mathbf{r}',\tau)}\right) G^A(\mathbf{r} \to \mathbf{r}',\tau) f^B(\mathbf{r},t) \tag{59}$$

But the Gaussian parts of the two Green's functions are the same, and so the ratio of the Green's functions is simply the ratio of the multiplicity factors, hence

$$f^B(\mathbf{r}',t+\tau) = \int d\mathbf{r}\left(\frac{e^{-\tau(E_L^B(\mathbf{r}) - E_T^B)}}{e^{-\tau(E_L^A(\mathbf{r}) - E_T^A)}}\right) G^A(\mathbf{r} \to \mathbf{r}',\tau) f^B(\mathbf{r},t) \tag{60}$$

Therefore the correct sampling of the expectation E^B is obtained by propagating the ensemble using the Green's function $G^A(\mathbf{r} \to \mathbf{r}',\tau)$ and weighting the kth move of a configuration by the factor

$$W_k = \prod_{i=1}^{k-1}\left(\frac{e^{-\tau(E_L^B(\mathbf{r}) - E_T^B)}}{e^{-\tau(E_L^A(\mathbf{r}) - E_T^A)}}\right)_i \tag{61}$$

where i labels the previous moves made by the configuration. The expectation E^B is therefore calculated as a weighted average of the local energy

$$E^B = \frac{< W_k M^B E_L^B >}{< W_k M^B >} \tag{62}$$

where $M^B = exp(-\tau(E_L^B(\mathbf{r}) - E_T))$. The correlated energy difference is therefore

$$\Delta E = \frac{< M^A E_L^A >}{< M^A >} - \frac{< W_k M^B E_L^B >}{< W_k M^B >} \tag{63}$$

In practice, the weight W_k is not accumulated over the complete past history of a configuration; rather, it has been found advantageous to define a chain length $l(0 \leq l \leq k)$ over which the weight is calculated. Thus after the kth move of a configuration the weight $W(l)$ is

$$W(l) = \prod_{i=k-l-1}^{k-1} \left(\frac{e^{-\tau(E_L^B(\mathbf{r}) - E_T^B)}}{e^{-\tau(E_L^A(\mathbf{r}) - E_T^A)}} \right)_i \tag{64}$$

and the energy difference $\Delta E(l)$ obtained using this value of $W(l)$ is

$$\Delta E(l) = \frac{< M^A E_L^A >}{< M^A >} - \frac{< W(l) M^B E_L^B >}{< W(l) M^B >} \tag{65}$$

Computationally, a single calculation evaluates $W(l)$ and $\Delta E(l)$ for a series of l values, and a graph of $\Delta E(l)$ vs l is plotted; as l increases from 0, the energy difference $\Delta E(l)$ approaches its asymptotic value.

The differential Green's function Monte Carlo method has been used to determine the dipole moment of LiH by means of a finite-field technique [14]: system B, which gives the expectation value E^B, described the molecule in the presence of an electric field of strength F directed along the internuclear axis, and system A, which gives the expectation E^A, described the molecule in the absence of an electric field. The energy in the presence of the electric field is given by the expression

$$E(F) = E(0) - \mu F - \frac{1}{2}\alpha F^2 - ... \tag{66}$$

and the differential Green's function Monte Carlo calculation evaluates the energy difference

$$\Delta E(F) = E(F) - E(0) \tag{67}$$

To calculate a value of the dipole moment μ free from low-order non-linear terms, four values of the electric field were considered:$\pm$F and $\pm$2F. Manipulation of the Taylor series then gives

$$\frac{1}{12}[\Delta E(2F) - \Delta E(-2F)] - \frac{2}{3}[\Delta E(F) - \Delta E(-F)] = -\mu F + \mathcal{O}(F^5) \tag{68}$$

The strength of the field F was determined by a compromise: a small field ensures a close similarity in the multiplicity patterns of the two systems, and so the reweighting factor $W(l)$ will not fluctuate greatly, but the small energy difference that results may be masked by the statistical variance inherent in the differential method itself; a large field will ensure a large and easily observable energy difference, but may result in the reweighting scheme (equation (64)) being unable to rescale fully the energy in system B. Prototye calculations were performed on a range of field strengths to determine the most suitable value of the field F and a value of $F = 0.0001$ a.u. was chosen, although the range of suitable field strengths extends at least an order of magnitude either side of this value. The results obtained after a total of 4050 moves per configuration in an ensemble of average size 100 configurations are shown in Table V. It is clear that the correlated energy difference is about an order of magnitude smaller than the error associated with the total energies of the two systems. Analysis of the results gave a dipole moment of 2.27 $\pm$ 0.03 a.u. [14], which agrees well with the experimental value(extrapolated to the equilibrium internuclear separation) of 2.29 a.u. [39].

TABLE V

Energy of system A (E(0))	Energy of system B (E(F))	Correlated energy difference ΔE
-8.0593(3)	-8.0595(3)	0.000227(3)

Table V. Energies (in Hartree) of the systems E(0) and E(F), and the correlated energy difference, for a field strength of 0.0001 a.u. The figures in parentheses indicate the error in the last decimal place.

8. References

1. Wells, B.H., 1987, in *Methods in Computational Chemistry Vol. 1* ed. Wilson, S., Plenum (New York).
2. Ceperley, D.M. and Alder, B.J., 1984, *J. Chem. Phys.* 81 5833.
3. Reynolds, P.J., Ceperley, D.M., Alder, B.J. and Lester, W.A. Jnr., 1982, *J. Chem. Phys.* 77 5593.
4. Moskowitz, J.W. and Kalos, M.H., 1981, *Int. J. Quant. Chem.* 20 1107.
5. Moskowitz, J.W., Schmidt, K.E., Lee, M.A. and Kalos, M.H., 1982 *J. Chem. Phys.* 76 1064.
6. Mentch, F. and Anderson, J.B., 1981, *J. Chem. Phys.* 74 6307
7. Moskowitz, J.W., Schmidt, K.E., Lee, M.A. and Kalos, M.H., 1982, *J. Chem. Phys.* 77 349.
8. Wells, B.H., 1984, *Inf. Quart. Comput. Sim. Cond. Phases* 14 14 (SERC Daresbury England)
9. Vrbik, J. and Rothstein, S.M., 1986, *Int. J. Quant. Chem.* 29 461.
10. Ball, R.G.J. and Wells, B.H., 1988, *J. Phys. B* 21 723.
11. Skinner, D.W., Moskowitz, J.W., Lee, M.A., Whitlock, P.A. and Schmidt, K.E., 1985, *J. Chem. Phys.* 83 4668.
12. Moskowitz, J.W. and Schmidt, K.E., 1986, *J. Chem. Phys.* 85 2868.
13. Ceperley, D.M., 1983, *J. Compt. Phys.* 51 404.
14. Wells, B.H., 1985, *Chem. Phys. Lett.* 115 89.
15. James, F., 1980, *Rep. Progr. Phys.* 43 1154.
16. Smith, E.B. and Wells, B.H., 1984, *Mol. Phys.* 52 701.
17. Jastrow, R., 1955, *Phys. Rev.* 98 1479.
18. Harrison, R.J. and Handy, N.C., 1985, *Chem. Phys. Lett.* 113 257.
19. Jelf, O., 1988, *Pt. II Thesis*, Oxford University.
20. Ferrante, G. and Geracitano, R., 1970, *Nuovo Cim. Lett.* 3 48.
21. Ball, R.G.J., 1988, *Pt. II Thesis*, Oxford University.
22. Frey, J.G., 1986, *Mol. Phys.* 57 1.
23. Bhatia, A.K. and Drachman, R.J., 1983, *Phys. Rev. A* 28 2523.
24. Burden, F.R., 1983, *J. Phys. B* 16 2289.
25. Mignaco J.A. and Roditti, I., 1981, *J. Phys. B* 14 L161.
26. Kolos, W., Roothaan, C.C.J. and Sack, R.A., 1960, *Rev. Mod. Phys.* 32 178.
27. Ramsden, J.J., 1987, *Proc. R. Soc. Lond.* A410 89.
28. Fojtik, A., Weller, H., Koch, U. and Henglein, A., 1984, *Ber. Bunsenges Phys. Chem.* 88 969.

29. Nozik, A.J., Williams, F., Nenadovic, M.T., Rajh, T. and Micic, O.J., 1985, *J. Phys. Chem.* 89 1985.
30. Rossetti, R., Ellison, J.L., Gibson, J.M. and Brus, L.E., 1984, *J. Chem. Phys.* 80 4464.
31. Brus, L.E., 1983, *J. Chem. Phys.* 79 5566.
32. Brus, L.E., 1984, *J. Chem. Phys.* 80 4403.
33. Schmidt, H. M. and Weller, H., 1986, *Chem. Phys. Lett.* 129 615.
34. Burr, J., 1987, *Pt. II Thesis*, Oxford University.
35. Harris, D., 1988, *Pt. II. Thesis*, Oxford University.
36. Burr, J., Harris, D. and Wells, B.H., 1988, to be submitted.
37. Flügge, S., 1971, *Practical Quantum Mechanics Vol. 1* Springer (Berlin).
38. Burr, J. and Wells, B.H., 1988, to be submitted.
39. Warton, L., Gold.and Klemperer, W., 1962, *J. Chem. Phys.* 37 2149.
40. Clementi, E. and Roetti, E., 1974, *At. Data Nucl. Data Tables* 14 185.
41. Hutchinson, K., 1988, *Pt. II Thesis*, Oxford University.

NUMERICAL DETERMINATION OF NON-RELATIVISTIC AND RELATIVISTIC PAIR CORRELATION

ANN-MARIE MÅRTENSSON-PENDRILL
Department of Physics
Chalmers University of Technology and University of Göteborg
S-412 96 Göteborg, Sweden

ABSTRACT: A numerical method is presented for obtaining atomic pair functions within the framework of many-body perturbation theory. The functions are obtained in a two-dimensional grid and finite differences are used for the second derivatives. Using an iterative scheme to solve the coupled one- and two-particle equations, correlation effects are treated to all orders. Calculated energies are shown both for He-like systems, where this is, in principle, an exact treatment, and for larger atoms. The extension to the relativistic case is possible using projection operators for positive energy states. Applications to the hyperfine structure, isotope shifts and weak interactions in atomic physics are discussed.

1. Introduction

The calculations of energy levels in simple atomic systems was a first crucial test of quantum mechanics. Although the basic equations describing an atomic system are now well known, the task of performing highly accurate calculations for large atoms still poses a challenge, in spite of the access to more and more powerful computers.

The developments in experimental techniques have given us extremely accurate results for many atomic levels. These results often contain indirect information which may be of interest outside atomic physics. E.g., the study of hyperfine structure can provide values for nuclear magnetic and electric moments. Isotope shifts (i.e. small shifts in energy levels between different isotopes) can tell about the differences in charge distribution. The study of parity non-conserving optical rotation in atoms can give values of weak-interaction parameters. However, in order to extract this information, a detailed understanding of the atomic structure is essential.

In this paper, we present a method based on many-body perturbation theory (MBPT) in combination with numerical solution of partial differential equations which describe the one- and two-particle excitations from the unperturbed wavefunction. By iterative solutions of these equations, the interaction between two electrons can in principle be treated exactly and very accurate results are, indeed, obtained for two-electron systems. The MBPT provides prescriptions for generalising the treatment to many-electrons. Although we do not explicitly include effects which involve the simultaneous excitations of more than two particles, the

M. Defranceschi and J. Delhalle (eds.), Numerical Determination of the Electronic Structure of Atoms, Diatomic and Polyatomic Molecules, 131–160.

use of the coupled-cluster formalism automatically includes also all *products* of single and double excitations.

In the next section we will describe the perturbation theory formalism used, and as an example we consider the exact equations for a two-electron system. In section 3, we present the numerical method used to solve the resulting differential equation. In Section 4, we present results obtained for He, Li and Ca^+.

For heavy atoms, a relativistic treatment must be used. The extension to the relativistic case involves non-trivial problems of both fundamental and practical character which are discussed in Section 5, where we present also second-order relativistic results for helium.

2. The Atomic Wavefunction

Our task is to obtain a wavefunction $\Psi(\mathbf{r}_1, \mathbf{r}_2, ..\mathbf{r}_N)$, which describes the distribution of the N electrons. For simplicity, we first assume that relativistic effcts can be neglected. The wave function then satisfies the Schrödinger equation. (Unless otherwise specified, we use atomic units, $e = m_e = \hbar = 4\pi\epsilon_0 = 1$.)

$$H\Psi = \Big(\sum_i(-\frac{\nabla_i^2}{2} - \frac{Z}{r_i}) + \frac{1}{2}\sum_{ij}\frac{1}{r_{ij}}\Big)\Psi = E\Psi$$

Obviously, the wavefunction involving the coordinates of all the electrons for a large system is too complicated to handle and the equation cannot be solved exactly. Instead, we start with an approximate hamiltonian, H_0, which usually is a sum of one-body operators, h_0,

$$H_0 = \sum_i h_{0,i} = \sum_i(-\frac{\nabla_i^2}{2} - \frac{Z}{r_i} + u_i) \qquad (1)$$

where u_i is an approximation of the interaction with all other electrons. The difference $V = H - H_0$ will then be treated as a perturbation. The one-body equation becomes

$$h_0\phi(\mathbf{r}_i) = \epsilon_i\phi(\mathbf{r}_i)$$

The approximate wavefunction can then be written as an antisymmetrised product (denoted by curly brackets) of one-particle wavefunctions

$$\Phi = \{\phi(\mathbf{r}_1)\phi(\mathbf{r}_2).....\}$$
$$H_0\Phi = E_0\Phi = (\sum_i \epsilon_i)\Phi$$

In the *central-field model* (CFM) the one-electron function is separated into radial, angular and spin parts as

$$\phi(\mathbf{r}) = \frac{P(r)}{r}Y_m^l(\theta,\varphi)\chi_\mu = \frac{P(r)}{r}|lm\mu\rangle$$

After expressing the Laplacian in spherical coordinates, the one-body equation then reduces to a one-dimensional second-order differential equation for P(r)

$$h_r^0(l)P(r) = \left(-\frac{1}{2}\left(\frac{d^2}{dr^2} - \frac{l(l+1)}{r^2}\right) - \frac{Z}{r} + u_i\right)P(r) = \epsilon P(r) \qquad (2)$$

which can be solved numerically.

The one-body potential u is often chosen to be the Hartree-Fock potential from the nearest closed-shell system

$$u^{HF}|i\rangle = \sum_c^{core}(\langle c|\frac{1}{r_{12}}|c\rangle|a\rangle - (\langle c|\frac{1}{r_{12}}|a\rangle|c\rangle)$$

The last, so-called *exchange* term makes the HF potential non-local and u^{HF} can thus not be used directly on the left-hand side of the differential equation. Instead, and iterative procedure is used, where an approximate local potential, u^{loc}, is used on the left-hand side, and the correction, $u^{HF} - u^{loc}$, which depends on the solution, is inserted on the right-hand side using the previously obtained solution.

The central-field model usually gives a good overall description of the atom and reproduces the qualitative features of atomic spectra. In combination with semi-empirical procedures, very good agreement with experiment can often be obtained. Nevertheless, it sometimes fails to give even the right sign of an interaction and some effects cannot be described within the CFM.

Several *ab initio* methods can be used to go beyond the CFM. For a two-electron system, it is possible to use Hylleraas-type wavefunctions that depend explicitly on the distance, r_{12}, between the electrons, and with the work of Pekeris (1958, 1962) and coworkers (Schiff *et al.* 1965, Accad *et al.* 1971), the non-relativistic two-electron problem is essentially solved. However, their method cannot easily be generalised to larger systems. A widely used approach, which can be used also for many-electron systems, is the method of *configuration interaction* (CI) where the wavefunctions are expanded in a basis set, which is usually analytic and discrete, and the coefficients are determined by minimisation of the energy expectation value

$$E = \min\left(\frac{\langle\Psi|H_0 + V|\Psi\rangle}{\langle\Psi|\Psi\rangle}\right)$$

In the *multi-configuration Hartree-Fock* (MCHF) approach, not only the coefficients, but also the basis functions are varied in the minimisation process, making it possible to obtain accurate results with relatively few configurations. In the present work we use *many-body perturbation theory* (MBPT) to improve the wavefunctions. In the pioneering work by Kelly (1962) the corrections were obtained by expansion in terms of excited orbitals, including the continuum. This aproach is used also in the work by Das and coworkers (e.g. Chang *et al.* 1968) and in the relativistic case discussed below e.g. by Vajed-Samii *et al.* (1979), Dzuba *et al.* (1985, 1987) and recently by Johnson *et al.* (1986, 1988) We have chosen a different approach, first used by Sternheimer (1950) for the one-particle case, and use numerical solutions of inhomogeneous differential equations to obtain summations over all excited orbitals.

2.1 Correcting the wavefunction using perturbation theory

In this work we obtain the corrections to the wave function within the framework of many-body perturbation theory and write

$$\Psi = \Phi + \delta\Psi, \quad H = H_0 + V, \quad E = E_0 + \Delta E$$

We usually work in P *intermediate normalisation* and require

$$\langle\Phi|\Psi\rangle = 1, \langle\Phi|\delta\Psi\rangle = 0 \tag{3}$$

The exact equation for Ψ can be written in the form

$$(E_0 + \Delta E - H_0 - V)|\Phi + \delta\Psi\rangle = 0$$

After some rearrangement, we get

$$(E_0 - H_0)|\Phi + \delta\Psi\rangle = (E_0 - H_0)|\delta\Psi\rangle = V|\Phi + \delta\Psi\rangle - \Delta E|\Phi + \delta\Psi\rangle \tag{4}$$

When $(E_0 - H_0)$ operates on the unperturbed wavefunction, it gives no contribution, and Φ can thus be omitted from the left-hand side. The energy correction ΔE can be found by using the intermediate normalisation condition and applying $\langle\Phi|$ from the left, which gives zero on the left-hand side. The right-hand side must thus be orthogonal to $\langle\Phi|$, giving

$$\Delta E = \langle\Phi|\frac{1}{r_{12}}|\Phi + \delta\Psi\rangle$$

Introducing the projection operators $P = |\Phi\rangle\langle\Phi|$ and Q=1-P we can write the equation for the correction $\delta\Psi$ as

$$(E_0 - H_0)|\delta\Psi\rangle = QV|\Phi + \delta\Psi\rangle - \Delta E|\delta\Psi\rangle \tag{5}$$

where Q makes the first term on the right-hand side orthogonal to the unperturbed wavefunction.

Before considering how this equation can be used in a general case, we study the application to a two-electron system.

2.2 A perturbative treatment of a helium-like system

A solution of the pair equation gives in principle the exact wave function for a two-electron system. The helium atom is therefore an important test of the equations and numerical methods used. It is also a relatively easy example of the application of many-body perturbation theory and understanding the pair equations for helium provides us with the building blocks needed to treat larger systems.

One possible starting point in the treatment of helium-like systems is to include only the interaction with the nucleus in the one-body hamiltonian and then treat the full interaction between the electrons as a perturbation:

$$h_0 = -\nabla^2/2 - Z/r \qquad V = 1/r_{12}$$

Consider two non-equivalent electrons in orbitals a and b. These may form either a singlet or a triplet state. The spin of the electrons does not enter explicitly in the wave equation, and it is sufficient to consider only the spatial part of the wave function, keeping in mind that the requirement that the total wavefuntion be anti-symmetric forces the spatial function to be symmetric for singlet states and anti-symmetric for triplet states. The approximate spatial function can be written as

$$\Phi = \{ab\} = (|ab\rangle \pm |ba\rangle)/\sqrt{2}$$

where the plus (minus) sign is associated with the singlet (triplet) state and the curly brackets denote the combination with the correct symmetry properties. In the case of two equivalent electrons (i.e. $a = b$) we get instead $\Phi = \{ab\} = |ab\rangle$.

The correction to this approximate wave function can be expressed in terms of a *pair function*, ρ_{ab}, giving a total wave function

$$\Psi = \Phi + \delta\Psi = \{ab + \rho_{ab}\}$$

The equation (4) for the the correction $\delta\Psi$ then becomes

$$(\epsilon_a + \epsilon_b - h_0(1) - h_0(2))\{\rho_{ab}\} = \frac{1}{r_{12}}\{ab + \rho_{ab}\} - \Delta E\{ab + \rho_{ab}\}$$
$$\Delta E = \langle\{ab\}|\frac{1}{r_{12}}|\{ab + \rho_{ab}\}\rangle = \langle ab|\frac{1}{r_{12}}|ab + \rho_{ab}\rangle \pm \langle ba|\frac{1}{r_{12}}|ab + \rho_{ab}\rangle \qquad (6)$$

where the total energy is $E = \epsilon_a + \epsilon_b + \Delta E$

In our work, we use instead the corresponding non-(anti-)symmetrised function, ρ_{ab}, which satisfies the equation

$$(\epsilon_a + \epsilon_b - h_0(1) - h_0(2))|\rho_{ab}\rangle =$$
$$\frac{1}{r_{12}}|ab + \rho_{ab}\rangle - \sum_{cd=ab,ba} |cd + \rho_{cd}\rangle\langle cd|\frac{1}{r_{12}}|ab + \rho_{ab}\rangle \qquad (7)$$

where the term $-|cd\rangle\langle cd|\frac{1}{r_{12}}|ab + \rho_{ab}\rangle$ makes the right-hand side orthogonal to cd (i.e. to ab and ba). By combining the equations for ρ_{ab} and ρ_{ba}, it is easy to show that the proper combination $\{\rho_{ab}\}$ satisfies the equation (6) above.

The correction ρ_{ab} defined in this way will contain not only double excitations, but also single excitations, where one electron is left in state a or b and we have a choice of treating these together with the pair excitations or removing also the single excitations from the right-hand side and treating them separately. The latter approach was used in the relativistic case, described in Section 5, where the single-particle equations can be solved exactly, whereas approximations are introduced in the solution of the pair equation.

The *first-order pair equation* is obtained by dropping all terms involving ρ_{ab} on the RHS of (7) , giving

$$
\begin{aligned}
(\epsilon_a + \epsilon_b - h_0(1) - h_0(2))|\rho_{ab}^{(1)}\rangle &= \frac{1}{r_{12}}|ab\rangle - \sum_{cd=ab,ba} |cd\rangle\langle cd|\frac{1}{r_{12}}|ab\rangle \\
&= \sum_{rs\neq ab,ba} |rs\rangle\langle rs|\frac{1}{r_{12}}|ab\rangle
\end{aligned} \tag{8}
$$

where we have used the closure relation $\sum_i^{all} |i\rangle\langle i| = 1$, to expand the right-hand side in a complete set of eigenfunctions to the single-particle hamiltonian. We now see that the first order correction to the wavefunction can be expressed as

$$
\rho_{ab}^{(1)} = \sum_{rs\neq ab,ba} \frac{|rs\rangle\langle rs|\frac{1}{r_{12}}|ab\rangle}{\epsilon_a + \epsilon_b - \epsilon_r - \epsilon_s} \tag{9}
$$

This first-order wavefunction correction can then be inserted on the RHS of (7) to obtain the next approximation to ρ_{ab} and we have an iterative procedure, where at each step the best ρ_{ab} available is inserted.

The exclusion of $rs = ab, ba$ from the RHS of (8) removes terms which would otherwise enter with zero energy denominator. However, the function obtained by adding any amount of $|ab\rangle$ or $|ba\rangle$ to $\rho_{ab}^{(1)}$ would still be a solution to the equation. If these homogeneous terms enter in the the pair function they should be removed by orthogonalising the solution. However, the presence of homogeneous terms may still destroy the numerical accuracy and a more stable procedure is obtained by shifting the energy denominators by adding a small correction, $\delta\epsilon\ \rho_{ab}$ both to the left- and right-hand sides and then solve the equation by iteration.

In the actual calculations described in Section 3 ρ_{ab} is divided into several terms depending on the l-values of the excited states. All these terms contribute both to the energy correction and to the first term on the RHS and thus the equations for all possible l-excitations are coupled.

2.3 Generalisation to a larger system

The correction $|\delta\Psi\rangle$, can in general be expressed in terms of single substitutions of the occupied orbital a by an unoccupied orbital r, Φ_a^r, double substitutions of orbitals a and b by orbitals r and s, Φ_{ab}^{rs}, ... from the unperturbed wavefunction. The total wavefunction can then be written in the form

$$
\Psi = \Phi + \sum_{ar} C_a^r \Phi_a^r + \sum_{\substack{a>b \\ r>s}} C_{ab}^{rs} \Phi_{ab}^{rs} + \ldots
$$

In the *configuration interaction (CI)* approach, the coefficients are determined by minimisation of the energy expectation value. Another possibility is to obtain the corrections using perturbation theory. We then introduce the *wave operator* Ω, which produces the exact wave function, $|\Psi\rangle$ when operating on the approximate wave function $|\Phi\rangle$. The equation for the wave function correction above can then be reformulated as an equation for Ω, noting that $H_0P = E_0P$

$$[\Omega, H_0]P = (V\Omega - \Omega PV\Omega)P \tag{10}$$

where $\Delta E = PV\Omega$.

This Bloch-type equation (Bloch 1958) can be used to generate the *Rayleigh-Schrödinger* perturbation expansion. To obtain the correction to the wave function to first-order in the perturbation V, we use P rather than Ω on the RHS of (10) (or, equivalently, neglect the $|\delta\Psi\rangle$ terms on the RHS of (4)). To obtain the second-order correction, we then use the first-order correction on the RHS If it is inserted also in the expression for $\Delta E = PV\Omega$, we will get some third-order terms, as well. However, as an alternative, used in this work, to the strict *order-by-order* expansion we can apply a *self-consistent* procedure, and use the best Ω available at every stage, until convergence has been obtained. Sometimes, it may be necessary to use weight factors, which make the new Ω contain a fraction of the old one, in order to improve the convergence properties.

Using *second quantisation* we can express Ω in terms of *creation operators* $a_i^\dagger$ and *annihilation operators* a_i for an electron in state i. These operators satisfy the anticommutation relations $a_i^\dagger a_j + a_j a_i^\dagger = \delta_{ij}$. We see that the single substitution Φ_a^r of orbital a by orbital r can be expressed as

$$\Phi_a^r = a_r^\dagger a_a \Phi$$

and a double substitution Φ_{ab}^{rs} as

$$\Phi_{ab}^{rs} = a_r^\dagger a_s^\dagger a_b a_a \Phi$$

Operating with H_0 on the unperturbed wavenfunction gives $E_0\Psi_0$ and operating on Φ_a^r and Φ_{ab}^{rs}, respectively, gives

$$H_0\Phi_a^r = (\epsilon_r - \epsilon_a + E_0)\Phi_a^r$$
$$H_0\Phi_{ab}^{rs} = (\epsilon_r + \epsilon_s - \epsilon_a - \epsilon_b + E_0)\Phi_{ab}^{rs}$$

We thus get an expression, e.g., for the commutator

$$[H_0, a_r^\dagger a_a] = \epsilon_r - \epsilon_a$$

and formally we can write the hamiltonian in second-quantised form:

$$H_0 = \sum_i^{all} \epsilon_i a_r^\dagger a_a$$

and a perturbing two-body potential as

$$V = \sum_{ijkl}^{all} a_i^\dagger a_j^\dagger a_l a_k \langle ij|V|kl\rangle$$

This form is very convenient for book-keeping of the various contributions, in particular in combination with the diagrammatic technique, described in detail, e.g. by Lindgren and Morrison (1982)

We can now obtain a first-order value for the coefficient C_{ab}^{rs} by collecting all terms of the first-order equation for Ω which contain the creation and annihilation operators $a_r^\dagger a_s^\dagger a_b a_a$

$$C_{ab}^{rs} = \frac{\langle rs|V|ab\rangle - \langle rs|v|ba\rangle}{\epsilon_r + \epsilon_s - \epsilon_a - \epsilon_b}$$

These coefficients can then be used on the RHS to create higher-order coefficients.

In this work we define instead *pair functions* ρ_{ab} given to first-order by (8) (with the summation over r and s restricted to excited states). The double substitutions to the wavefunction can then be written as

$$\sum_{a>b}^{occ} \sum_{r>s}^{exc} C_{ab}^{rs} \Phi_{ab}^{rs} = \sum_{a>b}^{occ} \Phi_{ab}^{\rho_{ab}}$$

When the pair corrections to the wavefunctions are inserted back in (10), they can give rise to single, double, triple and quadruple excitations, due to the term $QV\Omega$. The single excitations, although absent in first order if the Hartreee-Fock potential is used, enter in second and higher orders to modify the orbitals to approximate *Brueckner orbitals* (Lindgren *et al.*, 1976). These can then be used when higher-order pair functions are generated. Although practical limitations force us to truncate the expansion after double excitations, it is possible to include some of the triple and quadruple excitations by using the *coupled-cluster* approach and writing

$$\Omega = exp(S)$$

This form was introduced long ago in nuclear physics (See e.g. the review by Kümmel *et al.* 1978) and a similar formalism has been used in quantum chemistry. e.g. by Čížek (1966,1969). The cluster operator S is the *connected* part of the wave operator and it satisfies an equation similar to (10), but where only the connected parts are kept:

$$[S, H_0] = Q(V\Omega - \Omega PV\Omega)_{conn}$$

which can be separated into one-, two-, ... body equations (see e.g. Lindgren and Morrison, 1982, p406). When S is truncated after double excitations, the wave operator will still contain e.g. those triple excitations which are products of single and double excitations (or of three single excitations) as well as those quadruple excitations that are products of two independent pair excitations (*or* of one pair and two single excitations *or* of four single excitations). Sinanoğlu (1962,1964) found, in an analysis of the correlation effects in the ground state $1s^2 2s^2$ of beryllium that such products gave the most important quadruple excitations. We can also see e.g. that the wave function of two separated helium atoms will be given by the product of the wave functions for the two atoms if S is truncated after double excitations, but that this is not the case if Ω is truncated.

2.4 Correlation Effects on Additional Perturbations

Correlation energies are usually evaluated as a natural part of the calculation of pair functions, but the pair equation can be used to study also other properties in several ways. A perturbation h can be included formally in the one-body hamiltonian (for a two-body perturbation, an effective one-body interaction analogous to the Hartree-Fock potential can be used). The perturbation can also be together with V when the corrections to wavefunctions and energies are obtained, as described by Lindgren and Morrison (1982). It is also possible to use the general expression for a matrix element of h between states f and i

$$\frac{\langle \phi_f | \Omega^\dagger h \Omega | \phi_i \rangle}{\langle \phi_f | \Omega^\dagger \Omega | \phi_f \rangle^{1/2} \langle \phi_i | \Omega^\dagger \Omega | \phi_i \rangle^{1/2}} \tag{11}$$

This formula may be used also within an hermitean formulation of the perturbation theory, currently being developed by Lindgren and Mukherjee (1988), where the intermediate normalisation condition (3) is replaced by $P\Omega^\dagger \Omega P = P$.

By inclusion of the perturbation in the one-body potential h_0, all diagrams involving single excitations are formally included already at the unperturbed level, as well as diagrams involving factorisable two-particle excitation and chains of these (sometimes referred to as ground-state correlation). If, instead, the expression (11) were used directly, these chains would appear when connecting several powers of two-body clusters, $S_2^\dagger$ and S_2 from $\Omega^\dagger$ and Ω. Although, formally, h is treated together with h_0, in practice, the calculations are usually performed so that an extra set of orbitals $|\delta a\rangle$ including the perturbation to lowest order is obtained from the equation

$$(\epsilon_a - h_0)|\delta a\rangle = (h + v^h)|a\rangle - \delta\epsilon|a\rangle \tag{12}$$

where v^h is the correction to the potential due to the modification of the orbitals for the other electrons b, and $\delta\epsilon$ is the energy correction $\langle a|(h+v^h)|a\rangle$, making the RHS orthogonal to orbital a. If the Hartree-Fock potential from the core orbitals is used we get (using the exchange operator P_{12}, which interchanges electrons 1 and 2)

$$v^h = \sum_b^{core} \Big(\langle b | \frac{1}{r_{12}} (1 - P_{12}) | \delta b \rangle + \langle \delta b | \frac{1}{r_{12}} (1 - P_{12}) | b \rangle \Big) \tag{13}$$

This economical treatment was suggested by Sandars (1977) in connection with the study of weak interactions and has been used by several groups to study this and other problems (see e.g. Mårtensson-Pendrill 1985, Johnson *et al.* 1985,1986, Dzuba *et al.* 1985,1987). It is closely related to the *random-phase approximation* (see. e.g. Johnson *et al.* 1980) and to the method of summing all *core polarisation* effects (note, however, that this term is often used to refer instead to Brueckner-orbital like corrections) to all orders described by Garpman *et al.* (1976).

The orbital corrections (12) should then be included with the unperturbed orbitals when the pair equations are constructed and we would normally have to define a new set of pair functions including the perturbation. This would also be the case in the second approach, treating h tog ether with V, if pair correlation should be included to all orders. In lowest order this is not necessary, however, as found by Garpman *et al.* (1975), who follow this second approach in deriving the expressions for all third-order (two-orders in the electron-electron interaction, one in the hyperfine interaction, e.g.) contributions to the hyperfine structure in an alkali atom.

To obtain pair functions which include the perturbation appears to be the most convenient approach for studying two-particle perturbations, such as the mass polarisation or specific mass isotope shift, and the Breit interaction, where certain pair correlation effects cannot easily be included by using (11) (Lindroth *et al.* 1985).

For a one-particle perturbation, on the other hand it seems to be more straightforward to use the *all-order* pair functions directly in the evaluation of matrix elements using (11) rather than as input to obtain new pair functions which include the perturbation. However, is will be necessary to truncate the coupled-cluster expansion at some stage and it is possible to include longer chains of factorisable two-particle excitations by using the the modified one-electron orbitals, taking care to ensure that no double-counting occurs.

The expression (11) also has the advantage of giving a more symmetric treatment of the initial and final states, i and f in the evaluation of transition elements. In the study of parity non-conserving (PNC) weak interaction effects in atoms, where two external perturbations, the weak interaction and the applied photon field, must be included, it would be possible to include pair correlation to all orders by combining both approaches, calculating PNC pair functions and using these in the evaluation of PNC transition elements using (11).

3. Solving the pair equation

The pair function ρ_{ab}, describing the correlation between electrons a and b, involves two infinite summations over excited states. A convenient way of performing these summations is to solve the inhomogeneous two-particle differential equation, the *pair equation*, the function satisfies. This procedure was first applied by Musher and Schulman (1968) in a study of helium energies. About the same time a direct numerical solution of the eigenvalue equation for helium was performed by McKoy and Winter (1968) and by Winter *et al.* (1970). These pioneering works were followed by applications of the pair equation to calculations to second order in the electron-electron interaction for corrections to hyperfine structure in atoms (Schulman and Lee 1972, Garpman *et al.* 1975) and to correlation energies in larger systems (Morrison 1973). Our own calculations rely heavily on the solutions of the pair equation. By using pair functions to create approximate Brueckner orbitals (Lindgren *et al.* 1976) and by iterative solution of the pair equation (Mårtensson 1979) we have been able to include pair correlation effects to all orders. The approach has been extended to an open-shell coupled-cluster formalism (Lindgren and Salomonson 1980, Morrison and Salomonson 1980, Salomonson *et al.* 1980) and the procedure has proved to be capable of producing very accurate results also for atoms with many electrons (see e.g. Salomonson 1984, Mårtensson-Pendrill and Salomonson 1984, Lindgren 1985)

A computer program implementing the solution procedure described in this section, has been used in our group for over a decade. It has recently undergone major revision with intent of publication (Salomonson *et al.* 1988) and is now being tested for several problems. Among the new features included is the more accurate treatment of the cusp discussed in Section 3.2.3. It also includes a more complete treatment of single excitation effects, with an automatic coupling between the equation for single excitations, S_1 and double excitations, S_2, within the coupled-cluster approach and it can be used also for arbitrary one-body potentials.

The pair function ρ_{ab} for a larger system can be obtained as the solution to an equation analogous to (7) but with the right-hand side replaced by

$$\sum_{rs}^{exc} |rs\rangle\langle rs|\frac{1}{r_{12}}|ab\rangle = \frac{1}{r_{12}}|ab\rangle$$
$$-\sum_{c}^{core}(|c_1\rangle\langle c_1|\frac{1}{r_{12}}|ab\rangle + |c_2\rangle\langle c_2|\frac{1}{r_{12}}|ab\rangle) + \sum_{cd}^{core}|cd\rangle\langle cd|\frac{1}{r_{12}}|ab\rangle$$

where we have, again, used the closure relation to rewrite the RHS. The subscripts on the core orbital c in the second and third terms on the RHS denote the coordinate which is integrated over. The last three terms have the function of making the RHS orthogonal to all orbitals and will hereafter be referred to as 'orthogonality terms'. Instead of an infinite sum over excited states, we now have to evaluate a finite sum over the occupied orbitals. The pair equation is then written as

$$(\epsilon_a + \epsilon_b - h_0(\mathbf{r}_1) - h_0(\mathbf{r}_2))\rho_{ab}(\mathbf{r}_1, \mathbf{r}_2) = \frac{1}{r_{12}}|ab\rangle - \textit{orthogonality terms}$$

Below, we describe how this equation can be solved after a separation into radial and angular parts.

3.1 Obtaining the Radial Equation

In order to obtain an equation involving only r_1 and r_2, we use the common expansion of the electron-electron interaction

$$\frac{1}{r_{12}} = \sum_{kq}(-1)^q \frac{r_<^k}{r_>^{k+1}} \mathrm{C}_\mathrm{q}^\mathrm{k}(1) \cdot \mathrm{C}_\mathrm{-q}^\mathrm{k}(2)$$

where $r_>(r_<)$ is the larger (smaller) of the two values r_1 and r_2. In combination with the Wigner-Eckart theorem, this leads us to the following expansion of the pair function:

$$\rho_{ab}(\mathbf{r}_1,\mathbf{r}_2) = \sum_{kq}\sum_{l_r m_r l_s m_s} \frac{1}{r_1 r_2}\rho^k_{ab\to l_r l_s}(r_1,r_2)$$
$$\times (-1)^{l_r-m_r+l_s-m_s+k-q}\begin{pmatrix} l_r & k & l_a \\ -m_r & q & m_a \end{pmatrix}\begin{pmatrix} l_s & k & l_b \\ -m_s & -q & m_b \end{pmatrix}|l_r m_r\rangle|l_s m_s\rangle \tag{14}$$

The differential equation for the radial part of the pair function then becomes

$$(\epsilon_a+\epsilon_b-h^0_{r,1}(l_r)-h^0_{r,2}(l_s))\rho^k_{ab\to l_r l_s}(r_1,r_2)$$
$$= \frac{r_<^k}{r_>^{k+1}}P_a(r_1)P_b(r_2)(-1)^k\langle l_r\|\mathbf{C}^k\|l_a\rangle\langle l_s\|\mathbf{C}^k\|l_b\rangle - \textit{orthogonality terms} \tag{15}$$

where the radial part $h^0_{r,i}(l)$ of the one-body hamiltonian is given in (2)

Eq. (15) is a two-dimensional second-order differential equation and in Section 3.2 below we describe the numerical methods used to solve it.

Once a solution has been obtained, it can be inserted on the RHS to obtain higher-order contributions. E.g., inserting a pair function of the form (14) on the RHS of the all-orders pair equation gives a contribution from the term $\frac{1}{r_{12}}|\rho_{ab}\rangle$:

$$\sum_{k_1k_2l_tl_u}(2k+1)\begin{Bmatrix} l_r & l_a & k \\ k_1 & k_2 & l_t \end{Bmatrix}\begin{Bmatrix} l_s & l_b & k \\ k_1 & k_2 & l_u \end{Bmatrix}$$
$$\times\left((-1)^{k_2}\langle l_r\|\mathbf{C}^{k_2}\|l_t\rangle\langle l_s\|\mathbf{C}^{k_2}\|l_u\rangle\right)\left(\frac{r_<^{k_2}}{r_>^{k_2+1}}\right)\rho^{k_1}_{ab\to l_t l_u}(r_1,r_2)$$

which can be added to the RHS of (15). In the general case where we have more than two electrons, we would also get a number of terms involving other excited electron pairs. In addition there are similar terms arising from the energy correction. These can all be added as corrections to the right-hand side.

3.2 Numerical methods to solve the pair equation

Both the first-order and all-order pair equations can be written in the form

$$(\epsilon_1+\epsilon_2-h^0_r(l_1)-h^0_r(l_2))\rho(r_1,r_2) = g(r_1,r_2)$$

where $h^0_r(l)$ is given in (2), and we describe here how an equation of this type can be solved.

3.2.1 The finite difference method In the finite difference method the functions are represented by their values $\rho_{mn} = \rho(r_{1,m}, r_{2,n}), 1 \leq m,n \leq N$ in some, normally equidistant gridpoints. The second derivative of a function can be expanded in terms of differences as

$$\left(\frac{d^2 f}{dx^2}\right)_{x=x_0} = \frac{1}{h^2}\left(\delta_0^2 - \frac{1}{12}\delta_0^4 + \frac{1}{90}\delta_0^6 + \ldots\right) f$$

where $\delta_0 f = f(x_0+h/2) - f(x_0-h/2)$ and h is the distance between the gridpoints. As a first approximation to the second derivative, we can use the second difference. From the Taylor series expansion the difference error can be estimated as

$$\frac{1}{h^2}\delta_0^2 f - \left(\frac{d^2 f}{dx^2}\right)_0 = \frac{h^2}{12}\left(\frac{d^4 f}{dx^4}\right)_0 + \frac{h^4}{360}\left(\frac{d^4 f}{dx^4}\right)_0 + \ldots$$

To improve the approximation the fourth difference could also be included, which should remove the $\mathcal{O}(h^2)$ error. This assumes that the function and its derivatives are continuous. We will discuss this problem in more detail in Section 3.2.3 below. The energy obtained will then be dependent on h

$$E(h) = E(0) + C_4 h^4 + C_6 h^6 + \ldots$$

and this expression can be used to perform a Richardson extrapolation using at least two different grid sizes.

Close to the nuclues the atomic wavefunctions vary rapidly and many points are needed to describe this behaviour, whereas further out the grid points could be more sparsely distributed. A common way of obtaining such a distribution is to use a *logarithmic grid*, $r_n = \exp(x_n)$, where the equidistant points x_n are given by $x_n = x_{min} + (n-1)h$. The first-order derivatives which then appear can be eliminated (just as in the more familiar one-particle treatment) by using the transformation

$$\bar{\rho}(r_1, r_2) = \rho(r_1, r_2)/\sqrt{r_1 r_2}$$

The radial part of the pair equation can then be written in the form

$$\left[\frac{1}{2}\left(\frac{1}{r_1^2}\frac{\partial^2}{\partial x_1^2} + \frac{1}{r_2^2}\frac{\partial^2}{\partial x_2^2} - \frac{(l_1+\frac{1}{2})^2}{r_1^2} - \frac{(l_2+\frac{1}{2})^2}{r_2^2}\right) + \frac{Z}{r_1} + \frac{Z}{r_2} - u(r_1) - u(r_2) + \epsilon_1 + \epsilon_2\right]\bar{\rho}(r_1, r_2)$$

$$= g(r_1, r_2)/\sqrt{r_1 r_2} \equiv \bar{g}(r_1, r_2)$$

The boundary conditions require that $\bar{\rho}(r_1, r_2)$ vanishes when either variable is zero or infinity. The latter condition is aproximated by assuming that

$$\bar{\rho}_{mN} = \bar{\rho}_{Nn} = 0 \tag{16}$$

In the logarithmic grid, r_1 and r_2 never reach zero and the boundary condition for small r_1 and r_2 are replaced by the proper behaviour

$$\bar{\rho}(r_1, r_2)/\bar{\rho}(r_1', r_2) = (r_1/r_1')^{l_1+1/2} \tag{17}$$

for small r_1, r_1' and an analogous relation for small r_2. Inserting the second differnce approximation for the derivatives of $\bar{\rho}$ leads to a set of coupled linear equations for $\bar{\rho}_{mn}$

$$\frac{1}{r_m^2}(D_1(r_m)\bar{\rho}_{mn} - \bar{\rho}_{m-1n} - \bar{\rho}_{m+1n}) + \frac{1}{r_n^2}(D_2(r_n)\bar{\rho}_{mn} - \bar{\rho}_{mn-1} - \bar{\rho}_{mn+1}) = 2h^2\bar{g}_{mn}$$

where

$$D_i(r_j) = 2 + 2h^2\left(\frac{1}{2}(l_i + \frac{1}{2})^2 - Zr_j + r_j^2 u(r_j) - \epsilon_i\right)$$

In conventional treatments this system of N^2 equations is solved directly. Since the coefficient matrix is not symmetric and is not always diagonally dominant, it is not necessarily positively definite and relaxation methods sometimes fail to converge. The equation system must then be solved by Gaussian elimination, which puts heavy demands on computer time and storage even if the matrix contains mainly zeroes.

3.2.2 Transforming the finite-difference equations. An alternative method to solve the coupled equations for $\bar{\rho}_{mn}$, suggested by Lindgren (1975) is to express $\bar{\rho}_{m+1n}$ in terms of $\bar{\rho}_{mn}$, $\bar{\rho}_{m-1n}$ and $\bar{\rho}_{mn\pm1}$ as

$$\bar{\rho}_{m+1n} = \left(D_1(r_m) + \frac{r_m^2}{r_n^2}D_2(r_n)\right)\bar{\rho}_{mn} - \bar{\rho}_{m-1n} - \frac{r_m^2}{r_n^2}(\bar{\rho}_{mn-1} + \bar{\rho}_{mn+1}) + 2h^2 r_m^2 \bar{g}_{mn} \tag{18}$$

If the values of $\bar{\rho}_{mi}$ and $\bar{\rho}_{mi-1}$ are known for $1 \leq m \leq N$, the values $\bar{\rho}_{mi+1}$ can be obtained for $2 \leq m \leq N-1$ using this expression. The two remaining values $\bar{\rho}_{1i+1}$ and $\bar{\rho}_{Ni+1}$ can be obtained using implicit values for $\bar{\rho}_{0i}$ and $\bar{\rho}_{N+1i}$ obtained from the boundary conditions (16) and (17), respcetively. By recursion, we can then express all $\bar{\rho}_{mi}$ in terms of $\bar{\rho}_{1i}$ as

$$\bar{\rho}_{mn} = \sum_i C_{ni}^{(m)} \bar{\rho}_{1i} + B_n^{(m)} \tag{19}$$

by inserting this expression into (18), the recursion formulae for the coefficients B and C are obtained:

$$C_{ni}^{(m+1)} = \left(D_1(r_m) + \frac{r_m^2}{r_n^2}D_2(r_n)\right)C_{ni}^{(m)} - C_{ni}^{(m-1)} - \frac{r_m^2}{r_n^2}(C_{n+1i}^{(m)} + C_{n-1i}^{(m)}) \tag{20}$$

$$B_n^{(m+1)} = \left(D_1(r_m) + \frac{r_m^2}{r_n^2}D_2(r_n)\right)B_n^{(m)} - B_n^{(m-1)} - \frac{r_m^2}{r_n^2}(B_{n+1}^{(m)} + B_{n-1}^{(m)})$$

$$+2h^2r^2\bar{g}_{mn} \tag{21}$$

The definition (19) gives

$$C_{ni}^{(1)} = \delta_{ni} \qquad B_i^{(1)} = 0$$

and due to the boundary condition (17) we use implicitly

$$C_{ni}^{(0)} = \left(\frac{r_1}{r_2}\right)^{l_1+1/2} \delta_{ni} \qquad B_i^{(0)} = 0$$

By combining the boundary condition (16) for the values $\bar{\rho}_{Ni}$ with the definition (19), we can get a linear equation system for the N variables $\bar{\rho}_{1i}$ instead of the equation system for all the N^2 grid points.

However, the method does not work in this simple form. The B and C coefficients grow prohibitively large. From (20) and (21) we can see that the situation is especially bad for $m > n$ due to the (r_m^2/r_n^2) factors. For these points it is more advanteageous to express $\bar{\rho}_{mn}$ in terms of $\bar{\rho}_{i1}$ rather than $\bar{\rho}_{1i}$. The equation system – now with $2N - 1$ equations – is then obtained by requiring that the values $\bar{\rho}_{ii}$ and $\bar{\rho}_{i-1i}$ obtained from the two expressions be equal. This system is then solved by standard Gaussian elimination. The B and C coefficients still grow rather large and some numerical accuracy is lost, and the truncation errors introduced in this way do not have a simple h-dependence. The coefficients grow particularly fast in regions where relaxation methods converge well and the problem is alleviated by using a relaxation method for the outer points and matching the solutions obtained in the inner and outer regions.

Once the Gaussian elimination has been performed, repeated solutions with new right-hand sides for the same left-hand side can be obtained with relatively little effort. This feature is very useful since we want to add several solution-dependent corrections to the right-hand side. The first correction is obtained by noting that the values for $\bar{\rho}_{ii}$ and $\bar{\rho}_{ii-1}$ obtained by expressing them in terms of $\bar{\rho}_{i1}$ and $\bar{\rho}_{1i}$, respectively should have been equal, but are not quite equal due to the truncation error. By inserting the differences on the right-hand side of the equation and solving it again, we can improve the values of $\bar{\rho}_{i1}$ and $\bar{\rho}_{1i}$. A second correction performed is to add the difference $-\delta_0^4/12$ between the 3-point and the 5-point difference formula for the second derivative, which makes it possible to obtain a procedure with error $O(h^4)$ rather than $O(h^2)$. The appearence of homogeneous solutions may be prevented by adding an energy offset term both to the left- and right-hand sides of the equation, as discussed in 2.2. This is particularly important in the local potential case. If we use the Hartree-Fock potential, we must also add the difference from the approximate local potential used on the left-hand side, giving a term

$$\left((u^{HF} - u^{loc})_1 + (u^{HF} - u^{loc})_2\right) \bar{\rho}_{ab}$$

on the right-hand side.

3.2.3 Dealing with the Cusp at $r_i = r_j$ A problem one encounters in solving the radial pair equation is that the factor $r_<^k/r_>^{k+1}$ on the RHS has a cusp at $r_1 = r_2$, which destroys the accuracy of the numerical procedures used. This problem has to be remedied when evaluating Slater-type integrals, and we will see how this remedy can be applied also to the RHS of the pair equation.

Consider the integral of a function $f_i = f(x_i)$, defined on a grid of equidistant points, $x_i = x_0 + i \cdot h$. If the integrand and all its derivatives are continuous and the integrand vanishes at the limits, no numerical integration formula gives a better result than the simple trapeze formula

$$\int_{-\infty}^{\infty} f(x)dx = h \sum_i f_i$$

and this is the formula used in the evaluations of overlap integrals. The trapeze formula is, however, not adequate for the evaluation of Slater integrals since these involve $(r_<^k/r_>^{k+1})$, which has a cusp at $r_1 = r_2$. Then we use instead the approximation

$$\int_{x_i}^{x_{i+1}} f(x)dx = \frac{h}{24}(-f_{i-1} + 13f_i + 13f_{i+1} - f_{i+2}) + O(h^5)$$

This formula gives the total integral with an error $O(h^4)$ *if* the function f and its derivatives up to fourth order are continuous. In order not to loose the accuracy in the integration across the cusp, the integration is performed in two steps: First the integration over $r_1 < r_2$ is performed. In the evaluation for the last interval one point where r_1 is actually larger than r_2 is needed, but we use (r_1^k/r_2^{k+1}) also for this point, thus evaluating the integral of a function with *continuous* derivatives. Similarly, for the second part of the integral, $r_1 > r_2$, we use the form (r_2^k/r_1^{k+1}) also for the extra point with $r_1 < r_2$, needed for the first interval.

The problem of the cusp returns when the pair equation is solved. Using the five-point difference formula for the second derivatives should give an error $O(h^4)$, but, again, that assumes that the derivatives of the pair function up to fourth order are continuous, and since the RHS has a cusp, the *third* derivative of the pair function will have a discontinuity resulting, instead, in an error $O(h^2)$.

Maybe this is not so surprising — after all, evaluating the overlap with the RHS using the trapeze formula would give an $O(h^2)$ error and how could the differential equation know how to use a different formula? It is, however, possible to use the trapeze formula and still obtain an integral using the $O(h^4)$ formula given above by modifying the points next to the diagonal. Thus, if we define the RHS B_{ij} by

$$B_{ij} = \frac{r_<^k}{r_>^{k+1}} f_i g_j$$

for all points except the ones next to the diagonal, where we use instead

$$B_{i,i+1} = \left(\left(1 + \frac{1}{24}\right) \frac{r_i^k}{r_{i+1}^{k+1}} - \frac{1}{24} \frac{r_{i+1}^k}{r_i^{k+1}} \right) f_i g_{i+1}$$

with an analogous formula for $B_{i-1,i}$, then, using the trapeze formula for overlap integrals gives an $O(h^4)$ error. This error is, indeed, found also for the pair functions evaluated with the right-hand side defined in this way. This permits us to remove the $O(h^4)$ error using two rather than three different grid sizes in the Richardson extrapolation and also seems to make the final result more accurate as can be seen for the helium results shown in Table 1 below.

Table 1. Comparison between the results by various authors for the L-limits in the $(1s^2)^1S$ ground state of Helium .

L-Limit	Froese-Fischer (1973)	Bunge (1970)	Winter *et al.* (1970)	Mårtensson (1979)	Salomonson *et al.* (1988)
s^2	-2.87899	-2.879025	-2.87903	-2.87902	-2.8790288
p^2	-2.90015	-2.900507	-2.20051	-2.90050	-2.9005162
d^2	-2.90252	-2.902749	-2.90276	-2.90274	-2.9027668
f^2	-2.90291	-2.903307	-2.90331	-2.90328	-2.9033211
g^2	-2.90303	-2.903470	-2.90351	-2.90347	-2.9035186
h^2					-2.9036058
i^2					-2.9036500
k^2					-2.9036747
l^2					-2.9036897
Extrapolated					-2.9037248
'Exact' [a]					-2.90372435..

a) Schiff *et al.* (1965)

4. Results from Non-Relativistic Many-body Perturbation Theory

4.1 Helium Ground State Energies

Table 1 shows the results in different angular limits for the ground state of helium using the method described in Section 3 and compares them to results using the MCHF (Froese-Fischer, 1973) and CI (Bunge, 1970) methods and by solving the *eigenvalue equations* using the finite difference method (Winter *et al.*, 1970). In the s^2 limit only excitations $1s^2 \rightarrow s^2$ were considered and in our method only one equation, for $\rho_{1s^2\rightarrow s^2}$, was solved. In the p^2 limit also excitations into two p-orbitals were considered and the two coupled equations for $\rho_{1s^2\rightarrow s^2}$ and $\rho_{1s^2\rightarrow p^2}$ are solved together.

For higher l-values, the pair function becomes more and more localised to the area close to the diagonal, $r_1 = r_2$, reflecting the increased sharpness of the cusp of $(r_<^k/r_>^{k+1})$ for high k-values. For very high l-values very few grid points will be used to describe the pair function in areas where it is important and it seems that some numerical accuracy is lost, making it of little use to go beyond $l \approx 10$. To estimate the contributions of higher l-values, we perform an extrapolation using not only the $(l+\frac{1}{2})^{-4}$ dependence found by Schwartz (1962) for second-order energies but fit the values to an expression including also an $(l+\frac{1}{2})^{-6}$ term.

The older results by Mårtensson (1979) were obtained without using the cusp correction described in Sect. 3.2.3 and using at most 59 grid points in each direction, whereas the new results given in the last column where obtained using up to 90 points in each direction. The numerical accuracy was further improved by adding an energy offset term, $\delta\epsilon\, \rho_{ab}$ both to the left- and right- hand sides of the equation, so that it no longer has any homogeneous solutions $|ab\rangle$ or $|ba\rangle$ which could enter even if the right-hand side is made orthogonal to them, as discussed in Section 2.2. The results were obtained using hydrogenic orbitals as a starting point. Salomonson *et al.* (1988) have also performed the calculation starting from Hartree-Fock orbitals, and the results differ by at most one unit in the last digit shown, giving an indication of the numerical accuracy, which, although not comparable with that obtained by Pekeris and coworkers, should certainly be adequate for the study of larger atomic systems.

4.2 The $(1s2p)^{1,3}P$ States in Helium

As an example of the treatment of two non-equivalent electrons, we show in Table 2 the results obtained by Lindroth and Mårtensson-Pendrill (1984) for the $1s2p$ states in helium and also recent results by Salomonson *et al.* (1988) including the cusp correction. Following the procedure outlined in Section 2.2, we solved the equations for the functions $\rho_{1s2p\to ll'}$ and $\rho_{1s2p\to l'l}$ separately, rather than for the properly (anti-)symmetrised combinations $\{\rho_{1s2p\to ll'} \pm \rho_{2p1s\to ll'}\}$. A slight disadvantage of this approach is that the slower $(l+\frac{1}{2})^{-4}$ angular convergence for the singlet space, compared to $(l+\frac{1}{2})^{-6}$ for the triplet state (Schwartz 1962), will affect all excitations, and also that twice as many radial functions are coupled together, instead of treating two separate problems each with only half of the functions. For excitations from core orbitals, on the other hand, both approaches are comparable.

The pair functions used to evaluate these energies were also used to evaluate the expectation value $\langle \mathbf{p}_1 \cdot \mathbf{p}_2 \rangle$. This term is part of the *nuclear* kinetic energy

$$\mathbf{P}^2/2M = (\sum_i \mathbf{p}_i)^2/2M = \sum_i \mathbf{p^2}_i/2M + \sum_{i<j} \mathbf{p}_i \cdot \mathbf{p}_j/2M$$

where the momentum conservation has been used to express the nuclear momentum $\mathbf{P}$ in terms of the electronic momenta. The first term is the familiar *'reduced mass'* effect, which is proportional to the total energy. The second term can be described as a correlation of the electronic momenta through the motion of the nucleus and is referred to as *'mass polarisation'*. Between two isotopes of an element, this gives rise to the *'specific mass shift'* (SMS). Being a two-particle operator, it gives rise to large correlation effects and failure to treat these gives results far from experimental values. Lack of knowledge of the SMS is, in fact, a limitation in the extraction of changes in nuclear charge radii (which give rise to the 'field' or 'volume' isotope shift) from experimental isotope shifts, which have been obtained for several long chains of isotopes (see e.g. Thibault *et al.* 1981 for the case of Rb). A first calculation of correlation effects on the SMS for a many-electron atom was performed by Mårtensson and Salomonson (1982). Although this gave

Table 2. Angular limit convergence of the energy eigenvalue for the $(1s2p)^{1,3}P$ states in Helium

2^1P	Froese-Fischer (1973)	Davis and Chung (1982)	Lindroth and Mårtensson-Pendrill (1984)	Salomonson *et al.* (1988)
$sp - limit$	-2.122587	-2.1225935	-2.122592	-2.1225968
$pd - limit$	-2.123653	-2.1236858	-2.123686	-2.1236902
$df - limit$	-2.123766	-2.1237912	-2.123793	-2.1237970
$fg - limit$		-2.1238156	-2.123820	-2.1238232
$gh - limit$		-2.1238239	-2.123829	-2.1238327
$hi - limit$		-2.1238264	-2.123834	-2.1238370
$ik - limit$			-2.123835	-2.1238392
$kl - limit$				-2.1238405
Extrapolated				-2.123843
'Exact' [a]				-2.123843
2^3P				
$sp - limit$	-2.132367	-2.1323688	-2.132362	-2.1323714
$pd - limit$	-2.133054	-2.1331108	-2.133106	-2.1331149
$df - limit$	-2.133086	-2.1331508	-2.133147	-2.1331559
$fg - limit$		-2.1331562	-2.133153	-2.1331619
$gh - limit$		-2.1331572	-2.133154	-2.1331634
$hi - limit$			-2.133155	-2.1331638
$ik - limit$			-2.133155	-2.1331640
$kl - limit$				-2.1331641
Extrapolated				-2.133164
'Exact' [a]				-2.133164

a)Schiff *et al.* (1965), Accad *et al.* (1971)

Table 3. Convergence of the mass polarisation and of the normalisation integrals for the $(1s2p)^{1,3}P$ states in Helium (From Lindroth and Mårtensson-Pendrill 1984)

	$\langle \mathbf{p}_1 \cdot \mathbf{p}_2 \rangle$ $(10^{-2} a.u.)$		Normalisation Integral N^2	
	2^1P	2^3P	2^1P	2^3P
$sp - limit$	3.2710	-7.5157	1.9098	1.5566
$pd - limit$	4.5969	-6.4234	1.8682	1.5432
$df - limit$	4.5961	-6.4519	1.8637	1.5424
$fg - limit$	4.5994	-6.4552	1.8625	1.5422
$gh - limit$	4.6011	-6.4669	1.8621	1.5422
$hi - limit$	4.6020	-6.4562	1.8620	1.5422
$ik - limit$	4.6032	-6.4563	1.8617	1.5422
extrapolated	4.6043	-6.4564		
Araki *et al.*(1959) $pd - limit$	4.6122	-6.4636		
Davis and Chung (1982) $hi - limit$	4.6045	-6.4566		
'Exact' [a]	4.6045	-6.4572		

a)Schiff *et al.* (1965), Accad *et al.* (1971)

Table 4. Ionisation energies and contact hyperfine interaction parameters for the ground state of lithium (from Lindgren 1985) and of singly ionised calcium (from Mårtensson-Pendrill and Salomonson 1984) (The results are given in a.u.)

	Li 2^2S		Ca^+ 4^2S	
	Ionisation energy	*Contact parameter*	*Ionisation energy*	*Contact parameter*
Hartree-Fock	0.196304(1)	1.37222(1)	0.4150	15.69
Core-polarisation[a]		0.50914(1)		3.28
Brueckner orbitals	0.001834(2)	0.05466(2)	0.0207	2.84
Effect of B.O's on core pol.		0.01798(3)		0.43
Other third-order effects[b]		-0.00916(3)		-0.38
Relativistic corrections[c]	0.000016	0.0011	0.0017	0.95
Total	0.198154(3)	1.9459 (2)	0.4374	22.81
Experiment	0.198159[d]	1.9397[e]	0.4363[n]	22.8(4)[o]
Other calculations	0.19719[f]			23.03[p]
	0.19811[g,h]	1.937[g]		
	0.19694[i]	1.9148[i]		
	0.198(2)[j]	1.927[j]		
	0.198076[k]	1.939[l]		
	0.198158[m]			

a) Contribution from the potential correction (13) due to the hyperfine interaction, as discussed in Section 2.4

b) Including also some higher-order effects, since all-order pair functions were used in the evaluation

c) Obtained by using a relativistic treatment for all single-particle effects and estimating the effects for the correlation contributions

d) Moore (1952, 1971), converted to atomic units by using the reduced electron mass, corresponding to a nuclear mass of 6.5 amu.

e) Obtained using the nuclear magnetic moment $\mu_I = 3.2564$ (Lederer and Shirley 1978) and the experimental hyperfine structure from Schlecht and McColm (1966)

f) Weiss (1961) Configuration Interaction

g) Larsson (1968) h) Sims *et al.* (1976) Hylleraas-type wave functions

i) Nesbet (1967) 'Bethe-Goldstone equations' closely related to the all-orders pair functions but without couplings between different electron pairs and including excitations only up to $l = 3$

j) Chang *et al.* (1968) Perturbation theory up to third order.

k) Johnson *et al.* (1988a) *Relativistic* perturbation theory, up to third order

l) Garpman *et al.* Perturbation theory up to third order.

m) Johnson *et al.* (1988b) *Relativistic* perturbation theory, pair correlation to all orders. (Preliminary result)

n) Edlén and Risberg (1956)

o) Obtained from the value by Kelly *et al.* (1954), using $\mu_I(^{43}\text{Ca}) = 1.317642(7)$

p) Beigang and Timmermann (1982), semi-empirical value using the Fabry-Segrè formula a significant improvement of the results, we have found that, in general, it is not sufficient to include only the lowest-order correlation contributions to the SMS (see e.g. Lindroth *et al.* 1985). Inclusion of pair correlation effects to all orders would require the solution of pair equations including the SMS operator and we hope to develop such a procedure in the future. For two-electron systems, however, (11) does give the desired results, and Table 3 shows the values for the SMS in the 1s2p states in helium together with the values for the norm used in the denominator of (11). The largest contribution to the norm comes in the lowest angular limit, which includes terms which could be considered as potential corrections for the $2p$ electron due to the presence of the $1s$ electron.

4.3 MANY-ELECTRON SYSTEMS

4.3.1 The $1s^2 2s$ *Ground State of Lithium* For a three-electron system, an exact treatment is no longer possible, without including simultaneous excitations involving all three electrons. However, the helium-like $1s^2$ core, and the interaction between the valence electron $2s$ and one of the core electrons, as well as the coupling between the different pairs should be treated exactly when pair correlation is treated to all orders. To show the validity of this approximation we give in Table 4 the ionisation energies and hyperfine parameter for the 2^2S state in Li, from Lindgren (1985).

4.3.2 Singly Ionised Calcium To demonstrate that the method is not limited to few-electron cases we give in Table 4 also results for the 4^2S state in $^{43}Ca^+$, which where obtained as part of a calculation for neutral calcium (Salomonson 1984), providing, among other results, improved values of the quadrupole moment for $^{43}Ca^+$.

In the next section we describe the modifications necessary to treat heavy atomic systems.

5. Relativistic treatment

For a heavy atom it is essential to use a relativistic treatment. This modifies the kinetic energy term and the non-relativistic one-body hamiltonian (1) must then be replaced by

$$h_0 = (c\alpha \cdot \mathbf{p} + \beta mc^2) - \frac{Z}{r} + u \tag{22}$$

The energy eigenvalues, E, for this equation include the rest energy mc^2 and an electron with binding energy -ϵ will have $E = mc^2 + \epsilon$. The eigenfunctions to h_0 can be written in terms of two spatial components, F and G, each with two spin components

$$\phi_0 = \begin{pmatrix} F \\ G \end{pmatrix} = \frac{1}{r} \begin{pmatrix} P_{nlj}(r)|lsjm\rangle \\ iQ_{nlj}(r)|\bar{l}sjm\rangle \end{pmatrix}$$

The second form can only be used in the central-field model. We then have $\bar{l} = 2j - l$ and $|lsjm\rangle$ denotes a coupling of orbital and spin angular momenta, l and s, to a total angular momentum, j, with magnetic quantum number m.

In the non-relativistic limit, where $c(= \alpha^{-1}$ in atomic units) approaches infinity, the radial part of the upper component aproaches the radial part of a non-relativistic wavefunction. We can see that the size of the problem increases rapidly as we go to the relativistic treatment – where, in the non-relativistic case, we had one radial function for each l-value, we now have two possible j-values, $j = l \pm \frac{1}{2}$ (except for $l = 0$, where only $j = \frac{1}{2}$ is allowed), each with two radial components. However, in addition to these practical problems, there are problems of a more fundamental character related to the presence of *negative-energy solutions* to the one-particle Dirac equation (Dirac 1930).

Using the two-component form of the wavefunction, the one-body Dirac equation can be written as two coupled equations for F and G

$$(mc^2 - \frac{Z}{r} + u)F + c\sigma \cdot \mathbf{p}G = EF$$
$$(-mc^2 - \frac{Z}{r} + u)G + c\sigma{\cdot}\mathbf{p}F = EG$$

If F and G are solutions for a certain energy E, then it is obvious that a function with F and G interchanged will be a solution to an equation where E as well as the potential energy terms have changed signs. The states with negative energy can be considered as positron states. We note that an ordinary attractive central potential will have a set of both bound and continuum electron states, but only continuum positron states, since positrons are repelled from the nucleus.

The negative-energy states are not available for the electrons. When the one-body Dirac equation is solved numerically for atomic states, as done in the present work, this does not normally cause a problem, since the boundary conditions can be used to ensure that only bound states are obtained (Grant 1982). Another problem, commonly known as the *variational collapse*, (Schwarz and Wallmeier 1982) may, however, occur if the orbitals are expanded in a discrete basis set (which may contain unkown components of the negative energy solutions) and the eigenvalue is found by minimising the expectation value of h_0. Various methods have been suggested to circumvent this problem, as reviewed by Kutzelnigg (1984).

5.1 Many-electron case

The problems caused by the negative energy solutions become apparent if we consider the naïve relativistic generalisation, the so-called Dirac-Coulomb equation, of the two-electron equation discussed in Section 2.2. Consider e.g. the equation (8) for the first-order correction to the unperturbed wave function. The right-hand side can be rewritten as

$$\sum_{rs \neq ab,ba}^{pos,neg} |rs\rangle\langle rs|\frac{1}{r_{12}}|ab\rangle$$

Since the negative-energy states are part of the complete set of solutions to the equation they will automatically enter when the closure relation is used, and we can see that the right-hand side contains an infinite number of terms, degenerate with the combination $|ab\rangle$, where one electron is in the negative energy continuum and the other high up in the positive energy continuum. These solutions will be coupled by the two-electron interaction and an originally bound solution will 'dissolve into the continuum' when the perturbation is added. This problem was first pointed out by Brown and Ravenhall (1951) and has recently been emphasised, particularly by Sucher (1980,1984). However, even if a total collapse of the wavefunction into the continuum could be avoided, e.g., by the enforcement of suitable boundary conditions, the Dirac-Coulomb equation would still be incorrect, since it incorrectly treats the negative energy states in the same way as the positive energy ones — an excitation into a negative energy state should be allowed only when it is preceded by an excitation out of the same state, which corresponds to the creation of a virtual electron-positron pair requiring an excitation energy of about $2mc^2$.

In order to treat the negative energy states correctly, it is necessary to divide the two-particle interaction into several parts by using projection operators for positive and negative energy states and to treat each part individually. The distinction between positive and negative energy states is provided by the choice of one-body hamiltonian and a positive energy state in one potential will have a small admixture of negative energy states in another basis (Heully *et al.*, 1986a).

Formally, the one-particle projection operator, λ^+ for positive energy states can easily be written down in terms of eigenstates to the Dirac equation as

$$\lambda^+ = \sum_n^{pos} |n\rangle\langle n|$$

However, if a more or less complete set of solutions to the Dirac equation is available, it may also be used directly as a basis set for a configuration interaction type calculation or in an application of perturbation theory (Quiney *et al.* 1985). This type of approach has also been applied to calculations for heavy atoms (e.g. Vajed-Samii *et al.* 1979, Dzuba *et al.* 1985, 1987, Johnson and Sapirstein 1986 and Johnson *et al.* 1988) although only excitations into positive energy states have been considered. Such a calculation is *in principle* not much more complicated than the corresponding non-relativistic procedure, although *in practice* the demands on computing time and storage are very much larger.

In our non-relativistic work, using the inhomogeneous differential equation technique, only the occupied orbitals are obtained explicitly and the closure relation is used to provide the summation over all orbitals. The unwanted terms are then removed by performing a finite sum over occupied orbitals. In the relativistic case, there is an infinite number of unwanted negative-energy states, and we would like to find another expression for the projection operators. In the next section we show how this can be done.

5.2 Projection operators

We want to find a way to distinguish positive and negative energy solutions to the Dirac equation without having access to the energy eigenvalues and this can, indeed, be done by using the relations between the upper and lower components. The problem was investigated in some detail by Heully *et al.* (1986b) and methods based on relations between the components have been used also in quantum chemistry to avoid the 'variational collapse' (see e.g. Kutzelnigg 1984). It was found that operators $R_\pm$ satisfying

$$R_\pm = \pm\frac{1}{2mc^2}(c\sigma\cdot\mathbf{p} \mp [R_\pm, V] - R_\pm(c\sigma\cdot\mathbf{p})R_\pm) \tag{23}$$

are such that for positive energies $G = R_+F$, whereas for negative energies we have instead $F = R_-G$. The equation (23) is found by direct insertion in the Dirac equation of these relations. (Note also that (23) is of second order in the R operators and thus has two solutions - corresponding to positive and negative energies. The desired operator is the one that is in some sense 'small' so that the quadratic term on the RHS can be considered as a correction.) The two operators are related through $R_- = -R_+^\dagger$, as can be seen by comparing the equations for these operators.

We introduce the operator $\Omega_+^2 = (1 + R_+^\dagger R_+)^{-1}$ to maintain the norm of the wavefunctions. It is now easy to see that

$$\lambda_+ = \begin{pmatrix} \Omega_+^2 & \Omega_+^2 R_+^\dagger \\ R_+\Omega_+^2 & R_+\Omega_+^2 R_+^\dagger \end{pmatrix}$$

leaves a positive energy states unchanged while giving zero operating on a negative energy state and is thus our desired projection operator. A similar expression for λ_- can be found in an analogous way or by using $\lambda_+ + \lambda_- = 1$.

5.2.1 Approximations. As a first approximation to R_+ we use

$$R_+ \approx \frac{1}{2mc^2 - V} c\sigma\cdot\mathbf{p} \equiv R_0 \tag{24}$$

which differs from the exact operator in relative order α^2. (It would do so even without introducing V in the denominator, but not only the order in α is important — close to the nucleus V is certainly not small.) The approximate expression in (24) is very similar to the energy-dependent relation between F and G obtained from the Dirac equation (22). For a positive energy state with $E = \epsilon + mc^2$ we have

$$G = \frac{1}{2mc^2 + \epsilon - V} c\sigma\cdot\mathbf{p}F \tag{25}$$

and we find that

$$R_0F - G = \frac{\epsilon}{2mc^2(1 - V/2mc^2)} G = \mathcal{O}(\alpha)^2 G$$

The fact that V is large close to the nucleus actually improves the situation since *V is negative.* To estimate the error in R_0 we note that even for the extreme case of the *1s* state in uranium, $|\epsilon/2mc^2| \approx 0.13$. For the outer electrons of a neutral atom the approximation is much better. On the other hand R_0 is not adequate for discussions about the Brown-Ravenhall disease, where one electron has an energy of about $\epsilon \approx 2mc^2$.

5.3 THE RELATIVISTIC PAIR EQUATION

A two-particle Dirac function can be separated into four spatial components, each of which is a function of the coordinates for the two electrons, just like a non-relativistic pair functions. For any right-hand side the two-particle equation can be written in the general form

$$\begin{aligned}
&(\epsilon_a + \epsilon_b \qquad\qquad - u_1 - u_2)\rho^{LL} - c(\sigma\cdot\mathbf{p})_1\rho^{SL} - c(\sigma\cdot\mathbf{p})_2\rho^{LS} = P^{LL}\\
&(\epsilon_a + \epsilon_b + 2mc^2 - u_1 - u_2)\rho^{LS} - c(\sigma\cdot\mathbf{p})_1\rho^{SS} - c(\sigma\cdot\mathbf{p})_2\rho^{LL} = P^{LS}\\
&(\epsilon_a + \epsilon_b + 2mc^2 - u_1 - u_2)\rho^{SL} - c(\sigma\cdot\mathbf{p})_1\rho^{LL} - c(\sigma\cdot\mathbf{p})_2\rho^{SS} = P^{SL}\\
&(\epsilon_a + \epsilon_b + 4mc^2 - u_1 - u_2)\rho^{SS} - c(\sigma\cdot\mathbf{p})_1\rho^{LS} - c(\sigma\cdot\mathbf{p})_2\rho^{SL} = P^{SS}
\end{aligned} \tag{26}$$

where L and S denote large and small, respectively (for positive energies). A possible approach to solve this set of equations is to eliminate the 'mixed' small-large and large-small components in the same way as the small component G can be eliminated in the one-particle case by using (25). This gives us

$$\rho^{LS} = \frac{B_{ab}}{2mc}((\sigma\cdot\mathbf{p})_1\rho^{SS} + (\sigma\cdot\mathbf{p})_2\rho^{LL}) + \frac{B_{ab}}{2mc^2}P^{LS} \tag{27}$$

where B_{ab} is given by

$$B_{ab} = \left(1 + \frac{(\epsilon_a + \epsilon_b - u_1 - u_2)}{2mc^2}\right)^{-1} \tag{28}$$

Using the expression (27) for ρ^{LS} and the analogous equation for ρ^{SL}, the equation for the 'large-large' component can be rewritten as

$$\begin{aligned}
&\left(\epsilon_a + \epsilon_b - u_1 - u_2 - \frac{1}{2m}\Big((\sigma\cdot\mathbf{p})_1 B_{ab}(\sigma\cdot\mathbf{p})_1 + (\sigma\cdot\mathbf{p})_2 B_{ab}(\sigma\cdot\mathbf{p})_2\Big)\right)\rho^{LL}\\
&= P^{LL} + \tilde{R}_1^\dagger P^{SL} + \tilde{R}_2^\dagger P^{LS} + \frac{1}{2m}\Big((\sigma\cdot\mathbf{p})_1 B_{ab}(\sigma\cdot\mathbf{p})_2 + (\sigma\cdot\mathbf{p})_2 B_{ab}(\sigma\cdot\mathbf{p})_1\Big)\rho^{SS}
\end{aligned} \tag{29}$$

where we have introduced the operator

$$\tilde{R}_i = \frac{B_{ab}}{2mc}(\sigma\cdot\mathbf{p})_i$$

which differs from the R operator in (23) or (24) used in the projection operators only in relative order α^2. The equation for ρ^{SS} is analogous to (29), except for an extra $4mc^2$ which normally dominates the left-hand side. However, instead of using this equation for ρ^{SS}, we use the R operator to relate the upper and lower components of a pair function, *i.e.*

$$\rho^{SS} = R_1 R_2 \rho^{LL} \tag{30}$$

which is found to be an active ingredient in the antidote for the Brown-Ravenhall disease — with this expression used for ρ^{SS} the equation no longer has a combination of a positive and a negative energy state as a homogeneous solution.

5.3.1 Applying the projection operators. Although using the expression (30) prevents any disease of the left-hand side of the relativistic pair equation, the right-hand side may still contain illegal negative-energy components. The health of the right-hand side may be ensured by applying the positive-energy projection operators discussed in Section 5.2.

The application of $\Lambda_{12}^{++} = \lambda_1^+ \otimes \lambda_2^+$ to a right-hand side of the form in (26) gives

$$\bar{P}^{LL} = \Omega_1^2\Omega_2^2(P^{LL} + R_1^\dagger P^{SL} + R_2^\dagger P^{LS} + R_1^\dagger R_2^\dagger P^{SS})$$

and the other components of the projected right-hand side can be obtained as

$$\bar{P}^{SL} = R_1\bar{P}^{LL} \qquad \bar{P}^{LS} = R_2\bar{P}^{LL} \qquad \bar{P}^{SS} = R_1R_2\bar{P}^{LL}$$

Inserting these expressions into the first three terms of (29) gives

$$\begin{aligned}&\bar{P}^{LL} + \tilde{R}_1^\dagger\bar{P}^{SL} + \tilde{R}_2^\dagger\bar{P}^{LS}\\ &\quad = (1 + \tilde{R}_1^\dagger R_1 + \tilde{R}_2^\dagger R_2)\Omega_1^2\Omega_2^2(P^{LL} + R_1^\dagger P^{SL} + R_2^\dagger P^{LS} + R_1^\dagger R_2^\dagger P^{SS})\end{aligned} \tag{31}$$

By using the definition of Ω and the fact that $\tilde{R}_i^\dagger = R_i^\dagger(1 + O(\alpha^2))$ we find that

$$(1 + \tilde{R}_1^\dagger R_1 + \tilde{R}_2^\dagger R_2)\Omega_1^2\Omega_2^2 = 1 + O(\alpha^4)$$

and to lowest order in α^2 we need to consider only the terms in the last parenthesis in (31). For *diagonal interactions* (such as the Coulomb interaction between the electrons) the terms in the last bracket in the equation (31) are in ascending order of α^2, with the last term of order α^4, and we find that the RHS in equation (29) is changed only in order α^4 by the application of projection operators. For non-diagonal interactions, such as the Breit interaction, all four terms in the last parenthesis in (31) are of the same order and the effect of applying projection operators to the right-hand side enters already in order α^2.

We have thus found that by solving the equation (29) with ρ^{SS} given by (30) we solve, with an error O (α^4) a *no-virtual-pair pair equation*, where only excitations into the positive energy states are allowed. This equation is somewhat similar to a Schrödinger equation with relativistic corrections, although the presence of the B factors provides a welcome smoothing of the singularities close to the nucleus. The numerical procedure used to solve this equation for the large-large component is a non-trivial extension of the method presented in section 3, and has been described in detail by Lindroth (1988). Table 5 shows the results obtained for the second-order energies, starting from hydrogen-like orbitals, for the ground state of helium and of helium-like neon. It also shows the contributions from magnetic interactions, which contain one order of the Coulomb interaction and one order of the Gaunt (1929) interaction $-\alpha_1 \cdot \alpha_2/r_{12}$. The generalisation to larger systems can be done in the same way as in the non-relativistic case, and the development needed to treat also relativistic correlation effects to all orders is underway.

Table 5. Second-order correlation energies obtained with hydrogen-like orbitals for $Z = 2$ and $Z = 10$, given in units of $10^{-6}a.u.$ (From Lindroth 1988)

	$Z = 2$		$Z = 10$	
	Electrostatic[a]	*Magn.*	*Electrostatic*[a]	*Magn.*
$1s^2 \to s^2$	-125356.0(22.3)	-68.3	-125896	-1706
$p_{1/2}^2$	-8828.3 (3.4)	-22.0	-8747	-542
$p_{3/2}^2$	-17664.1 (-0.7)	-23.6	-17681	-587
$d_{3/2}^2$	-1561.0 (1.4)	-6.3	-1529	-151
$d_{5/2}^2$	-2343.5 (0.1)	-5.1	-2343	-125
$f_{5/2}^2$	-461.2 (0.6)	-2.9	-446	-69
$f_{7/2}^2$	-615.7 (0.1)	-2.4	-613	-57
$g_{7/2}^2$	-180.1 (0.4)	-1.7	-172	-39
$g_{9/2}^2$	-225.5 (0.1)	-1.4	-223	-33
$h_{9/2}^2$	-83.8 (0.2)	-1.1	-79	-24
$h_{11/2}^2$	-100.8 (0.1)	-0.9	-99	-21
⋮	⋮	⋮	⋮	⋮
Total[b]	-157681 (15)	-146	-158065(-402)	-3531
Splines[c]	-157681.5		-158083	
Pekeris[d]	-157681.3(14.9)		-158038(-372)	
All orders				
$1s^2 \to s^2$	-129049.5	-54.7	-126441	-1615
Contributions				
from higher orders	-3693.5 (1.4)	13.6	-545 (9)	91

a) The values in parentheses are the relativistic corrections obtained by subtracting the results obtained by solving the non-relativistic pair equation using the same grid-size and extrapolation procedures.

b) The final results were obtained by including excitations up to $l = 10$ and performing an extrapolation to estimate the contributions from higher l-values.

c) Second-order energy obtained by Johnson and Sapirstein (1986) by using basis splines to expand the orbitals

d) Pekeris (1958) used his 'exact' non-relativistic wavefunctions to evaluate the relativistic corrections to lowest order. The second-order result was extracted by fitting the results for *Z=1-10* to a Z-expansion.

The development of methods to treat relativistic correlation effects was stimulated by the need for atomic calculations to extract the weak interaction parameters from experiments on atomic parity non-conservation (PNC) (see. e.g. Bouchiat and Pottier 1986 and Commins 1987). Impressive progress has been made over the last few years and two groups have presented calculations of PNC effects including correlation. Dzuba *et al.* (1987) have performed calculations including all second- and selected higher-order pair correlation effects for the PNC transitions studied in Cs and Tl and also for several other properties. Johnson *et al.* (1988) have performed a complete calculation of the most important second-order effects, including excitations both to high l-values and from deep-lying core orbitals. By using iterative procedures it should be possible to treat the effect of pair correlation to all orders also for PNC transitions — it is a giant undertaking, but we think that the importance of the results justifies the effort.

Acknowledgements

The numerical methods described have been developed in Prof. Ingvar Lindgren's research group in Göteborg and I would like to express my gratitude for happy and fruitful collaboration to him and to the other past and present coworkers in this project: John Morrison, Johannes Lindgren, Sten Salomonson, Jean-Louis Heully, Eva Lindroth and Per Öster.

The work has been supported by the Swedish Natural Science Research Council. Further, I would like to thank the Institute for Theoretical Physics, University of California, Santa Barbara, where this manuscript was produced using TeX, for hospitality and support (by National Science Foundation under Grant No. PHY82-17853, supplemented by funds from the National Aeronautics and Space Administration) and for a very stimulating environment.

References

Accad Y, Pekeris C L and Schiff B 1971 *Phys. Rev. A* **4** 516
Araki G, Mano K and Otha M 1959 *Phys. Rev.* **115** 1222
Beigang R and Timmermann A 1982 *Phys. Rev. A* **25** 1496
Bloch C 1958 *Nucl. Phys.* **6** 329
Bouchiat M A and Pottier L 1986 *Science* **234** 1203
Brown G E and Ravenhall D G 1951 *Proc. Roy. Soc. A* **208** 552
Bunge 1970 *Theor. Chim. Acta* (Berlin) **16** 126
Chang E S, Pu R T and Das T P 1968 *Phys. Rev.* **174** 1
Čížek J 1966 *J. Chem. Phys.* **45** 4256
1969 *Adv. Chem. Phys.* **14** 35
Commins E D 1987 *Physica Scripta* **35** 468
Davis B F and Chung K T 1982 *Phys. Rev. A* **25** 1328
Dirac P A M 1930 *Proc. Roy. Soc. A* **126** 360
Dzuba V A, Flambaum V V, Silvestrov P G and Sushkov O P 1985 *J. Phys. B* **17** 1953-6
1987 *Physica Scripta* **35** 69, *J. Phys. B* **20** 1399
Edlén B and Risberg P 1956 *Ark. Fys.* **10** 553
Froese-Fischer C 1973 *J. Comput. Phys.* **13** 502
Garpman S, Lindgren I, Lindgren J and Morrison J 1975 *Phys. Rev. A* **11** 758
1976 *Z. Physik A* **276** 167
Gaunt J A 1929 *Proc. Roy. Soc. A* **122** 513

Grant I P 1982 *Phys. Rev. A* **25** 1230-2
Heully J-L, Lindgren I, Lindroth E and Mårtensson-Pendrill A-M 1986a *Phys. Rev. A* **33** 4426-9
Heully J-L, Lindgren I, Lindroth E, Lundqvist S and Mårtensson-Pendrill A-M 1986b *J. Phys. B* **19** 2799
Johnson W R, Guo D S, Idrees M and Sapirstein J 1985 *Phys. Rev. A* **32** 2093
1986 *Phys. Rev. A* **34** 1043
Johnson W R, Blundell S A and Sapirstein J 1988a *Phys. Rev. A* in print
Johnson W R, Blundell S A, Liu Z W and Sapirstein J 1988b to be published
Johnson W R, Lin C D, Cheng K T and Lee C M 1980 *Physica Scripta* **21** 409
Johnson W R and Sapirstein J 1986 *Phys. Rev. Lett.* **57** 1126
Kelly F M, Kuhn H and Pery A 1954 *Proc. Roy. Soc. A* **67** 450
Kelly H P 1963 *Phys.Rev.* **131** 684
Kümmel H, Lührmann K H and Zabolitzky J G 1078 *Phys Rep.* **36C** 1
Kutzelnigg W 1984 *Int. J. Quantum Chem.* **25** 107-29
Larsson S 1968 *Phys. Rev.* **169** 49
Lederer C M and Shirley V S 1978 *Table of Isotopes* (Wiley, New York)
Lindgren I 1975 *private communication*
Lindgren I 1985 *Phys. Rev. A* **31** 1273-86
Lindgren I, Lindgren J and Mårtensson A-M 1976 *Z. Physik A* **279** 113-25
Lindgren I and Morrison J 1982 *Atomic Many-Body Theory* Springer Series in Chemical Physics, vol 13
Lindgren I and Mukherjee D 1988 to be published
Lindgren I and Salomonson S 1980, *Physica Scripta* **21** 335
Lindroth E 1988 *Phys. Rev. A* **37** 316
Lindroth E and Mårtensson-Pendrill A-M 1984 *Z. Physik A* **316** 265-73
Lindroth E, Mårtensson-Pendrill A-M and Salomonson S 1985 *Phys. Rev. A* **30** 712
Mårtensson A-M and Salomonson S 1982 *J. Phys. B* **15** 2115
Mårtensson A-M 1979 *J. Phys. B* **12** 3995
Mårtensson-Pendrill A-M 1985 *J. de Physique (Paris)* **46** 1949-59
1985 *Phys. Rev. Lett.* **54** 1153-5
Mårtensson-Pendrill A-M and Salomonson S 1984 *Phys. Rev. A* **30** 712
McKoy V and Winter N W 1968 *J.Chem. Phys.* **48** 5514-23
Moore C E 1952, 1971 *Atomic Energy Levels* Natl. Bur. Stand. (U.S) Circ. No. 467 (U.S. GPO, Washington, D.C)
Morrison J C 1973 *J. Phys. B* **6** 2205-12
Morrison J C and Salomonson S 1980 *Physica Scripta* **21** 343
Musher J I and Schulman J M 1968 *Phys. Rev.* **173** 93-107
Nesbet R K 1967 in *Proceedings of the International Colloquium on Magnetic Hyperfine Structure of Atoms and Molecules, Paris, 1966,* edited by R Lefebvre and C Moser (CNRS, Paris)
Pekeris C L 1958 *Phys. Rev.* **112** 1649
1962, *Phys. Rev.* **126** 143-5 , 1470-6, *ibid.* **127** 509-11
Quiney H M, Grant I P and Wilson S 1985 *J. Phys. B* **18** 2805-15
Salomonson S 1984 *Z. Physik A* **316** 135-48
Salomonson S, Lindgren I and Mårtensson A-M 1980 *Physica Scripta* **21** 351
Salomonson S, Öster P, Lindgren I and Mårtensson-Pendrill A-M 1988, to be published
Sandars P G H 1977 *J. Phys. B* **10** 2983
Schiff B, Lifson H, Pekeris C L and Rabinowitz P 1965 *Phys. Rev.* **140** A1104
Schlecht R G and McColm D W 1966 *Phys. Rev.* **142** 11

Schulman J M and Lee W S 1972 *Phys. Rev. A* **5** 13-21
Schwartz C 1962, *Phys. Rev.* **126** 1015-9
Schwarz W H E and Wallmeier H 1982 *Mol. Phys.* **46** 1045-61
Sims S, Hagstrom S A and Rumble, Jr. P R 1976 *Phys. Rev. A* **14** 576
Sinanoğlu O 1962 *J. Chem. Phys.* **36** 706, 3198
1964 *Adv. Chem. Phys.* **6** 315
Sternheimer 1950 *Phys.Rev.* **80** 102
Sucher J 1980 *Phys. Rev. A* **22** 348-62
1984 *Int. J. Quantum. Chem.* **25** 3-21
Thibault C *et al.* 1981 *Phys. Rev. C* **23** 2720
Vajed-Samii M, Ray S N, Das T P and Andriessen J 1979 *Phys. Rev. A* **20** 1787
Weiss A W 1961 *Phys. Rev.* **122** 1826
Winter N W, Laferrière A and McKoy V 1970 *Phys. Rev. A* **2** 49-60

FULLY NUMERICAL CALCULATIONS FOR DIATOMIC SYSTEMS

P. PYYKKÖ
Department of Chemistry, University of Helsinki
Et. Hesperiankatu 4, SF-00100 Helsinki, Finland

Abstract. The fully numerical, finite-difference successive overrelaxation approach developed in the author's group for molecular Schrödinger (or second order Dirac) equations is outlined. The latest numerical results are summarized, including Q(Li) from LiF, the dipole moment of the molecular ion ArH^+ and He---He interatomic potentials.

Introduction

The Hartree-Fock equations

$$F\Psi_a = (T + V_n + V_c + V_x)\Psi_a = \varepsilon_a \Psi_a + \sum_{a \neq b} \varepsilon_{ab} \Psi_b \tag{1}$$

($T = -\nabla^2/2m$, n = nuclear, c = electronic Coulomb, x = exchange) for molecules are commonly solved in quantum chemistry by the "linear combination of atomic orbitals (LCAO)" approach, using either Slater or Gaussian AO's ($\exp(-\zeta r)$ or $\exp(-\alpha r^2)$, respectively). This procedure always contains a basis-set truncation error.

While numerical methods are for molecules computationally much more expensive, they converge rapidly, as a function of the grid, and provide definitive benchmarks for LCAO calculations. They also can be used for calculating "difficult" molecular properties, like electric field gradients, where LCAO methods are unable to provide unambiguous answers.

Our non-relativistic [1-17] and relativistic [18-22a] fully numerical, two-dimensional ("2D") calculations have recently been reviewed [23]. Therefore we

M. Defranceschi and J. Delhalle (eds.), Numerical Determination of the Electronic Structure of Atoms, Diatomic and Polyatomic Molecules, 161–175.

here only outline the method and bring the list of references [1-23] up to date. The molecules treated are listed in Table 1 below. References [24, 25] use these data in conjunction with basis-set methods.

As to work by other groups, by far the most fertile approach has been the "Partial Wave (PWSCF)" one by McCullough [26], recently reviewed in reference [27]. While we solve the functions f_a in

$$\Psi_a (\zeta,\eta,\varphi) = f_a (\zeta,\eta) \exp(im_a\varphi) \tag{2}$$

$$\zeta = (r_1 + r_2)/R, \qquad 1 \leq \zeta \leq \infty \tag{3}$$

$$\eta = (r_1 - r_2)/R, \qquad -1 \leq \eta \leq 1$$

using 2D numerical methods, McCullough introduces the expansion

$$\Psi_a (\zeta,\eta,\varphi) = \exp(im_a\varphi) \sum_{l=|m|}^{l_{max}} P_l^m(\eta)\, f_{alm}(\zeta) \tag{4}$$

and solves the 1D numerical problem for each f_{alm}. Even for a molecule like Cr_2, he finds $l_{max} = 20$ sufficient [28]. The method has been extended to coupled Hartree-Fock studies of polarizabilities [29], coupled-cluster treatments of correlation effects [30] and spin-polarized calculations of magnetic hyperfine constants [31].

Local-exchange 2D results for diatomics, using iterative matrix methods, have been reported by Becke [32,33].

The main contender of the finite-difference method is the finite-element one. Its first applications to molecules (H_2^+ [34,35], H_2 [36], N_2 etc...[38]), atomic collisions [39] or vibrational problems [40,41] are promising.

A further approach is to avoid the differentiation in the T operator altogether by using fast Fourier transforms (FFT) to carry out the multiplication by T in p space and the multiplication by V in r space. This approach [42,43] has been used for some time for vibrational problems. Its applications [44] on the first 3D calculations on a non-linear molecule (H_3^+) are competitive with the best LCAO calculations.

Momentum-space results for linear molecules , up to H_3, have been published earlier by the Paris group [45,46]. For technical improvements, see references [46-48]

A " pseudospectral" method, combining LCAO basis sets and grids, was presented by Friesner [49]. Becke [50] proposed a multicenter numerical integration scheme for polyatomic molecules and obtained a five- or six-figure accuracy using a few thousand integration points per atom.

For references to 3D calculations in nuclear and solid-state physics, see reference [23].

Method

Instead of the variables (ζ,η) we actually now use the variables (μ,ν) giving a quadratic distribution of points near the nuclei:

$$\zeta = \cosh \mu \quad , \quad 0 \leq \mu < \infty \tag{5}$$

$$\eta = \cos \nu \quad , \quad 0 \leq \nu \leq \pi.$$

An equidistant grid of the order of $n_\nu \times n_\mu = 100 \times 100$ points is used. The run is started with a coarse grid, followed by Stirling interpolation to denser one(s). We never return, however, to the coarser ones, as in the multigrid methods [51].

The eigenvalues, ε_a, are obtained as Rayleigh quotients

$$\varepsilon_a = \langle a \,|\, F \,|\, a \rangle / \langle a \,|\, a \rangle \tag{6}$$

The eigenfunctions Ψ_a, the Coulomb potentials V_c

$$V_c = \sum_{b \neq a} V_c^{\,b} \tag{7}$$

and the exchange potential

$$V_x = \sum_{b \neq a} V_x^{ab} \tag{8}$$

are obtained by solving eq. 1, eq. 9 or eq. 10, respectively, using the successive overrelaxation method (SOR), (Eq. 11).

$$\nabla^2 V_c^{\,b} = -\,4\pi\,\Psi_b^{\,2} \tag{9}$$

$$\nabla^2 V_x^{\,ab} = 4\pi\,\Psi_a \Psi_b \tag{10}$$

Seven-point formulae are used for the derivation and five-point ones for the integration (along one dimension). In this method, the discretized equations (1, 9,10) are solved for one point at a time. The overrelaxation step

$$\Psi_{final}^{\;(k+1)} = (1-\omega)\,\Psi^{(k)} + \omega\,\Psi^{(k+1)} \tag{11}$$

with $1 < \omega < 2$ accelerates the convergence. In our work, typical values are around 1.8. For the sweep sequences, see reference [23]. The method is a "Seidel" one; new points are immediately used as old ones.

Results

The obtained results consist of total and orbital energies, electric multipole moments, electric field gradients and other ground-state properties, such as deformation densities (between atoms and molecules). Electric polarizabilities can be calculated by a finite-field method [23]. So far we have not considered other excited states of many-electron molecules, than the lowest, $X\ ^2\Sigma^+$ and $A\ ^2\Pi_r$ states of BeH [14]. For a summary of the systems treated so far, see Table 1.

The 2D HF of MC SCF results can be improved by adding to them correlation contributions and vibrational averages from various LCAO methods, such as many-body perturbation theory with up to quadrupole substitutions ("MBPT (4)") [24] or the " Complete Active Space SCF (CASSCF)"[5,6,9] approach. This has so far been tried on the electric dipole moment of LiH [5], the bond lengths of ArH^+[13]

and BCl [25], and the electric field gradient, q(Li) in LiH [6] and LiF [24], or q(N) in NO^+ and N_2 [9].

The latter comparison permits a determination of the nuclear quadrupole moment, Q, from the experimental nuclear quadrupole coupling constant, eqQ/h. For Li, a serious discrepancy is found betwen our molecular values for LiH and LiF (which agree with themselves) and the nuclear, Coulomb scattering value [53] (see Table 2). Nuclear many-body calculations seem to be too inaccurate to resolve the issue [54].

In the case of the nitrogen quadrupole moment, later calculations confirm our result.

A particularly interesting molecular dipole moment [13] is that of ArH^+; this is the first experimentally known dipole moment (with respect to the center of mass) for a molecular ion. The 2D value confirms the earlier, LCAO results, suggesting that its deviation from experiment is due to experimental inaccuracies.

Due to the distribution of grid points in 2D calculations, large bond lengths are technically difficult while small ones should be easy. This suggests the method might be helpful for studying short-range interatomic potentials. As a preliminary example we show in Table 3 some results for the He---He interatomic potential [55]. It is seen that the basis set superposition error (BSSE), compared to our 2D results, decreases in the order: Wells and Wilson [60], Senff and Burton [56] and Liu and McLean [57]. The correlation contribution between the Monte Carlo results [58] and the HF ones becomes smaller, in relative terms, when R approaches zero.

Table 1: Molecules treated by our 2D method or by the Kassel FEM method, providing an even higher accuracy. The methods are labelled: HF= Hartree-Fock, HFS= Hartree-Fock- Slater, DF= Dirac-Fock, DS= Dirac-Slater. Tests on atoms are included.

Method	Heaviest Atom	Molecule	Reference(s)
HF	H	H^-	2
		H_2^+	1, 6
		H_2	2, 11, 23, 36
	He	He	2
		HeH^{2+}	1
		HeH^+	2, 11
		He_2	10, this work
	Li	LiH	3, 6, 23
		Li_2	10
		$LiHe^+$	11
	Be	BeH^+	11
		BeH	14
		Be_2	10
	B	BH	3
	C	CH^+	11
		CH	14
		C_2	11
	N	NH	14
		N_2	9, 10
		NO^+	9, 10
		CN^-	10

Table 1 (cont.)

Method	Heaviest Atom	Molecule	Reference(s)
HF	O	OH	14
		OH^-	10
		CO	10
		BeO	11
	F	HF	10
		LiF	11
		BF	10
		F_2	11
	Ne	Ne	10
		NeH^+	11
		HeNe	11
	Na	NaH	11
		LiNa	11
		NaF	13
	Mg	MgH	14
		MgH^+	11
		MgO	13
	Al	AlF	12
	Si	SiH	14
		SiO	12
	P	PN	12
		P_2	12
	S	BeS	13
		CS	12
		SH^-	12
	Cl	HCl	15
		BCl	12, 25

Table 1 (cont.)

Method	Heaviest Atom	Molecule	Reference(s)
HF	Ar	Ar	12
		ArH^+	13
MCSCF	H	H_2	5
	Li	LiH	5
	C	C^{2+}	5
HFS	Li	LiH	8, 23
	B	B_2	4
		BH	8, 23, 37
	C	C_2	4
	N	N_2	4, 37
	O	CO	4, 16, 37
		O_2	4
	F	F_2	4
		HF	8, 23
		BF	4
	Na	NaH	8, 23
	Al	AlH	8, 23
	Cl	HCl	8, 23
	Cr	Cr_2	7
DF	H	H_2^+	18, 20, 21
		H_2	19, 21
	He	HeH^+	19

Table 1 (cont.)

Method	Heaviest Atom	Molecule	Reference(s)
DS	Li	LiH	21
		Li_2	21
	B	BH	21
	C	CH^+	21

Table 2: Values of Q(^{7}Li).

Source and method	Q/10^{-28} m^2	Reference
LiH, LCAO	-0.0366(3)[a]	52
LiH, 2D + corr.	-0.0406	6
LiF, 2D + corr.	-0.04055(80)	24
Coulomb scattering	-0.O370(8)	53
Nuclear Theory	-0.0342[b]	54
"	-0.0419[c]	"
"	-0.0371[d]	"

[a] As shown in ref. [6], this result suffers from serious basis-set truncation errors

[b] "Best" wavefunction, WF4

[c] "Second best", WF3

[d] "Poorest", WF1

Table 3: The He---He intermolecular potential (in 10^{-6} hartree). The 2D value at R = 5.6 a.u. is a $n_{\nu} \times n_{\mu} = 81 \times 177$ one from ref. 10; the other 2D results are new small-grid ($n_{\nu} \times n_{\mu} = 41 \times 89$) ones [55].

R /a.u.	Hartree-Fock				Correlated	Reference
	2D	Literature	BSE	Reference		
5.6	29.2	29.1	-0.1	60	-28.722	57
		29.29	0.09	59		
		29.165	0.035	57		
4.0	1357.	1357.9	0.04	60	988.	57
		1357.94	0.0	56		
		1355.37	-2.6	57		
3.0	13516.8	13518.4		60	12033.7	58
		13517.92		56		
		13518.9		57		
2.0	120749.	120753.7		60	113370.	58
		120759.93		56		
1.5	342724.	342727.6		60	332510.	58

References

[1] L. Laaksonen, P. Pyykkö and D. Sundholm, Two-dimensional fully numerical solutions of molecular Schrödinger equations. I. One-electron molecules, *Int. J. Quantum Chem.* **23**, 309-317(1983)

[2] L. Laaksonen, P. Pyykkö and D. Sundholm, Two-dimensional fully numerical solutions of molecular Schrödinger equations. II. Solution of the Poisson equation and results for singlet states of H_2 and HeH^+, *Int. J. Quantum Chem.* **23**, 319-323(1983)

[3] L. Laaksonen, P. Pyykkö and D. Sundholm, Two-dimensional fully numerical solutions of molecular Hartree-Fock equations: LiH and BH, *Chem. Phys. Lett.* **96**, 1-3(1983)

[4] L. Laaksonen, D. Sundholm and P. Pyykkö, Two-dimensional, fully numerical molecular calculations. IV. Hartree-Fock-Slater results on second-row diatomic molecules, *Int. J. Quantum Chem.* **27**, 601-612(1985)

[5] L. Laaksonen, D. Sundholm and P. Pyykkö, Two-dimensional, fully numerical MC SCF calculations on H_2 and LiH: The dipole moment of LiH, *Chem. Phys.Lett.* **105**, 573-576(1984)

[6] D. Sundholm, P. Pyykkö, L. Laaksonen and A.J. Sadlej, Nuclear quadrupole moment of lithium from combined fully numerical and discrete basis-set calculations on LiH, *Chem. Phys. Lett.* **112**, 1-9 (1984)

[7] D. Sundholm, P. Pyykkö and L. Laaksonen, Fully numerical HFS calculations on Cr_2: Basis-set truncation error on the bond length and interaction on the semicore orbitals, *Finn. Chem. Lett.,* 51-55 (1985)

[8] D. Sundholm, P. Pyykkö and L. Laaksonen, Two-dimensional, fully numerical molecular calculations. VIII. Electric field gradients of diatomic hydrides LiH- ClH at the HFS level, *Mol. Phys.* **55**, 627-635 (1985)

[9] D. Sundholm, P. Pyykkö, L. Laaksonen and A.J. Sadlej, Nuclear quadrupole moment of nitrogen from combined fully numerical and discrete basis-set calculation on NO^+ and N_2, *Chem. Phys.* **101**, 219-225 (1986)

[10]D. Sundholm, P. Pyykkö and L. Laaksonen, Two-dimensional, fully numerical molecular calculations. X. Hartree-Fock results for He_2, Li_2, Be_2, HF, OH^-, N_2, CO, BF, NO^+ and CN^-, *Mol. Phys.* **56**, 1411-1418 (1985)

[11]P. Pyykkö, D. Sundholm and L. Laaksonen, Two-dimensional, fully numerical molecular calculations. XI. Hartree-Fock results for BeH^+, $LiHe^+$, CH^+, NeH^+, C_2, BeO, LiF, NaH, MgH^+, HeNe, LiNa, and F_2, *Mol. Phys.* **60**, 597-604 (1987)

[12]P. Pyykkö, G.H.F. Diercksen, F. Müller-Plathe and L. Laaksonen, Fully numerical Hartree-Fock calculations on the third-row diatomics AlF, SiO, PN, CS, BCl, SH^- and P_2, *Chem. Phys. Lett* . **134**, 575-578 (1987)

[13]P. Pyykkö, G.H.F. Diercksen, F. Müller-Plathe and L. Laaksonen, Fully-numerical Hartree-Fock molecular calculations on NaF, MgO, BeS and ArH^+. The dipole moment of ArH^+, *Chem. Phys. Lett.* **141**, 535-539 (1987)

[14]L. Laaksonen, F. Müller-Plathe and G.H.F. Diercksen, On the basis-set truncation error: fully-numerical RHF calculations on open-shell hydrides, to be published. (References 1-14 form the Parts 1-14 of our "2D series").

[15]L. Laaksonen, I.P. Grant and S. Wilson, The Dirac equation in the algebraic approximation. VI. Molecular self-consistent field studies using basis-sets of Gaussian-type functions, *J. Phys. B: At. Mol. Phys.* (1988)
[16]E.J. Baerends, P. Vernooijs, A. Rozendaal, P.M. Boerrigter, M. Krijn, D. Feil and D.Sundholm, Basis set effects on the electron density and spectroscopic properties of CO, J. *Mol. Struc.(Theochem)* **133**, 147-159 (1985)
[17]L. Laaksonen, Fully numerical, One and Two-Dimensional Solutions of Molecular Schrödinger and Dirac Equations: The Lone-Pair Problem, Techn. Dr. Thesis, Åbo Akademi, Finland (1983)
[18]L. Laaksonen and I.P. Grant, Two-dimensional fully numerical solutions of molecular Dirac equations. One-electron molecules, *Chem. Phys. Lett.* **109**, 485-487 (1984)
[19]L. Laaksonen and I.P. Grant, Two-dimensional fully numerical solutions of molecular Dirac equations. Results for ground singlet states of H_2 and HeH^+, *Chem. Phys. Lett.* **112**, 157-159 (1984)
[20]D. Sundholm, P. Pyykkö and L. Laaksonen, Two-dimensional, fully-numerical solutions of second order Dirac equations for diatomic molecules. Part 3, *Phys. Scripta* **36**, 400-402 (1987)
[21]D. Sundholm, Two-dimensional fully numerical solutions of molecular Dirac equations. IV. Dirac-Slater calculations on LiH, Li_2, BH and CH^+ (to be published)
[22]a) D. Sundholm, Applications of Fully numerical, Two-dimensional Self-Consistent Methods on Diatomic Molecules, Ph. D. Thesis, University of Helsinki, Finland (1985)
b) D. Sundholm, A block preconditioned conjugate gradient method for solving high-order finite element matrix equations, to be published.
[23]L. Laaksonen, P. Pyykkö and D. Sundholm, Fully numerical Hartree-Fock methods for molecules, *Comp. Phys. Rep.* **4**, 313-344 (1986)
[24]G.H.F. Diercksen, A.J. Sadlej, D. Sundholm and P. Pyykkö, Towards an accurate determination of the nuclear quadrupole moment of Li from molecular data: LiF, *Chem. Phys. Lett*. **143**, 163-168 (1988)
[25]F. Müller-Plathe and G.H.F.Diercksen, Properties of the BCl molecule studied by fourth-order many-body pertrubation theory, *Int. J. Quantum Chem.* **S21**, 595-602 (1987)
[26]E.A. McCullough, Jr., Seminumerical SCF calculations on small diatomic molecules, *Chem. Phys. Lett* . **24**, 55-58 (1974)
[27]E.A. McCullough, Jr, Numerical Hartree-Fock methods for diatomic molecules: A partial-wave expansion approach, *Comp. Phys. Rep.* **4**, 265-312 (1986)
[28]K. W. Richman and E.A. McCullough, Jr., Numerical Hartree-Fock calculations on diatomic chromium, *J. Chem. Phys.* **87**, 5050-5051 (1987)
[29]L. Adamowicz and R.J. Bartlett, Numerical coupled Hartree-Fock study of the total (electronic and nuclear) parallel polarizability and hyperpolarizability for the FH, H_2^+, HD^+ and D_2^+ molecules, *J. Chem. Phys.* **84**, 4988-4991 (1986)
[30]L. Adamowicz and R.J. Bartlett, Coupled cluster calculations with numerical orbitals for excited states of polar anions, *J. Chem. Phys.* **83**, 6268-6274 (1985)
[31]K.W. Richman, Z.-G. Shi and E.A. McCullough, Jr, Calculations of the hyperfine splitting constants for diatomic molecules using numerical wavefunctions, *Chem. Phys. Lett.* **141**, 186-192 (1987)
[32]A.D. Becke, Numerical Hartree-Fock Slater calculations on diatomic molecules, *J. Chem. Phys.* **76**, 6037-6045 (1982)

[33]A.D. Becke, Completely numerical calculations on diatomic molecules in the local-density approximation, *Phys. Rev.* A**33**, 2786-2788 (1986)
[34]W.K. Ford and F.S. Levin, Channel-coupling theory of molecular structure. Finite-element method solution for H_2^+, *Phys. Rev.* A**29**, 43-51 (1984)
[35]W. Schulze and D. Kolb, H_2^+ correlation diagram from finite element calculations, *Chem. Phys. Lett*. **122**, 271-275 (1985)
[36]D. Heimann, D. Kolb and B. Fricke, H_2 solved by finite element method, *Chem. Phys. Lett.* **137**, 180-182 (1987)
[37]D. Heinemann, B. Fricke and D. Kolb, Accurate Hartree-Fock-Slater calculations of small diatomic molecules with the finite element method, *Chem. Phys. Lett* (1988)
[38]D. Heinemann, Die Methode der finiten Elemente in der Quanten-mechanik-Lösung der Hartree-Fock-Slater Gleichungen, Thesis, Kassel (1987)
[39]M. Mishra, J. Linderberg and Y. Öhrn, Characterization of adiabatic states in triatomic collisions, *Chem. Phys. Lett.* **111**, 439-444 (1984)
[40]N. Sato and S. Iwata, Application of finite element method to the two-dimensional Schrödinger equation (to be published)
[41]R. Jaquet, Investigations with the finite element method on the Cyber 205. I. The collinear X + YZ (X, Y, Z = H, D, Mu) reaction, *Theor. Chim. Acta* **71**, 425-448 (1987)
[42]M.D. Feit, J.A. Fleck Jr. and A. Steiger, Solution of the Schrödinger equation by a spectral method, *J. Comp. Phys.* **47**, 412-433 (1982)
[43]D. Kosloff and R. Kosloff, A Fourier method solution for the time-dependent Schrödinger equation as a tool in molecular dynamics, *J. Comp. Phys.* **52**, 35-53 (1983)
[44]S.A. Alexander, R.L. Coldwell and H.J. Monkhorst, Polyatomic SCF calculations with numerical orbitals. II. Methods to reduce integration and truncation error, *J. Comp. Phys.* (in press)
[45]M. Defranceschi, M. Suard and G. Berthier, Numerical solution of Hartree-Fock equations for a polyatomic molecule: Linear H_3 in momentum space, *Int. J. Quantum Chem.* **25**, 863-867 (1984)
[46]M. Defranceschi, Theoretical investigations of the momentum densities for molecular hydrogen, *Chem. Phys.* **115**, 349-358 (1987)
[47]J. Delhalle and M. Defranceschi, Toward fully numerical evaluation of momentum space Hartree-Fock wave functions. Numerical experiments on the He atom, *Int. J. Quantum Chem.* S**21**, 425-433 (1987)
[48]W. Rodriguez and Y. Ishikawa, to be published
[49]R.A. Friesner, Solution of the Hartree-Fock equations by a pseudo-spectral method: Application to diatomic molecules, *J. Chem. Phys.* **85**, 1462-1468 (1986)
[50]A.D. Becke, A multicenter numerical integration scheme for polyatomic molecules, *J. Chem. Phys.* **88** (1988)
[51]F.F. Grinstein, H. Rabitz and A. Askar, The multigrid method for accelerated solution of the discretized Schrödinger equation, *J. Comp. Phys.* **51**, 423-443 (1983)
[52]S. Green, Quadrupole moment of Li, *Phys. Rev.* A**4**, 251-253 (1971)
[53]A. Weller, P. Egelhof, R. Caplar, O. Karban, D. Krämer, K.-H. Möbius, Z. Moroz, K. Rusek, E. Steffens, G. Tungate, K. Blatt, I. Koenig and D. Frick, Electromagnetic excitation of aligned ^{7}Li nuclei, *Phys. Rev. Lett* . **55**, 480-483 (1985)

[54]T. Mertelmeier and H.M. Hofmann, Consistent cluster model description of the electromagnetic properties of lithium and beryllium nuclei, *Nucl. Phys.* A**459**, 387-416 (1986)
[55]P. Aaltonen and P. Pyykkö, unpublished results
[56]U.E. Senff and P.G. Burton, A CEPA2 investigation of the He-He and He-Li^+ potential functions, *Mol. Phys.* **58**, 637-645 (1986)
[57]B. Liu and A.D. McLean, Accurate calculation of the attractive interaction of two ground state helium atoms, *J. Chem. Phys.* 59, 4557-4558 (1973)
[58]D.M. Ceperley and H. Partridge, The He_2 potential at small distances, *J. Chem. Phys.* **84**, 820-821 (1986)
[59]H. Tatewaki, K. Tanaka, Y. Ohno and T. Nakamura, The interaction potentials for He-He and He-Li^+, *Mol. Phys.* **53**, 233-240 (1984)
[60]B.H. Wells and S. Wilson, Universal basis sets of elliptical functions. Applications to simple diatomic molecules, *J. Phys. B: At. Mol. Phys.* **19**, 17-32 (1986)

VERY ACCURATE CALCULATIONS FOR DIATOMIC, NEUTRAL AND ANIONIC SYSTEMS WITH NUMERICAL ORBITALS.

Ludwik Adamowicz
Department of Chemistry
University of Arizona
Tucson, Arizona 85721, U.S.A.

ABSTRACT. The computational strategy and numerical results are presented for our recently developed method, which combines numerical orbital technique for diatomic molecules with the many body perturbation theory (MBPT) and the coupled cluster (CC) method. This composite produces the most accurate results ever obtained for diatomic systems. We also show applications of the present technique to study ground and excited states of anions of polar diatomic molecules.

1. INTRODUCTION

Following the development of many-body perturbation theory (MBPT) by Brueckner and Goldstone [1,2], Kelly made the initial applications to correlation energies and other properties of atoms [3]. By exploiting the spherical atomic symmetry, which permits separation of the radial equation from the angular part, these calculations were performed purely numerically.

These applications included the second and third orders of perturbation theory with some consideration of higher-order diagrams via denominator shifts. Kelly and co-workers [4] attempted to use one-center expansion techniques for some molecules (H_2, HF, H_2O) to retain the same numerical approach as for atoms, but this was generally unsatisfactory due to the extreme difficulty in overcoming the singularities of the multicenter Coulomb potential.

To rectify the multicenter problem, MBPT/CC could be performed using the conventional Slater- or Gaussian-type basis sets employed in molecular theory. However, due to basis-set limitations, the result is subject to basis-set incompleteness error, which was virtually eliminated in Kelly's initial atomic calculations. This follows since the asymptotic dependence of correlated calculations such as CCSD (CC with single and double excitations) scales as n^6 for n-basis functions, and CCSDT-1 as n^7. This places a severe restriction on the

M. Defranceschi and J. Delhalle (eds.), Numerical Determination of the Electronic Structure of Atoms, Diatomic and Polyatomic Molecules, 177–200.

number of basis functions that may be employed in molecular calculations, and, consequently, it is seldom possible to extrapolate to the basis-set limit for molecules.

2. METHODOLOGY

2.1 Numerical MCSCF for diatomic systems

A many-electron wave function can be written exactly as an infinite configuration (CI) expansion

$$\psi = \sum_I C_I \Phi_I \quad (1)$$

where the Φ_I are symmetry adapted linear combinations of Slater determinants constructed from a complete set of orthonormal orbitals. For a truncated CI expansion, the best wave function (in variational sense) is obtained when both the CI coefficients and the functional forms of the orbitals are varied to minimized the energy. This defines the multiconfiguration Hartree-Fock approximation.

Formal variation of the orbitals yields a complicated set of coupled operator equations (Fock equations). Numerical approach is common for atoms [4], and recent years have witnessed development of several alternative numerical approaches for diatomic molecules [5,6,7]. All these approaches make use of a natural coordinate system for diatomic molecules, prolate spheroidal coordinates:

$$(2) \quad \begin{aligned} \zeta &= (r_a + r_b)/R & 1 \le \zeta < \infty \\ \eta &= (r_a - r_b)/R & -1 \le \eta \le 1 \\ \phi & & 0 \le \phi \le 2\pi \end{aligned}$$

where r_a and r_b are the scalar distance of the point from the corresponding nucleus and ϕ is the angle of rotation about the internuclear axis and R is the internuclear separation. The principle advantage of this coordinate system is that the Coulombic singularities at both nuclei can be made to coincide with coordinate origin. If this is not the case, one can always encounter some difficulty representing the electron-nucleus attraction operator or the orbital behavior or both at a non-origin nucleus, which must be done to high accuracy since the electron-nucleus attraction dominates the potential energy.

Assuming the separation of the spacial and spin components of the spinorbital, and furthermore the separation of the ζ-dependent part from the function of η and ϕ coordinates, the numerical orbital for diatomic system can be written as:

$$(3) \qquad \psi_p^m(\zeta,\eta,\phi) = \chi_p^{m_p}(\zeta,\eta) \frac{e^{im\phi}}{(2\pi)^{1/2}}$$

with a two-dimensional component to be determined from the variational principle. m_p is a positive integer or zero and m can take on the value $\pm m_p$. Shells with $m_p = 0$ are called 'sigma'; m_p, 'pi'; etc. Sigma shells contain only one distinct orbital (two distinct spin orbitals); all other shells contain two (four).

While in the approach developed by Pyykkö et al. [6] and Becke [7], $\chi_p^{m_p}(\zeta,\eta)$ is represented as a single function, we have taken a different approach [5], in which $\chi_p^{m_p}$ is represented as a finite sum of products of one-dimensional functions of η and ϕ separately. The numerical orbital has the following form

$$(4) \quad \psi_p^m(\zeta,\eta,\phi) = [\sum_{l=m_p}^{l_p} X_{l,p}^{m_p}(\zeta)P_l^{m_p}(\eta)] \cdot \frac{e^{im\phi}}{(2\pi)^{1/2}} = \sum_{l=m_p}^{l_p} X_{l,p}^{m_p}(\zeta)\, Y_l^m(\eta\phi)$$

where the $P_l^{m_p}$ are associated Legendre functions and Y_l^m are usual spherical harmonics. We refer to eq.(4) as a partial wave expansion since it resembles the partial-wave expansions of scattering theory. Being of a product form, which involves η-dependent predetermined functions, $P_l^{m_p}(\eta)$ and a one-dimension numerical part, our ansatz is an intermediate one between a fully numerical two-dimensional approach taken by Pyykkö et al. and Becke and a basis set approach, in which a set of predetermined functions in both variables, ζ and η, is used. The advantage of a partial wave representation (also two-dimensional representation) is that, provided that orbitals are mathematically well behaved, the error in determining them can be established and, more importantly, controlled. Thus, for all practical purposes, by the adjustment of the summation limit, l_p, and the size of the numerical grid for ζ we can assure a desired accuracy.

The numerical procedure for determining the numerical components in the partial-wave expansions have been developed by McCullough [5a, 5b] for the single configuration Hartree-Fock case. The extension for the multiconfiguration wave function followed [5d,5e].

The MCSCF total energy can be written as a linear combination of one- and two-electron integrals

$$(5) \quad E = \sum_{ij} a_{ij} \langle i|h|j\rangle + \sum_{ijkl} b_{ijkl} \langle ij|1/r_{12}|kl\rangle$$

where a's and b's depend on the orbital occupancies, spin and spatial symmetry restrictions, and the CI coefficients. Given some initial estimates of these, the Fock equation for each orbital can be constructed ($\sum_j F_{ij} \; |j\rangle = \sum_j \epsilon_{ij} \; |j\rangle$). Variation of orbital **i** generates the contributions to its Fock equation listed in Table I. Each contribution has the general form of a one-electron operator acting on an orbital, where the operator contributed by a two-electron integral itself depends on the orbitals. We evaluate all the operators as well as their effects on any orbital except **i** using previous estimates of the orbitals. The Fock equation for orbital **i** can now be solved numerically subject to orthonormality constrains yielding an improved estimate of orbital **i**. The orthonormality conditions are introduced to the Fock equations through conventional Lagrange multipliers.

Table I. Fock Operator Contributions

$$\langle i|h|i\rangle \rightarrow 2h|i\rangle$$

$$\langle i|h|j\rangle \rightarrow h|j\rangle$$

$$\langle i|\langle i|1/r_{12}|i\rangle|i\rangle \rightarrow 4\langle i|1/r_{12}|i\rangle|i\rangle$$

$$\langle i|\langle i|1/r_{12}|i\rangle|l\rangle \rightarrow 2\langle i|1/r_{12}|l\rangle|i\rangle + \langle i|1/r_{12}|i\rangle|l\rangle$$

$$\langle i|\langle i|1/r_{12}|k\rangle|l\rangle \rightarrow \langle i|1/r_{12}|k\rangle|l\rangle + \langle i|1/r_{12}|l\rangle|k\rangle$$

$$\langle i|\langle j|1/r_{12}|k\rangle|i\rangle \rightarrow 2\langle j|1/r_{12}|k\rangle|i\rangle$$

$$\langle i|\langle j|1/r_{12}|k\rangle|l\rangle \rightarrow \langle j|1/r_{12}|k\rangle|l\rangle$$

Repetition of this process for all orbitals defines an iterative scheme for solution of the Fock equations for fixed CI coefficients. With new estimates of the orbitals the Hamiltonian can be constructed and diagonalized, yielding new values of the CI coefficients and so on. There is an additional step involving orbital rotations. This is necessary to alleviate the undesired constraint which results from selecting a particular order of solving Fock equations. This is a well known procedure in the first-order MCSCF method which utilizes Fock equations to determine orbitals. The rotation of **i** and **j** orbitals

$$(6)\quad |i\rangle \rightarrow N\,(|i\rangle + \theta|j\rangle)$$

$$|j\rangle \rightarrow N\,(|j\rangle - \theta|i\rangle)$$

$$N = (1 + \theta^2)^{-\frac{1}{2}}$$

is performed by expanding the energy in a power series of θ, dropping the third and higher-order terms and then finding the minimum of the resulting quadratic function with respect to θ. The complete iterative scheme then consists of the following steps: solution of the Fock equations, diagonalization of the Hamiltonian, and rotations.

The MCSCF Fock equations are coupled, integro-differential equations. The procedure for solving these equations has been developed by McCullough [5a,5b], and was recently summarized in a review paper [5f]. Multiplying the Fock equation by spherical harmonics and utilizing well known relations for these functions, we can reduce the three-dimensional equation to a set of coupled one-dimensional equations for radial numerical functions, $X(\varsigma)$. A conventional approach to numerical integration of differential equations is based upon approximating a continuous function using values at a set of equally-spaced discrete points. Equally spaced points in ς would be quite inefficient for the MCSCF problem because the shell function changes very rapidly at small ς, but slowly and smoothly at long range. Consequently, it is convenient to introduce a new radial variable, ρ, which is defined by

$$(7)\quad \rho = \ln\,[R(Z_a + Z_b)\,(\varsigma-1)/2]$$

The scaling factor $R(Z_a + Z_b)$ is introduced to retain the same accuracy of the numerical representation regardless of the charges of the nuclei.

2.2 Basis set MBPT/CC methods

The basis-set MBPT (also referred to as Moller-Plesset perturbation theory) approach was accomplished and presented in a series of papers by Bartlett and co-workers [8a] which has since led to extensive numbers of MBPT applications. In general these applications explicitely include contributions from diagrams of higher-order than Kelly considered for the atomic case - typically all diagrams through the fourth-order perturbation theory [8b]. Furthermore, infinite-order coupled-cluster [9] generalizations of MBPT have been implemented; these are CC methods which include T_2 operators (CCD), all diagrams resulting from T_1 and T_2 cluster operators (CCSD) [10a] and their products, partial contributions of T_3 (CCSDT-1) [10b], and recently all T_1, T_2 and T_3 diagrams (CCSDT) [10c].

The CC ansatz is built upon a wavefunction of the form:

$$(8) \quad \Psi_{exact} = \Psi_{CC} = e^{\hat{T}}\Phi(0)$$

where $\Phi(0)$ is some single determinant reference function, often but not necessarily the SCF solution. The excitation operator, $\hat{T}$, is:

$$(9) \quad \hat{T} = \hat{T}_1 + \hat{T}_2 + \hat{T}_3 \; . \; . \; . \; .$$

$$(10a) \quad \hat{T}_1 = \sum_{i} t_i^a \; a_a^\dagger a_i$$

$$(10b) \quad \hat{T}_2 = \sum_{\substack{i>j \\ a>b}} t_{ij}^{ab} \; a_a^\dagger a_i \; a_b^\dagger a_j$$

The amplitudes $\{t_i^a, t_{ij}^{ab} \ldots\}$ are to be determined by the CC equations, and a,b,c,... represent excited orbitals (and associated second-quantized operators) and i,j,k,... occupied orbitals (operators). The CC coefficients are obtained by solving the system of nonlinear equations, which result from the projection of the Schrödinger equation for the CC wave function onto the reference function, singly excited determinants, doubly excited determinants, etc. In the case of the CCSD method, the CC coefficients, t_i^a and t_{ij}^{ab} result from the following set of projected Schrödinger equations:

$$(11a) \quad \langle\Phi(0), \; \langle{}_i^a|, \; \langle{}_{ij}^{ab}|, \; (H-E_{CCSD})|\exp(\hat{T}_1+\hat{T}_2)\Phi(0)\rangle = 0$$

with the exponential wavefunction expanded as:

$$(11b) \quad \exp(\hat{T}_1+\hat{T}_2) = 1 + \hat{T}_1 + \hat{T}_2 + \tfrac{1}{2}\hat{T}_1^2 + \hat{T}_1\hat{T}_2 + \tfrac{1}{2}\hat{T}_2^2 + \tfrac{1}{2}\hat{T}_1^2\hat{T}_2 + \tfrac{1}{3!}\hat{T}_1^3 + \tfrac{1}{4!}\hat{T}_1^4 + \ldots$$

In essence, the single reference CC method offers a procedure for finding solutions of the Schrödinger equation not only for the ground state but for any bound state, provided that the reference function (in currently implemented version - single determinant) constitutes a good approximation to the state under consideration. Our model calculations for the LiH molecule [11] have shown that this is really the case even for highly excited states. We have employed the direct CC method to study higher excited states of polar diatomic anions and I shall review these calculations in the following sections. If in the state under consideration a leading determinant cannot be distinguished, the multi-determinantal single-reference CC method needs to be applied.

A glossary of some shorthand notations which are used in Tables to indicate different versions of the MBPT and CC methods are as follows:

MBPT D(2)	- Second-order correlation energy.
MBPT D(3)	- Sum of correlation corrections up to third-order.
MBPT SDQ(4)	- Sum of correlation corrections up to fourth-order. Fourth-order contribution is calculated only with single, double and quadruple excitations.
MBPT SDTQ(4)	- Complete sum of correlation corrections up to fourth-order.
E(2),E(3),E(4)	- individual, second-, third- and fourth-order MBPT contributions.
CCD	- CC calculation with double excitations only ($\hat{T} = \hat{T}_2$).
CCSD	- CC calculation with single and double excitations ($\hat{T} = \hat{T}_1 + \hat{T}_2$).
CCSD-T(CCSD)	- CC calculation with triple excitations ($\hat{T}_3$) included only in the last CCSD iteration (see ref. 10b).
CCSDT-1	- CC calculation with triple excitation ($\hat{T}_3$) included in a linear fashion (see ref. 10b).

2.3 Numerical orbitals with MBPT/CC methods

A procedure for solving the CC equations numerically has been accomplished for atoms [12], and an analogous approach would be highly desirable for diatomic molecules. However, computational difficulties would make a similar procedure almost prohibitive for molecular systems. Therefore, we have chosen an intermediate approach, which combines the MCSCF generation of numerical orbitals and basis-set CC method.

To define correlation orbitals, we employ the idea of the Bethe-Goldstone (BG) hierarchy of equations introduced to the electronic structure calculations by Nesbet [13] in the late sixties. The basis principle of the BG method is closely related to the CC partition of the electron correlation into cluster contributions. The first step in the procedure is the evaluation of the reference function, which for a closed shell system is usually a restricted Hartree-Fock (HF) wavefunction (for open-shell systems it is a restricted or unrestricted HF wavefunction). In the next step the electronic correlation is separated into two-, three-, four-, etc. electron clusters and electrons for each cluster are correlated in an independent variational MCSCF calculation, while the rest remain in the HF state. This can be followed by formation of super-clusters (clusters of clusters) with correlation developed between them. The cluster structure is not determined by a mathematical formalism (e.g. a formal separation of the total Hamiltonian in the perturbation theory), but is decided based on the physical nature of the electronic correlation.

The BG cluster approach allows for a variety of different strategies. One possibility is to correlate electrons in one cluster while the other electrons are not in their HF state but are partially correlated by short configuration expansions. This alternative eliminates the isolation of the clusters already at the first level of the BG cluster partitioning. In essence, the BG cluster hierarchy is complete and in the limit should provide a basis set in which the exact solution of the Schrödinger equation can be expressed. In this respect, the union of the BG cluster fragmentation and the CC method seems to be a natural consequence of the physical concept which lies behind both approaches.

The procedure combining the numerical orbital technique for diatomic systems with the CC method can be summarized in the following four steps:

1. Generation of the numerical reference functions (e.g. numerical HF calculation).

2. Correlation partitioned into clusters and an independent MCSCF calculation for each cluster resulting in a set of correlation numerical orbitals, which are orthogonal to the reference function orbitals and to orbitals from the same cluster but not orthogonal to correlation orbitals of other clusters.

3. A union of correlation numerical orbitals is formed. To remove possible linear dependencies, the overlap is constructed and diagonalized. Vectors, whose eigenvalues are below the assumed threshold are discarded. For the rest, the numerical construction of orthonormal orbitals is performed as follows:

$$(12) \qquad \psi_\alpha = \sum_p c_{p\alpha}\, \psi_p = \sum_{l=m_p}^{l} \Big[\sum_p c_{p\alpha}\, X_{l,p}^{m_p}(\zeta) \Big]\, Y_l^m(\eta,\phi)$$

$$= \sum_{l=m_\alpha}^{l_\alpha} X_{l,\alpha}^{m_\alpha}(\zeta)\, Y_l^m(\eta,\phi)$$

where $c_{p\alpha}$ are linear coefficients resulted from the orthogonalization and $X_{l,\alpha}^{m_\alpha}$ are new numerical functions obtained by the point-by-point summations.

4. All one- and two-electron integrals are calculated for the ψ_α orbital set, which includes the reference function orbitals and correlation orbitals. Next, MBPT/CC calculations are performed by using the PROPAGATOR program system [14].

The purpose of the numerical orthogonalization of orbitals is not only to remove linear dependencies, but also to reduce an enhancement of the numerical error which originates in numerical integrations. With non-orthogonal numerical orbitals the MO expansion coefficients, which result from the diagonalization of the Fock matrix (it is convenient to work with canonical orbitals in the MBPT/CC program), are usually much larger. The larger the coefficients are, the less accurate molecular integrals become after the four-index transformation. Lately, we have tried to avoid the integral-transformation step by constructing canonical numerical orbitals using eq. 12, and recalculating molecular integrals, which can be directly used in the MBPT/CC calculations. This step is particularly useful in conjunction with the procedure for the reduction of the correlation orbital space [28], which will be discussed later.

3. NUMERICAL RESULTS

3.1 Neutral systems

The present procedure has been applied to several diatomic systems. The first system studied was the H_2 molecule. Of course, since the BG electron-pair wave-function in this case is exactly equivalent to the full solution of the Schrödinger equation, the converged CCSD energy should be exactly the variational MCSCF result in the same basis set. The number of correlation shells in the H2 calculation was equal to 45 distributed as 18σ, 15π and 12δ. The MBPT/CC results are given in Table II [15].

For the LiH molecule calculations have been performed using eighty-two numerical correlation orbitals (distributed as: 30σ, 24π, 16δ, 12d) [16]. These orbitals resulted from independent MCSCF calculations for $(1\sigma)^2$, $(2\sigma)^2$, $(1\sigma 2\sigma)$ (singlet- and triplet-coupled) electron pairs (two-electron BG clusters). The present value for the CC correlation energy obtained with the highest level of the theory applied, CCSDT-1, reproduces 99% of the experimental value, which is also the case for the H_2 molecule. For comparison CC gaussian geminal (GG) results [17] and recent variational and Monte Carlo results [18, 19] are cited. The CC method is not rigorously variational for the four-electron system, although it may be shown to provide upper-bounds up to some order in perturbation theory and, except for pathological cases, in practice show upper-bound character in comparison to full CI [20] (i.e. the best possible basis set result). It is worthwhile to notice a considerable difference between our second-order MBPT result (-0.07088 a.u.) and the one obtained with GG expansions (-0.07217 a.u.). We attribute this to the BG cluster generation of the numerical orbitals being an infinite prescription while the GG approach optimized their geminals at second-order level.

Table II. MBPT/CC correlation energies for the H_2 molecule (r_{H-H} = 1.4 a.u), the LiH molecule (r_{Li-H} = 3.015), and the FH molecule (r_{F-H} = 1.7328 a.u.)

	H_2	LiH	FH
Numerical Hartree-Fock	-1.133630	-7.987352	-100.070807
MBPT			
E(2)	-0.03334	-0.07088	-0.36788
E(3)	-0.00507	-0.00835	+0.00205
E(4)	-0.00148	-0.00210a	-0.01189[a]
CC			
CCD	-0.04037	-0.08152	
CCSD	-0.04051	-0.08200	-0.36728
CCSD+T(CCSD)	-	-	-0.37622[a]
CCSDT-1	-	-0.08216[a]	-
Experimental correlation energy		-0.0832±0.0001[b]	-0.381[c]
Literature results	-0.04085[d]	-0.08169[e]	
		-0.08238±0.00026[f]	

[a] Triple excitation contribution calculated with reduced virtual space.

[b] Bunge, ref. [25].

[c] Bender and Davidson, ref. [26].

[d] Kolos and Rychlewski, ref. [27]. Highly accurate variational treatment.

[e] Handy et al., ref. [18]. Multireference CI.

[f] Harrison and Handy, ref. [19]. Monte Carlo calculation.

For the next system, the FH molecule, a slightly different BG cluster structure has been chosen [28]. For core electrons the pair approach has been retained, but all four electrons in the π shell have been correlated simultaneously. Subsequently, the numerical MCSCF calculations have been performed for the following nine clusters:

$(1\sigma^2)$, $2\sigma^2)$, $(3\sigma^2)$, $(1\pi^4)$, $(1\sigma 2\sigma)$, $(1\sigma 1\pi)$, $(2\sigma 1\pi)$, $(3\sigma 1\pi)$

The total number of numerical correlation shells generated in the procedure is 119, distributed as 45σ, 40π, 26δ and 8ϕ. This number has been reduced to 75 (28σ, 23π, 7δ, 7ϕ) after removing linear dependencies. The results of MBPT/CC calculations done with the resulting set of orbitals are presented in Table II. Again, the procedure yields almost 99% of the experimentally estimated value for the total electronic correlation energy. It is worthwhile to mention that all calculations have been performed on smaller computers, namely a Digital Equipment Corporation VAX11/750 (for numerical orbital generation) and a Floating Point Systems FPS-164 (for MBPT/CC calculations), which is an attractive feature. In the next section we shall discuss an application of the present procedure to study ground and excited electron affinities of polar diatomic anions.

Once the extended basis set of numerical orbitals is generated via the BG portioning and the MCSCF cluster calculations a question arises as to how much the size of the set can be reduced through an orbital contraction procedure without a significant energy loss. By answering this question we would like to address a more general problem of the effectiveness of the orbital description of the electronic correlation. Moreover, the number of numerical orbitals generated with the present procedure can significantly exceed the capabilities of the existing implementation of the MBPT/CC computer program, especially for larger systems with a large number of BG clusters. Therefore, an orbital contraction procedure is desirable.

Recently, we have proposed the OVOS (optimized virtual orbital space) method [28] and it has been employed for the present purpose. The method is based on the second-order Hylleraas functional, which constitutes an upper bound to the second-order MBPT energy E(2) for a trial first-order wave function $\Phi^{(1)}$.

$$(13) \qquad E(2) \leq \langle\Phi^{(1)}|H_0 - E(0)|\Phi^{(1)}\rangle + 2\langle\Phi^{(1)}|V - E(1)|\Phi^{(0)}\rangle \equiv J(2)$$

where the unperturbed Hamiltonian H_0 is defined as a sum of Fock operators, and the correlation perturbation operator is equal to $V = H - H_0$. The virtual space of correlation numerical orbitals is divided into an active and nonactive part and the active virtual orbitals are rotated with nonactive orbitals to produce the minimal value of J(2) determined for the active space. A unitary orbital transformation is used. It is represented in an exponential form with an antisymmetric matrix **R**

$$(14) \qquad \mathbf{U}^{+}\mathbf{U} = 1$$

$$(15) \qquad \mathbf{U} = \exp(\mathbf{R})$$

$$(16) \qquad \mathbf{R} = -\mathbf{R}^{+}$$

A similar technique is commonly used in the MCSCF orbital optimization.

The orbital rotation matrix **R** is obtained in an iterative procedure which utilizes the Newton-Raphson quadratic formula

$$(17) \qquad \mathbf{R} = -\mathbf{G} \cdot \mathbf{H}^{-1},$$

where gradient, **G**, and hessian, **H**, matrices are defined as

$$(18) \qquad \mathbf{G} = \left. \frac{\partial J(2)}{\partial R} \right|_{R = 0}$$

$$(19) \qquad \mathbf{H} = \left. \frac{\partial^2 J(2)}{\partial R^2} \right|_{R = 0}$$

An important advantage of using the second-order Hylleraas functional is that only exchange integrals of the type <occ(1)occ(2)|vir(1)vir(2)> are involved in evaluating the gradient and hessian matrices. This makes the procedure feasible even for large numbers of basis functions. The reduction of the numerical orbital set is performed in the same fashion as it would be done for any other kind of basis set, e.g. Gaussian orbitals. There is, however, a difference in utilizing the contracts over numerical orbitals in high-level correlated calculations in comparison with conventional analytical orbitals. Namely, once the contraction coefficients are determined, one can combine numerical orbitals together in point-by-point summations (see. eq.12) and evaluate new numerical functions. Consequently, there is no need for even a partial but still costly four-index integral transformation. Instead, this is replaced by the reevaluation of a new integral set over contracted numerical orbitals. Those orbitals can be directly used in subsequent correlated calculations. The numerical contraction becomes especially beneficial for higher reduction levels of the virtual space.

The reduction scheme has been applied in the calculations for the FH molecule. The set of 75 numerical virtual orbitals (we should rather say shells) resulting from the BG generation procedure has been reduced via the OVOS procedure to several different sizes ranging from 56 to 19. The results are presented in Table III. Upon inspection of the second-order energy it is clear that the initial set contains an excessive number of functions. Even 50% reduction (from 75 to 37) reduces E(2) by only 4.[illegible]%, while a reduction by 25% (from 75 to 56) causes little change in E(2) (about 1%).

Table III. FH molecule. Second-order energy for different levels of OVOS reduction (a.u.).

Number of numerical orbitals	Symmetry distribution (σ, π, δ, ϕ)	Second-order energy
75	28/23/17/7	-0.36788
62	22/20/14/6	-0.36617
56	20/18/13/5	-0.36465
50	19/15/11/5	-0.36265
37	14/12/ 8/3	-0.35535
25	9/ 8/ 6/3	-0.33772
19	7/ 6/ 4/2	-0.32141
10	4/ 3/ 2/1	-0.26591

In Table V we present MBPT/CC results obtained with different number of OVOS orbitals. It is interesting that the third-order energy, however small, is very sensitive to the orbital reduction. In contrast, the fourth-order SDQ contribution remains almost unchanged. The contribution of the triple excitations is relatively stable at both the MBPT and CC level.

Since E(2) is a byproduct of the OVOS optimization technique, it is possible to use OVOS only for the higher-order correlation corrections augmenting those effects by the exact E(2). By doing so, most of the error in OVOS should be eliminated as long as E(2) constitutes the dominant part of the correlation correction. The last two lines demonstrate the accuracy of the hybrid method. Even for the reduction of 25% of the available space (19 orbitals), the CCSD+T(CCSD) result is essentially identical, and CCSD slightly better than the 56 orbital result. Obviously being better is an accident that occurs since E(2) is proportionately larger for the largest orbital space artificially enhancing the D(2)+Δ[CC] results slightly. However, from a pragmatic viewpoint, the hybrid methods should be exceptionally accurate for most examples.

Table IV. FH molecule. MBPT and CC results with different OVOS's.

		Number and angular distribution of OVOS orbitals (σ,π,δ,ϕ)			
		75 (28/23/27/ 7)	56 (20/18/13/ 5)	37 (14/12/ 8/ 3)	19 (7/ 6/ 4/ 2)
MBPT	D(2)	-0.36788	-0.36465	-0.35535	-0.32141
	D(3)	+0.00205	+0.00150	+0.00023	-0.00069
	SDQ(4)	-0.00204	-0.00206	-0.00216	-0.00220
	SDTQ(4)	-	-0.01190	-0.01132	-0.00868
CC	CCSD	-0.36728	-0.36461	-0.35669	-0.32391
	CCSD+T(CCSD)	-	-0.37354	-0.36507	-0.32993
	D(2)+Δ[CCSD]	-0.36728	-0.36784	-0.36922	-0.37038
	D(2)+ Δ[CCSD+T(CCSD)]	-	-0.37677	-0.37760	-0.37640

The OVOS procedure also allows us to study the convergence of the second-order energy as a function of the orbital symmetry quantum number m. In the course of this study several reductions have been performed with simultaneous suppression of the number of orbitals and elimination of consecutive angular momentum components. Results are presented in Table V. The energy series as a function of m values allows us to estimate the limit for the second-order energy using [1/1] Padé approximant. This value is 0.36973 a.u. If one takes the second-order contribution from orbitals with m>3 of 0.00356 a.u. and adds it to our best CCSD+T(CCSD) result of -0.37622 a.u. one gets -0.37978 a.u. for the correlation energy (99.7% of the experimental value). This procedure is justified by a close agreement between the E(2) and CCSD.

Table V. Angular contributions to the second-order correlation energy (a.u.).

Angular distribution ($\sigma/\pi/\delta/\phi$)	Second-order energy
22/ 0/ 0/ 0	-0.07623
22/20/ 0/ 0	-0.27965
22/20/14/ 0	-0.35183
22/20/14/ 6	-0.36617
Padé [1/1]	-0.36973

We plan to apply the OVOS technique to transition metal diatomics.

3.2 Ground and excited states of diatomic anions

It was theoretically proven [29] and experimentally verified [30] that polar molecules are able to trap an extra electron within the attractive field generated by their dipoles. Recent years witnessed extensive theoretical and experimental studies of this intriguing phenomena [31].

To the first approximation anions of polar molecules may be understood as the binding of an electron by a molecular dipole fields. It has been shown that an infinite number of bound-electron states exist in the Born-Oppenheimer approximation for any system with a long range dipole field exceeding 1.625 Debye. The results are drastically altered when one considers corrections to the Born-Oppenheimer electron affinity, leading to only a few lowest anionic states to remain bound. For a dipole-bound state to remain bound it is necessary to possess electron affinity as large as one-tenth to one-twentieth of the molecular rotational constant, as was theoretically established by Crawford and Garrett using a model of a rotating dipole and an electron [32].

Our primary goal in our studies of polar-diatomic anions was to predict vertical electron affinities especially for electronic attachment which undergoes with the formation of the electronic excited states. The anions considered so far are: LiH^-, LiF^-, BeO^-, $NaCl^-$, NaF^- and $LiCl^-$ [33-38]. The Crawford and Garrett's criterion indicates that the stationary dipole approximation should be sufficient to study the ground and first excited states of these anions, or even second excited states in the case of heavier systems. The LiF^- second-excited state is an interesting intermediate case. Our calculations [34,37] indicate that there is significant difference between the prediction for the electron affinity based on the Born-Oppenheimer approximation and a model calculation by Gerrett [39], where an electron is collided with a rotated dipole. This case would be a possible medium for experimental verification of the theory.

One of the most difficult problems faced in electron affinity calculations is the evaluation of the total energy for the anion and the neutral systems with precisely the same accuracy. This would involve calculations which incorporate correlation effects common to the anion and neutral in precisely the same way. Such calculation must also account for all important orbital relaxation effects, as well as all important correlation effects unique to the anion.

The electrophilic reactions under study can be formally written as:

$$\text{neutral} + \bar{e} \longrightarrow \text{anion (ground state)}$$

$$\text{neutral} + \bar{e} \longrightarrow \text{anion (excited state)}$$

For the LiH molecule this would mean the following processes:

$$\text{LiH}(1\sigma^2 2\sigma^2) + \bar{e} \longrightarrow \text{LiH}^- \text{ g.s. } (1\sigma^2 2\sigma^2 3\sigma^1)$$

$$\text{LiH}(1\sigma^2 2\sigma^2) + \bar{e} \longrightarrow \text{LiH}^- \text{ Ix.s. } (1\sigma^2 2\sigma^2 3\sigma^0 4\sigma^1)$$

The reference functions for LiH and LiH^- g.s. are numerical RHF determinants, while for the LiH^- Ix.s. it is a Slater determinant constructed out two LiH core orbitals plus a Koopmans orbital 4σ obtained numerically in a frozen-core calculation in which optimized orbital was kept orthogonal to previously determined 3σ Koopmans orbital unoccupied but present in the calculation.

In the next step we portioned the electron correlation for the neutral and anionic systems into the following BG cluster structure:

LiH	:	$(1\sigma^2)$, $(2\sigma^2)$, $(1\sigma 2\sigma)$
LiH^- (g.s.)	:	$(1\sigma^2)$, $(2\sigma^2)$, $(1\sigma 2\sigma)$, $(1\sigma 3\sigma)$, $(2\sigma 3\sigma)$, $2\sigma^2 3\sigma)$
LiH^- (Ix.s.)	:	$(1\sigma^2)$, $(2\sigma^2)$, $(1\sigma 2\sigma)$, $(1\sigma 4\sigma)$, $(2\sigma 4\sigma)$, $(2\sigma^2 4\sigma)$

and MCSCF numerical orbitals were generated to correlate electrons in each cluster. The reason for considering the three-electron clusters: $(2\sigma^2 3\sigma)$ and $(2\sigma^2 4\sigma)$ was to allow the singly occupied orbital to adjust to the dipole field change produced by the correlation of the core electrons. In the case of cluster correlation for LiH^- (Ix.s.) variational MCSCF optimizations were performed for the second CI root, and involve both single and double excited determinants.

The final step involves constructing combined basis sets of numerical orbitals and the MBPT/CC calculations performed with one- and two-electron integrals generated for numerical functions. The basis sets used for the calculation of the neutral LiH were different for calculation of the ground and excited state electron affinities by virtue of including correlation orbitals of clusters for the anionic system, which do not exist for the neutral. One can find the details of the basis-set construction procedure in ref. [35]. The combined correlating basis set for the LiH-LiH^- g.s. electron attachment consists of 46 numerical shells distributed as 22 of σ symmetry, 14 of π symmetry and 10 of δ symmetry. The final result (CCSDT-1) for the ground state vertical electron affinity is 0.01101 a.u., and agrees well with the full CI result of 0.01097 from Liu et al. [40] interpolated for internuclear distance of 3.015 a.u.. For the LiH-LiH^- (Ix.s.) electron attachment, final MBPT/CC calculations were performed with 47 numerical shells, one more orbital of σ symmetry than in the ground state electron affinity calculation. Since the calculation for the LiH^- x.s. anion is done with the reference function which is not an UHF solution, the electron affinity could only be determined at CCSD level - the only method which in the current implementation does not demand the orbitals to be HF-canonical. However, the CCSDT-1 result, which is also included in Table VI, should also be very close to the true value, even though the current CCSDT-1 implementation is not invariant upon a unitary transformation of correlation orbitals. A direct CC calculation with the reference determinant $|1\sigma^2 2\sigma^2 4\sigma|$ converges well and the final result for the excited state electron affinity is 0.000103 a.u., slightly larger than our previous MCSCF result of 0.000095 a.u. obtained in relatively small numerical basis set [33].

Determinations of electron affinities of the BeO molecule appeared to be more difficult than for LiH due to a significant mobility of the electronic charge for BeO along the bond axis due to the well known 2s-2p degeneracy. The BeO reference function for the correlation calculation was the $^1\Sigma$ determinant, $|1\sigma^2 2\sigma^2 3\sigma^2 4\sigma^2 1\pi^4|$. The dipole moment calculated for this function, 7.52 D, is larger than the experimental estimate of 7.1±0.3 D [41]. It is known from previous low-level correlated calculations [42] that the electron correlation significantly lowers the dipole moment. However, a decrease down to 6.40 D might be an exaggeration caused by limiting the correlation effects to only second order in energy. The summary of numerical results obtained for BeO electron affinities are presented in Table VII.

Table VI. Electron affinities for LiH and BeO molecules in a.u.

LiH-LiH⁻ g.s.	LiH	LiH^-	Electron affinity
Koopmans Theorem			0.00783
numerical HF	-7.988753	-7.996429	0.00908
MBPT D(2)	-8.055729	-8.066045	0.01032
MBPT D(3)	-8.065140	-8.075813	0.01067
MBPT SDTQ(4)	-8.067254	-8.078099	0.01084
CCSD	-8.067905	-8.078658	0.01075
CCSDT-1	-8.068057	-8.079068	0.01101
$LiH-LiH^-$ Ix.s.			
Koopmans theorem			0.000086
CCSD	-8.067853	-8.067957	0.000103
CCSDT-1	-8.068005	-8.068112	0.000106

Table VII. Electron affinities for the BeO in a.u.

$BeO-BeO^-$ g.s.	BeO	BeO^-	Electron affinity
Koopmans theorem			0.064
Numerical HF	-89.453259	-89.530600	0.077
CCSD	-89.751454	-89.828139	0.077
BeO-BeO x.s.			
Koopmans theorem			0.00131
CCSD	-89.749964	-89.750819	0.00086

In the case of the BeO^- x.s. one sees a strong departure from the Koopmans electron affinity which, however does not appear for the ground-state case. This seems to suggest that correlation-related change in the dipole moment of the neutral BeO core for the excited state attachment undergoes unperturbed by the diffused additional electron and it is significantly limited for ground state process.

Our studies on electron affinities of larger polar diatomics, NaF, LiCl and NaCl [38] have been conducted using an alternative technique, which combines a numerical HF orbitals with Slater orbitals. The Slater correlation basis sets were carefully chosen based on Slater orbital representation of the HF numerical orbitals generated by our mapping procedure [45]. Our primary goal in these calculation was to determine first excited state electron affinities, which in light of recent progress in the area of photoelectron spectroscopy [44] may soon become an interesting experimental medium. The calculations involving numerical HF orbitals and Slater functionions followed the same scheme as used for all-numerical basis sets. Our previous study indicated that in the formation of the first excited state there are two opposite factors at work. The first is the reduction of the electron affinity due to the decrease of the correlation-related decrease of the molecular dipole moment. The second is a positive contribution due to the correlation of the extra electron with the core electrons. For LiCl and NaCl, the second effect dominates slightly, while for NaF, the two effects exactly cancel (see Table VIII).

Table VIII. Koopmans and CCSD electron affinities for first excited state electron attachment. In a.u.

	Koopmans theorem	CCSD
NaF g.s. + $\bar{e}$ ⟶ NaF^- Ix.s.	0.00044	0.00044
LiCl g.s. + $\bar{e}$ ⟶ $LiCl^-$ Ix.s.	0.00031	0.00033
NaCl g.s. + $\bar{e}$ ⟶ $NaCl^-$ Ix.s.	0.00073	0.00077

4. SUMMARY

The present paper considers our recent computational results, which have been obtained by combining the partial-wave numerical orbital technology for diatomic molecules with the MBPT and CC methods. This union allows the calculation of ground and some excited states of diatomic systems with very high accuracy. Moreover, the present composite method offers partial relief from the basis set problems in the sense that numerical functions even as a basis set follow the nuances of orbitals better than normal Gaussian or Slater basis sets. However, without direct numerical solutions of the correlated CC equations, there remains a basis set problem in some incompleteness in covering the correlation space.

The electron affinity calculations demonstrate the general chemical applicability of the present method; but at the same time, it reveals how complex the computational process becomes when the basis set, instead of being arbitrarily chosen, is optimized for particular state and geometrical structure, which seems to be an unavoidable condition in maintaining a very high accuracy standard.

There is an optimistic conclusion emerging from our very accurate calculations for small diatomic molecules. This is that one does not have to resort to explicitely correlated basis functions to generate over 99% of the molecular correlation energy, which is quite a popular belief among quantum chemists. This can be accomplished within an orbital approximation, provided that orbitals maintain sufficient flexibility and are carefully optimized. An area in which we hope to apply our procedure is transition-metal diatomic, neutral and ionic systems which suffer in traditional approaches from basis-set limitations. In general, the current approach should contribute to diminishing the effect of this restriction in the ab-initio results.

The big question, however, is: can this approach be generalized to polyatomic molecules? Certainly, the prolate spherical coordinate system is unique for two-center systems, and therefore the partial wave technology only works for this case. There is, however, a method which we have proposed and applied to the SCF procedure [45], as well as to solving first-order pair equations [46]. It is based on the realization that in the basis set approach, one can construct the Fock operator not by distributing tape-located two-electron integrals multiplied by appropriate density matrix elements, but by constructing numerically represented Coulomb and exchanged potentials. This results in the reduction of the N^4-dependence of the computational time on the number of basis functions for the ordinary SCF scheme, to at most N^2-dependence. It also avoids the construction and manipulation of large integral files even when the calculation is done with a basis set. The substance of the method is to find a suitable representation for basis functions, which can be analytical, numerical, or combined analytical and numerical (the partial wave representation is the last case), and to be able to fold the total electronic density and, consequently, the Coulomb potential, as well as individual molecular orbital contributions, to the exchange potential in a form of simple functions.

For the diatomic case, the Slater orbital can be transformed into the partial wave representation:

$$(20) \qquad \psi^{m}_{nl}(\zeta,\eta,\phi) = \sum_{L=|m|}^{L_p} X^{m}_{nl,L}(\zeta) \; Y^{m}_{L}(\eta,\phi)$$

and then the molecular orbital representation can be simply generated by using eq. 12. Now, employing the Neumann expansion of $1/r_{12}$ in spheroidal coordinates and integrating over angular coordinates we can reduced the Coulomb contribution of the (p,q) Fock matrix element to a finite sum of one dimensional integrals (see ref.45 for details)

$$(21) \qquad \langle p|J_i|q\rangle = \sum_s \int_1^\infty d\zeta \; \rho_{pq,s}(\zeta) \; V_{ii,s}(\zeta)$$

where index s results from the Neumann expansion. The expression under the integral sign factors into the density function $\rho_{pq,s}$, which depends only on basis functions p and q, and into the potential function $V_{ii,s}$, which depends only on occupied orbital i and, therefore, is identical for all Fock matrix elements.

Can this strategy be implemented for polyatomic molecules? The answer is yes, and we can give a brief sketch of one approach. Several existing polyatomic integral programs [47] employ a potential-charge-density formulation and multidimensional numerical quadrature to evaluate Coulomb and exchange integrals. In these programs, the precise analytic form of the basis functions is used only to simplify construction of the potentials. One could, however, determine the potentials by numerical solution of Poisson's equation

$$(22) \qquad \nabla^2 V = -\rho$$

for a purely numerical charge density. Furthermore, with some non-iterative numerical methods for the solution of eg.(22) much of the computational work need not be repeated for different ρ. Whether or not this or any other implementation for polyatomics could achieve acceptable computational efficiency is another question, but the possibility seems worthy of investigation.

References:

[1] K. A. Brueckner, Phys.Rev. **97**, 1353 (1955) and **100**, 36 (1955).

[2] J. Goldstone, Proc.Roy.Soc.London, Ser.A **239**, 267 (1957).

[3] H. P. Kelly, Phys.Rev. **131**, 684 (1963) and 136, B896 (1964); H. P. Kelly, Adv.Chem.Phys. **14**, 129 (1969); H. P. Kelly, Phys.Rev.Lett. **23**, 455 (1969).

[4] C. F. Fisher, *The Hartree-Fock Method for Atoms*, Wiley, New York, 1977.

[5] a) E. A. McCullough, Jr., Chem.Phys.Lett. **24**, 55 (1974); b) E. A. McCullough, Jr., J.Chem.Phys. **62**, 3991 (1975); c) P. A. Christiansen and E. A. McCullough, Jr., ibid. **67**, 1877 (1977); d) L. Adamowicz and E. A. McCullough, Jr., ibid. **75**, 2475 (1981); e) E. A. McCullough, Jr. J.Phys.Chem. **86**, 2178 (1982); f) E. A. McCullough, Comp.Phys.Rep. **4**, 265 (1986).

[6] L. Laaksonen, P. Pyykkö and D. Sundholm, Chem.Phys.Lett. **96**, 1 (1983); L. Laaksonen, D. Sundholm and P. Pyykkö, Chem.Phys.Lett. **105**, 573 (1984).

[7] A. D. Becke, J.Chem.Phys. **76**, 6037 (1983).

[8] a) R. J. Bartlett and D. M. Silver, Phys.Rev. A**10**, 1927 (1974), and **13**, 1 (1976), and Int.J.Quantum Chem.Symp. **8**, 271 (1974), and Chem.Phys.Lett. **29**, 199 (1974); R. J. Bartlett and I. Shavitt, Chem.Phys.Lett. **50**, 190 (1977) and **57**, 157 (1878); R.J. Bartlett, G. D. Purvis and I. Shavitt, J.Chem.Phys. **71**, 281 (1979); R. J. Bartlett and G. D. Purvis, Int.J.Quant.Chem. **14**, 561 (1978), and Phys.Scr. **21**, 255 (1980); b) See R. J. Bartlett, Annu.Rev.Phys. Chem. **32**, 359 (1981) for a review.

[9] F. Coester, Nucl.Phys. **1**, 421 (1958); F. Coester and H. Kummel, Nucl.Phys. **17**, 477 (1960); J. Cizek, J.Chem.Phys. **45**, 4256 (1966), and Adv.Chem.Phys. **14**, 35 (1969).

[10] a) G. D. Purvis and R. J. Bartlett, J.Chem.Phys. **76**, 1980 (1982); b) Y. S. Lee, S. A. Kucharski and R. J. Bartlett, J.Chem.Phys. **81**, 5906 (1984); c) J. Noga and R. J. Bartlett, J.Chem.Phys. **86**, 7041 (1987).

[11] L. Adamowicz and R. J. Bartlett, Int.J.Quantum Chem.Symp. **19**, 217 (1986).

[12] I. Lingren and S. Salomonson, Phys. Scr. **21**, 335 (1980).

[13] R. K. Nesbet, Adv.Chem.Phys. **14**, 1 (1969).

[14] PROPAGATOR program system consists of GRNFNC that does SCF calculation and integral transformation of G. D. Purvis, and CC/MBPT for coupled-cluster and many-body perturbation theory calculations, written by R. J. Bartlett, G. D. Purvis, Y. Lee, S. J. Cole, and R. Harrison.

[15] L. Adamowicz, R. J. Bartlett and E. A. McCullough Jr., Phys.Rev. Lett. **54**,426 (1985).

[16] L. Adamowicz and R. J. Bartlett, in *Applied Quantum Chemistry*, V. H. Smith ed., Reidel Publishing Company, 1986.

[17] B. Jeziorski, H. J. Monkhorst, K. Szalewicz and J. G. Zabolitsky, J.Chem.Phys. **81**, 368 (1984).

[18] N. C. Handy, R. J. Harrison, P. J. Knowles and H. F. Schaefer, J.Chem.Phys. **88**, 4852 (1984).

[19] R. J. Harrison and N. C. Handy, Chem.Phys.Lett. **113**, 257 (1985).

[20] R. J. Bartlett, H. Sekino and G. D. Purvis, Chem.Phys.Lett. **98**, 66 (1983).

[25] C. F. Bunge, Phys.Rev. A **14**, 2496 (1977).

[26] C. F. Bender and Davidson, Phys.Rev. **183**, 23 (1969).

[27] W. Kolos and J. Rychlewski, Acta.Phys.Pol. A **53**, 281 (1978).

[28] L. Adamowicz and R. J. Bartlett, Phys.Rev. A **37**, 1 (1988).

[29] R. F. Wallis, R. Herman and H. W. Milnes, J.Mol.Spec. **4**, 51 (1960).

[30] L. Carlsten, L. R. Peterson and W. C. Lineberger, Chem.Phys.Lett. **37**, 5 (1976).

[31] J. Simons and K. D. Jordan, Chem.Rev. **87**, 535 (1987).

[32] O. H. Crawford and W. R. Garret, J.Chem.Phys. **66**, 4968 (1977).

[33] L. Adamowicz and E. A. McCullough, Jr.,J.Chem.Phys. **88**, 2045 (1984).

[34] L. Adamowicz and E. A, McCullough, Jr.,Chem.Phys.Lett. **107**, 72 (1984).

[35] L. Adamowicz and R. J. Bartlett, J.Chem.Phys. **83**, 6268 (1985).

[36] L. Adamowicz, J. C. Ellenbogen and E. A. McCullough, Jr., Int.J. Quantum Chemistry, **30**, 617 (1986).

[37] L. Adamowicz and R. J. Bartlett, Chem.Phys.Lett. **129**, 159 (1986).

[38] L. Adamowicz and R. J. Bartlett, J.Chem.Phys. **88**, 313 (1988).

[39] W. R. Gerrett, J.Chem.Phys. **77**, 3666 (1982).

[40] B. Liu, K. O-Okata and K. Kirby-Docken, J.Chem.Phys. **67**, 1850 (1977).

[41] M. Yoshimine, J.Phys.Soc.Jpn. **25**, 1100 (1968).

[42] Y. Yoshioka and K. D. Jordan, J.Chem.Phys. **75**, 5899 (1980).

[43] L. Adamowicz and E. A. McCullough, Int.J.Quantum.Chem. **24**, 19 (1983).

[44] S. M. Burnett, A. E. Stevens, C. S. Feigerle and W. C. Lineberger, Chem.Phys. Lett. **100**, 124 (1983).

[45] L. Adamowicz and E. A. McCullough, Int.J.Quantum Chem. **26**, 373 (1984).

[46] L. Adamowicz and R. J. Bartlett, Int.J.Quantum Chem. **26**, 213 (1984).

[47] for example A. C. Wahl and R. H. Land, Int.J.Quantum Chem.Symp. **1**, 375 (1967).

The Development of an Efficient Numerical Orbital Algorithm for Polyatomic Systems: A Review of the Various Options

H.J. Monkhorst and S.A. Alexander
Quantum Theory Project
Departments of Physics and Chemistry
University of Florida
Gainesville, FL 32611

1. Introduction

In most ab initio calculations, atomic and molecular orbitals are described by an expansion of simple basis functions e.g. Gaussians or Slaters. The nonlinear parameters in these basis functions are chosen using a variety of methods including optimized atomic calculations and various tempering schemes [1]. When only low to moderate accuracy is required (greater than 1 millihartree), an appropriate set of parameters can be easily selected and the desired calculation quickly completed. As the demand for higher and higher accuracy has grown, however, the limitation of these methods has become increasingly apparent. In particular, there is still no systematic way to increase the size of the basis set or to reliably estimate the basis set truncation error despite a number of recent studies [2].

One way to avoid these problems is to eliminate any sort of basis set expansion and calculate the orbitals directly i.e. numerically. This approach was first applied to atoms [3] and then later to diatomics [4-8]. Because these calculations can be systematically improved and their error computed, numerical orbital methods have been able to accurately determine the energies and properties of many "difficult" systems. These successes have created a demand for similar methods which can be applied to polyatomic systems. Such methods, however, are still in their infancy. In this

M. Defranceschi and J. Delhalle (eds.), Numerical Determination of the Electronic Structure of Atoms, Diatomic and Polyatomic Molecules, 201–207.

short review we describe the differences between the various algorithms being developed today as well as their advantages and disadvantages.

2. Position space and Momentum space

Although efficient algorithms for calculating numerical orbitals have been developed for both atoms and diatomics, none of these methods can be applied to general polyatomics because of the many electron-nuclear and electron-electron singularities in the Hartree Fock equations. For closed shell molecules, the restricted version of these equations have the form

$$\varepsilon_i \phi_i(r) = -\frac{\nabla^2}{2}\phi_i(r) - \sum_{a=1}^{\#\,\mathrm{Atoms}} \frac{Z_a}{|r - \mathbf{R}_a|}\phi_i(r) + 2\sum_j \int d^3r' \frac{\phi_j^*(r')\phi_j(r')}{|r - r'|}\phi_i(r) - \sum_j \int d^3r' \frac{\phi_j^*(r')\phi_i(r')}{|r - r'|}\phi_j(r) \tag{1}$$

where ε_i is the i'th orbital energy, $\mathbf{R}_a$ is the a'th nuclear position. For atoms and diatomics these singularities can be eliminated by choosing a proper coordinate system. No such system exists, however, for general polyatomics. Several methods have been proposed which would sidestep this problem but none has been widely adopted. Hinze and Chemie, for example, proposed one scheme in which each molecular orbital was decomposed into a sum of functions located at different centers [9]. This approach lead to a complicated series of integrations over spheres and was later abandoned. Adamowicz has recently suggested [8] that such calculations could instead be performed using Poisson's equation.

Although the Coulomb singularities cannot be eliminated in the position space formulation of the Hartree-Fock equations, Navaza and Tsoucaris showed in 1981 [10] that the momentum space version of these equations had only one, removable singularity at the origin

$$\varepsilon_i \phi_i(\mathbf{p}) = \frac{p^2}{2}\phi_i(\mathbf{p}) - \int d^3q\, \phi_i(\mathbf{p}-\mathbf{q}) \sum_{a=1}^{\#\,\mathrm{Atoms}} Z_a\, e^{i\,\mathbf{q}\cdot\mathbf{R}_a} / (2\pi^2 q^2) \, +$$

$$2 \sum_j \int d^3q \int d^3s\, \phi_j^*(\mathbf{s})\, \phi_j(\mathbf{s}+\mathbf{q})\, \phi_i(\mathbf{p}-\mathbf{q}) / (2\pi^2 q^2) \, -$$

$$\sum_j \int d^3q \int d^3s\, \phi_j^*(\mathbf{s})\, \phi_i(\mathbf{s}+\mathbf{q})\, \phi_j(\mathbf{p}-\mathbf{q}) / (2\pi^2 q^2) \qquad (2)$$

This approach has proved quite promising and a number of algorithms have been developed to evaluate equation 2. Although most of these methods currently require large amounts of CPU time, some results are already competitive with the best LCAO calculations on small molecules.

3. Evaluating the convolutions

When evaluating equation 2, the most time consuming part of any algorithm is the calculation of the convolutions. One approach is to evaluate these quantities entirely in momentum space. This method is potentially quite powerful since a fixed number of grid points can be judiciously placed so as to most accurately represent each orbital. Such calculations scale like M^2 where M is the total number of points and they typically require interpolations since the point **p-q** is usually not on the same grid as the points **p** and **q**. This approach has been used in a number of calculations [10-18]. Although no attempt has been made to push for highly accurate results, several algorithmic improvements are currently being developed to improve the accuracy and reduce the amount of CPU time required.

Another approach to computing the convolutions is based on a property of Fourier transforms

$$\int d^3q\, f(\mathbf{q})\, g(\mathbf{p}-\mathbf{q}) = (2\pi)^{3/2} \int d^3r\, f(\mathbf{r})\, g(\mathbf{r})\, e^{i\,\mathbf{p}\cdot\mathbf{r}} \qquad (3)$$

Here the functions f(**p**) and g(**p**) are transformed from momentum to position space where the functions f(**r**) and g(**r**) are multiplied together at each point on the grid. The product is then transformed back to momentum space. This method does not require interpolations and it also manipulates the entire orbital rather than individual grid points. Since each numerical orbital is implicitly or explicitly described by a spline, this last feature helps minimize the error introduced by the discontinuities in this representation. In addition, the fast Fourier transform (FFT) scales like $N \log_2 N$ where N is the total number of grid points. If N~M, this method will be much faster than those which scale like M^2. The use of the FFT is not without its disadvantages, however. It requires a Cartesian grid with equidistant points and in those calculations which have been performed with this approach a large number of points were needed to obtain accurate energies [19,20]. A Cartesian grid also complicates the removal of the singularity at the origin. In fact as Ref. [20] shows, the accuracy of this method depends heavily on the skill with which this is done.

One possible hybrid of these two approaches is to use the slow Fourier transform to calculate the convolutions. This method scales as M^2 and allows a much greater flexibility in the placement of points than does the FFT. Like the FFT it evaluates the entire orbital and needs no interpolations. To the best of our knowledge, however, this approach has not yet been implemented.

4. Choosing a grid

As we saw above, integration grids are usually not chosen so as to obtain the most accurate representation of an orbital with the fewest number of points but rather for other reasons. In Ref. [19,20] a set of equally spaced points on a Cartesian grid was used because the FFT required it. In Ref. [10-18] a spherical grid was used primarily because it automatically canceled the singularity at the origin. With the success of the method in Ref. [20], removing the Coulomb singularity should become less of a reason for selecting a particular grid. One criteria should be how well a grid can represent the wavefunction of a general polyatomic. Another criteria should be how well a grid handles the problem of truncation error. This error is caused by calculating an orbital over a finite rather than an infinite region. One could, of course, simply increase the size of the grid to reduce this problem but the number of grid points would also have to be increased to keep the integration error constant. For methods which scale like M^2, a number of solutions are available. In Ref. [18] for example, truncation error was reduced simply by including a small number of grid points in the asymptotic region. If the method scales like $N \log_2 N$, however, the problem is much more difficult since most changes in the structure of the grid are incompatible with the use of the FFT. One novel solution is outlined in Ref. [21]. In this procedure each orbital is written as the sum of a numerical part and an analytic part. Since the analytic part is defined over all space, it contains information about the orbital outside the grid and thus decreases truncation error. A very different solution to this problem has been proposed by Ishikawa and his coworkers [22,23]. They have shown that numerical orbitals can be computed on grids which cover the entire space. Such an approach directly reduces truncation error and the initial results are encouraging.

5. Conclusions

Developing an efficient numerical orbital algorithm for polyatomic systems has not been as easy as creating similar programs for atoms or diatomics. One reason is that unlike atoms and diatomics no natural choice of a coordinate system exists. Another reason is the explicit three-dimensional nature of these systems. This extra dimension means that over an order of magnitude more CPU time and memory are needed than in similar two-dimensional diatomic calculations and much more than in one-dimensional atomic calculations. In spite of these difficulties much progress has been made. It has now been firmly established that numerical orbitals can be computed for general polyatomic systems and that in certain situations their results are already competitive with large basis set calculations. Since many of the computational barriers which remain are exacerbated by the various choices each algorithm has explicitly or implicitly made, we propose that future work consider two numerical experiments:

1) On the same size grid what are the relative sizes of M and N needed to achieve the same accuracy and how do the CPU times compare? This test would tell us what the tradeoffs are between computational speed with the FFT and increased flexibility with the other methods.

2) Can a spherical grid be used to efficiently describe general polyatomic systems? We believe that the valence orbitals of these systems will need to be accurately described both near the origin (since these functions are strongly peaked there) and far from the origin (because of the oscillations caused by the atomic positions). In such a situation the straightforward use of a spherical grid would invariably lead to an oversupply of points near the origin. We believe that a Cartesian grid is more appropriate for such cases. Comparing the number of spherical and Cartesian grid points needed to accurately calculate the properties of some simple polyatomics would test the relative effectiveness of each grid.

These tests together with those of the mixed numerical/analytic orbital idea and the use of integration methods to cover the entire space should provide a much clearer view of how to more efficiently calculate numerical orbitals for general polyatomic systems.

Acknowledgements

We wish to thank Dr. J. Delhalle and Dr. M. Defranchechi for organizing such a stimulating conference. This work was supported with a PACER fellowship from Control Data.

References

[1] Two excellent reviews of the LCAO problem have been given by S. Wilson, Methods in Computational Molecular Physics G.H.F Diercksen and S.Wilson Eds. (Reidel, Dordrecht, 1983) and by E.R. Davidson and D. Feller,Chemical Reviews 86,681 (1986).

[2] A method to generate systematic basis sets has recently been proposed by L. Adamowicz and R.D. Bartlett, Phys. Rev. A 37, 1 (1988); the question of basis set error has been examined in

detail by B. Klahn and J.D. Morgan III, J. Chem. Phys. 81,410 (1984)

[3] C. Froese Fischer, The Hartree-Fock Method for Atoms (Wiley, New York,1977) and references therein.

[4] E.A. McCullough Jr., Comp. Phys. Rep. 4,265 (1986) and references therein.

[5] L. Laaksonen, P. Pyykko and D. Sundholm, Comp. Phys. Rep. 4,315 (1986) and references therein.

[6] A.D. Becke, J. Chem. Phys. 76,6037 (1983).

[7] P. Pyykko, this proceedings.

[8] L. Adamowicz, this proceedings.

[9] J. Hinze and F.F. Chemie, Int. J. Quant. Chem. S15,69 (1981).

[10] J. Navaza and G. Tsoucaris, Phys. Rev. A 24,683 (1981).

[11] M. Defranceschi, M. Suard and G. Berthier, Comptes Rendus 299,9 (1984).

[12] M. Defranceschi, M. Suard and G. Berthier, Int. J. Quantum Chem. 25,863 (1984).

[13] M. Defranceschi, M. Suard and G. Berthier, Comptes Rendus Acad. Sci. serie II 299,9 (1984).

[14] G. Berthier, M. Defranceschi, J. Navaza, M. Suard and G. Tsoucaris, Theochem 120,343 (1985).

[15] M. Defranceschi and J. Delhalle, Comptes Rendus Acad. Sci. serie II 301,1405 (1985).

[16] M. Defranceschi and J. Delhalle, Phys. Rev. B 34,5862 (1986).

[17] J. Delhalle and M. Defranceschi, Int. J. Quantum Chem. S21,425 (1987).

[18] G. Berthier, M. Defranceschi and J. Delhalle, this proceedings.

[19] S.A. Alexander and H.J. Monkhorst, Int. J. Quantum Chem. 32,361 (1987).

[20] S.A. Alexander, R.L. Coldwell and H.J. Monkhorst, to appear J. Comp. Phys.

[21] S.A. Alexander and H.J. Monkhorst, this proceedings.

[22] W. Rodriguez and Y. Ishikawa, to appear Chem. Phys. Lett.

[23] W. Rodriguez and Y. Ishikawa, to appear Int. J. Quant. Chem. Symp. 22.

ELECTRONIC STRUCTURE THEORY IN MOMENTUM SPACE

G. BERTHIER[+], M. DEFRANCESCHI[*] and J. DELHALLE[#]

[+]*Laboratoire de Biochimie Théorique de l'I.B.P.C.*
13, rue Pierre et Marie Curie, F-75005 Paris, France.

[*]*Laboratoire de Chimie, Ecole Normale Supérieure*
1, rue Maurice Arnoux, F-92120 Montrouge, France.

[#] *Laboratoire de Chimie Théorique Appliquée, Facultés Universitaires*
Notre-Dame de la Paix, 61, rue de Bruxelles, B-5000 Namur, Belgium.

" A Joseph FOURIER, Préfet de l'Isère de 1801 à 1815 "

ABSTRACT. A momentum space MO theory is developed, using the Fourier transforms of various effective Fock Hamiltonians defined for molecules and polymers in position space. As proved by the examples of H_2 and H_3, the molecular orbitals of polyatomic molecules can be computed without resorting to expansions in terms of basis functions, by solving the momentum space Hartree-Fock equations numerically. A special attention is paid to the integration scheme involving p, θ and φ polar coordinates.

1. Momentum Space Formulation of the Electronic Schrödinger Equation.

Formulating and perhaps solving the electronic Schrödinger equation in terms of momentum coordinates **p** instead of the usual space coordinates **r** was a possibility recognized in the early days of Quantum Mechanics. Because the potential terms occurring in the electronic Hamiltonian involve the electron variables **r** rather than **p**, the most convenient way is to write the Schrödinger equation in an integral form through Fourier transformations involving $\mathbf{p} \rightarrow \mathbf{q}$ shifts in momentum space, *i.e.* for an electron subjected to a local potential V(**r**),

M. Defranceschi and J. Delhalle (eds.), Numerical Determination of the Electronic Structure of Atoms, Diatomic and Polyatomic Molecules, 209–238.

$$(\frac{1}{2}\mathbf{p}^2 - E)\,\Phi(\mathbf{p}) = -\int d\mathbf{q}\,\Phi(\mathbf{q})\,V'(\mathbf{p}-\mathbf{q})\ , \tag{1}$$

with, in atomic units,

$$V'(\mathbf{p}) = (2\pi)^{-3/2}\int d\mathbf{r}\,e^{-i\mathbf{p}.\mathbf{r}}\,V(\mathbf{r})\ . \tag{2}$$

The hydrogenoid atom in momentum space was treated by several authors, including Hylleraas [1] in 1932, and its discrete and continuous spectra in terms of Z can be found in Bethe and Salpeter textbook [2]. A vigorous attempt to extend the theory to many-electron many-nuclei systems in the frame of the Born-Oppenheimer approximation was made by McWeeny and Coulson in 1949. These authors investigated the cases of the helium atom [3] and H_2^+ molecule [4], and tried to apply to the hydrogen molecule their iterative calculation method based on the Svartholm treatment of nuclear forces [5]. They faced "formidable difficulties" in considering more complicated systems and concluded pessimistically as to the possibility of generalizing their work.

For a long time, the problem of direct momentum space calculations remained as it was in the fifties. People interested in momentum space distributions in connection with experiment had to be content with theoretical results coming from Fourier transformations of wave functions first computed in the usual position space, as Duncanson and Coulson did before for atomic Compton profiles [6], or nowadays Brion and his associates for (e,2e) molecular spectroscopy [7]. The situation changed in 1981, when two crystallographers, Navaza and Tsoucaris, decided to treat by Fourier transformations, not the Schrödinger equation itself, but one of its most popular approximate forms for electron systems, namely the Hartree-Fock equations. The form of these equations was known before, in connection with electron-scattering problems[8], but their advantage for Quantum Chemistry calculations was not yet recognized.

The work by Navaza and Tsoucaris on the H_2 molecule [9] proved the feasibility of direct numerical MO computations, *i.e.* without atomic basis functions. For chemists interested in polyatomic molecules, this method was quite attractive because it was not limited to systems whose geometry determines the coordinates to be used for integrating the position space equations, as for example polar coordinates for AH_n molecules [10] because they have approximate spherical symmetry and/or spheroidal coordinates for diatomic molecules [11]. During the last years, we have contributed to demonstrate that direct momentum space calculations are in principle feasible for molecules of any size by studying hydrogen systems of increasing complexity : the H_2 ground state at the SCF and MC-SCF level [12], an H_3 open-shell system [13] and a chain of H atoms including an infinite number of electrons and nuclei [14]. More complex systems are also studied in connection with experiments : the form factors of the Li_2 molecule [15], the triple differential cross-sections of the helium atom in direct relation with (e,2e) measurements [16]. Systematic studies of Compton profiles of polymers are now in progress.

The Fourier transformation method enables us to immediately write the momentum space equations as soon as the SCF theory used to describe the system under consideration allows us to build one or several effective Fock Hamiltonians for the orbitals to be determined. This includes a rather large variety of situations:

♦ Closed-shell systems as defined in the standard Hartree-Fock theory [17,18].

♦ Unrestricted monodeterminantal treatments using different orbitals for different spins for open-shell systems (free radicals, triplet states, etc.) [19,20].

♦ Roothaan open-shell treatments involving a closed-shell subsystem and outer unpaired electrons interacting through two-index integrals of Coulomb and exchange type only [21].

♦ MC-SCF treatments written in terms of coupled Fock equations [22]. The simplest examples are the two-configuration SCF theory [23] used in s^2p^n-p^{n+2} atomic mixing[24], or bonding-antibonding molecular problems [25], and more generally the Clementi-Veillard electron-pair MC-SCF theory [26].

♦ SCF treatments for infinite chains having translational symmetry [27,28].

In this contribution we emphasize the principles of the momentum space approach for quantum chemistry calculations of molecules and polymers. To avoid unnecessary complications, but without loss of generality, we consider in details the case of the H_3 molecule (section 2) and, in section 3, that of the infinite chain of hydrogen atoms $(\text{-H-})_\infty$. However, whenever appropriate we quote more general results on other systems to demonstrate that, even though in its infancy, the momentum space approach is not restricted to such simple cases and has a real potential for wider applications.

2. Hartree-Fock Equations in Momentum Space for H_3.

In this section we consider the trihydrogen molecule H_3 in its linear symmetrical conformation as a typical example illustrating the transformation of the Hartree-Fock equations in momentum space. This system studied in connection with the saddle-point of the H+ H_2 reaction is a sort of quantum-mechanical benchmark, for which we have very precise calculations by *ab initio* LCAO-SCF-CI methods [29] as well as quantum Monte Carlo results [30]. As already pointed out, this particularly simple example is selected to give insight into the method , but without loss of generality. As a matter of fact, the complication of the momentum equations does not increase with the size of the chemical system, only the numerical effort scales accordingly. The equations for the more general cases can be found elsewhere [31].

2.1. EQUATIONS.

The restricted Hartree-Fock equations for the symmetrical H_3 are particularly simple because there is only one unpaired electron in a molecular orbital automatically orthogonal to the orbital assigned to the other two electrons [32]. So, they do not contain any off-diagonal Lagrange parameter between the doubly and singly occupied orbitals. More generally, if off-diagonal parameters of this type would be necessary, it should be possible to evaluate them at each iteration from an appropriate combination (with a minus sign) of the closed-shell and open-shell Fock Hamiltonian and integrate it, using the orbitals obtained at the previous iteration cycle. This process, currently used in standard Hartree-Fock calculations for atoms, would require inverse Fourier transformations.

In position space, the equations giving the doubly and singly occupied MO's φ_1 ($\in \sigma_g{}^+$) and φ_2 ($\in \sigma_u{}^-$) of the linear symmetrical H_3 read :

$$(\frac{1}{2}\nabla^2 - \sum_a \frac{Z_a}{|r-R_a|} + 2J_1 - K_1 + J_2 - \frac{K_2}{2})\,\varphi_1 - e_1\varphi_1 = 0\,, \tag{3a}$$

$$(\frac{1}{2}\nabla^2 - \sum_a \frac{Z_a}{|r-R_a|} + 2J_1 - K_1)\varphi_2 - e_2\varphi_2 = 0\,. \tag{3b}$$

(in atomic units and with obvious notations for the nuclear terms as well as Coulomb and exchange operators). Performing the Fourier transformations on the various potential terms, we get the momentum space Hartree-Fock equations for H_3 in integral form :

$$(\frac{1}{2}\mathbf{p}^2 - e_1)\,\varphi_1(\mathbf{p}) = \frac{1}{2\pi^2}\int \frac{d\mathbf{q}}{q^2}\,[\,F(\mathbf{q}) - W^*_{11}(\mathbf{q}) - W^*_{22}(\mathbf{q})\,]\,\varphi_1(\mathbf{p}-\mathbf{q})$$

$$+ \frac{1}{4\pi^2}\int \frac{d\mathbf{q}}{q^2}\,W^*_{12}(\mathbf{q})\,\varphi_2(\mathbf{p}-\mathbf{q}), \tag{4a}$$

and

$$(\frac{1}{2}\mathbf{p}^2 - e_2)\,\varphi_2(\mathbf{p}) = \frac{1}{2\pi^2}\int \frac{d\mathbf{q}}{q^2}\,[\,F(\mathbf{q}) - 2\,W^*_{11}(\mathbf{q})\,]\,\varphi_2(\mathbf{p}-\mathbf{q})$$

$$+ \frac{1}{4\pi^2}\int \frac{d\mathbf{q}}{q^2}\,W^*_{21}(\mathbf{q})\,\varphi_1(\mathbf{p}-\mathbf{q}) \tag{4b}$$

where $F(\mathbf{q})$ is the so-called form factor,

$$F(\mathbf{q}) = \sum_a Z_a\, e^{i\mathbf{q}.\mathbf{R}_a} \tag{5}$$

corresponding to the nuclear attraction potential ($a = H_1, H_2, H_3$; R_{H1} or $R_{H3} = \pm R$, $R_{H2} = 0$). The quantities W come from the Fourier transformation of the Coulomb and exchange terms according to the rules :

$$J_{n'}(\mathbf{r})\,\varphi_n(\mathbf{r}) \quad \text{----FT---->} \frac{1}{2\pi^2}\int \frac{d\mathbf{q}}{q^2} W^*_{n'n'}(\mathbf{q})\,\varphi_n(\mathbf{p}-\mathbf{q})\,, \tag{6a}$$

$$K_{n'}(\mathbf{r})\,\varphi_n(\mathbf{r}) \quad \text{----FT---->} \frac{1}{2\pi^2}\int \frac{d\mathbf{q}}{q^2} W^*_{nn'}(\mathbf{q})\,\varphi_{n'}(\mathbf{p}-\mathbf{q})\,, \tag{6b}$$

$$W_{mn}(\mathbf{q}) = \int d\mathbf{r}\,\varphi^*_m(\mathbf{r})\,\varphi_n(\mathbf{r})\,e^{i\mathbf{q}.\mathbf{r}} = \int d\mathbf{p}\,\varphi^*_m(\mathbf{p})\,\varphi_n(\mathbf{p}-\mathbf{q})\,. \tag{6c}$$

Because of Fourier transform properties, symmetric and antisymmetric position space orbitals φ_1 and φ_2 become real and purely imaginary, respectively, in momentum space. Also important to notice, Fourier transformation preserves the point group symmetry of the molecule, but for the total electronic system be at rest, the momentum density $\rho(\mathbf{p})$ has to obey inversion symmetry [33,34]. This important aspect of momentum space density is treated with some details in the Appendix. As shown below, taking into account these symmetry properties simplifies the calculations.

As for their configuration space counterparts, the above equations have to be solved iteratively. One starts with trial one-electron functions $\varphi_1^{(0)}$ and $\varphi_2^{(0)}$ giving an initial guess for the orbital energies, $e_1^{(0)}$ and $e_2^{(0)}$, as well as the interaction terms W. These orbital energies are obtained by numerical integration according to the following expressions :

$$e_1^{(0)} = \int d\mathbf{p}\, \varphi_1^{(0)*}(\mathbf{p}) \frac{p^2}{2} \varphi_1^{(0)}(\mathbf{p}) - \frac{1}{2\pi^2} \int \frac{d\mathbf{q}}{q^2} \left[F(\mathbf{q}) - W_{11}^{(0)*}(\mathbf{q}) - W_{22}^{(0)*}(\mathbf{q}) \right] W_{11}^{(0)}(\mathbf{q})$$

$$+ \frac{1}{4\pi^2} \int \frac{d\mathbf{q}}{q^2} W_{12}^{(0)*}(\mathbf{q})\, W_{12}^{(0)}(\mathbf{q}), \tag{7a}$$

$$e_2^{(0)} = \int d\mathbf{p}\, \varphi_2^{(0)*}(\mathbf{p}) \frac{p^2}{2} \varphi_2^{(0)}(\mathbf{p}) - \frac{1}{2\pi^2} \int \frac{d\mathbf{q}}{q^2} \left[F(\mathbf{q}) - 2W_{11}^{(0)*}(\mathbf{q}) \right] W_{22}^{(0)}(\mathbf{q})$$

$$+ \frac{1}{2\pi^2} \int \frac{d\mathbf{q}}{q^2} W_{21}^{(0)*}(\mathbf{q})\, W_{21}^{(0)}(\mathbf{q}). \tag{7b}$$

The quantities $W^{(0)}$ are computed according to Eq.(6c) using $\varphi_1^{(0)}$ and $\varphi_2^{(0)}$.

New approximations, $\varphi_1^{(i+1)}$ and $\varphi_2^{(i+1)}$, to the true solutions are generated iteratively by introducing the quantities of the previous iteration, the i-th iteration, into the following expressions :

$$\varphi_1^{(i+1)}(\mathbf{p}) = \left[\frac{p^2}{2} - e_1^{(i)} \right]^{-1} \left\{ \frac{1}{2\pi^2} \int \frac{d\mathbf{q}}{q^2} \left[F(\mathbf{q}) - W_{11}^{(i)*}(\mathbf{q}) - W_{22}^{(i)*}(\mathbf{q}) \right] \varphi_1^{(i)}(\mathbf{p}-\mathbf{q})\right.$$

$$\left. + \frac{1}{4\pi^2} \int \frac{d\mathbf{q}}{q^2} W_{12}^{(i)*}(\mathbf{q})\, \varphi_2^{(i)}(\mathbf{p}-\mathbf{q}) \right\}, \tag{8a}$$

$$\varphi_2^{(i+1)}(\mathbf{p}) = [\frac{p^2}{2} - e_2^{(i)}]^{-1} \{ \frac{1}{2\pi^2}\int \frac{d\mathbf{q}}{q^2}[\, F(\mathbf{q}) - 2W_{11}^{(i)*}(\mathbf{q})\,]\, \varphi_2^{(i)}(\mathbf{p}-\mathbf{q})$$

$$+ \frac{1}{4\pi^2}\int \frac{d\mathbf{q}}{q^2} W_{21}^{(i)*}(\mathbf{q})\, \varphi_1^{(i)}(\mathbf{p}-\mathbf{q}) \} . \tag{8b}$$

The fact that we are interested in bound states, $e_k^{(i)} < 0$, prevents the denominator from diverging.

The total energy is obtained in a similar way by summing its various kinetic, nuclear attraction and Coulomb-exchange terms, as given by integrating their momentum space counterpart over **q**. For instance, one has

$$J_{nn'} = \frac{1}{2\pi^2}\int \frac{d\mathbf{q}}{q^2} W_{nn}^{*}(\mathbf{q})\, W_{n'n'}(\mathbf{q}), \tag{9a}$$

and,

$$K_{nn'} = \frac{1}{2\pi^2}\int \frac{d\mathbf{q}}{q^2} W_{nn'}^{*}(\mathbf{q})\, W_{nn'}(\mathbf{q}) . \tag{9b}$$

The total energy of H_3 molecule is :

$$E = 2\int d\mathbf{p}\ \varphi_1^{*}(\mathbf{p})[\frac{p^2}{2}]\ \varphi_1(\mathbf{p}) + \int d\mathbf{p}\ \varphi_2^{*}(\mathbf{p})[\frac{p^2}{2}]\ \varphi_2(\mathbf{p})$$

$$- \frac{1}{2\pi^2}\int \frac{d\mathbf{q}}{q^2} F(\mathbf{q})\, [\, 2\, W_{11}^{*}(\mathbf{q}) + W_{22}^{*}(\mathbf{q})\,]$$

$$+ \frac{1}{2\pi^2}\int \frac{d\mathbf{q}}{q^2}[\, W_{11}^{*}(\mathbf{q})\, W_{11}(\mathbf{q}) + W_{22}^{*}(\mathbf{q})\, W_{22}(\mathbf{q}) - W_{12}^{*}(\mathbf{q})\, W_{21}^{*}(\mathbf{q})\,] . \tag{10}$$

As shown below, we can use a polar grid and integrate with respect to the φ, θ and p coordinates successively by a technique of Gauss-Legendre type. The terms W, (Eq.6c), imply a shift (**p**-**q**) of the φ functions, thus requiring the evaluation of these φ functions at points different from those defined by the integration grid. Presently, interpolation by one-dimensional splines is used in our group. As trial functions for the H_3 molecule, we have chosen the Fourier transforms of the corresponding Hückel molecular orbitals built from the 1s atomic functions of the three hydrogen atoms :

$$\varphi_1^{(0)}(\mathbf{p}) = \frac{1 + C\cos(\mathbf{p}.\mathbf{R})}{N_1\,[\,p^2 + \alpha^2\,]^2}, \tag{11a}$$

$$\varphi_2^{(0)}(\mathbf{p}) = \frac{i\sin(\mathbf{p}.\mathbf{R})}{N_2\,[\,p^2 + \alpha^2\,]^2}, \tag{11b}$$

where α denotes the orbital exponent and C the coefficient of the middle-hydrogen atomic function contributing to φ_1.

2.2. COMPUTATIONAL ASPECTS.

Several attempts have been made to solve numerically the momentum space equations.

The oldest of all proposed by Fock [35] consists in an hypergeometrical transformation of the ordinary 3D momentum space into a 4D hypershere. This choice replaces the p radial variable by a variable α defined as $p = \tan(\alpha/2)$; its integration range is $0 \leq \alpha \leq \pi$ instead of the infinite range in the case of p, $0 \leq p \leq \infty$. Several authors have considered this transformation in the past : Monkhorst [36] and Novosadov [37] have used it to study its properties and advantages in the context of momentum space Hartree-Fock equations. Following the pioneering work of Shibuya and Wulfman [38], Novosadov [39] has also applied successfully this technique to treat the molecule H_2^+. More recently, Ishikawa, starting with the Fock transformation, has published a fully numerical treatment for H_2^+ [40].

The fast Fourier transform (FFT) is another approach for fully numerical treatments. It was first outlined by Navaza and Tsoucaris [9] and it is now considered in a systematic fashion by Monkhorst and Alexander [41,42].

When McWeeny investigated the solution of the momentum space integral equation for the hydrogen molecule ion [4] he found that, already in the calculation of the first iterated correction to the ground state LCAO molecular orbital, the resulting expressions were too complicated to be integrated in closed form. Thus he concluded that momentum space was not promising. However, inspection of these expressions points to quite interesting features for a pure numerical treatment directly in momentum space.

As a consequence of the Fourier transformations applied to the nuclear attraction terms, the Coulomb singularities originating from the nuclear framework are transferred to infinity in momentum space (when two particles are interacting their kinetic energy raises as a consequence of the increase of their momentum). This is in contrast with the ordinary position space situation where for integration purposes we have to find an appropriate coordinate system, if any, isolating these singularities at origins (*i.e.*, where we are limited to geometries corresponding to exactly solvable one-electron problems). The transformed equations have a single singularity, $\mathbf{q}^{-2}$, at the origin, suggesting the use of the polar momentum space coordinates p, θ, φ. This idea is strengthened by the observation that the largest amplitudes of the **p**-space functions [12-14] are essentially centered around the origin of the **p** space.

In our applications, we have adopted the polar momentum space coordinate system in conformity with the existence of a molecular axis in the position space, which reminds us of the positions of the nuclei, even if they are immaterial in momentum space. The reminiscence of the nuclear framework is of great help for understanding quantum mechanics in momentum space [43]. The range of integration is $0 \leq p \leq \infty$; $0 \leq \theta \leq \pi$, and $0 \leq \varphi \leq 2\pi$; it is divided in concentric spherical shells around the origin. Since momentum space wave functions have their maximum amplitude near the origin, the largest part of the numerical efforts can be spent in the inner shells.

The volume where the $\varphi(\mathbf{p})$ functions take appreciable values does not very much depend on the molecular size so that an integration sphere with a radius R of a few atomic units seems reasonably appropriate. However, the integration process cannot be simply stopped at the boundary of spheres. A recent attempt to evaluate the error as a function of the sphere radii [44] has shown that its incidence largely depends on the nature of the properties. In fact, it was calculated that for the one-electron energy of the helium atom to be accurate to 9 decimal digits, one would need to consider radii of integration as large as 2500 au! Simply ignoring the tail of the wave function in the calculation of H_2 [45] leads to an orbital energy 7% higher than the Hartree-Fock limit for a radius of integration equal to 3.5 au, and 3% higher for a radius equal to 5.5 au. Thus, an accuracy comparable to that currently obtained with LCAO-SCF values is hardly obtainable by brute force numerical integration.

At the present stage of the development of the momentum space approach, it seems unlikely that purely numerical schemes will be practical and there is a need for asymptotic forms to continue the numerical results from the inner to the outer regions . In our applications, we have added known functional forms to describe the tail (*i.e.* outside the inner regions) of the $\varphi(\mathbf{p})$ functions. In the case of the hydrogen systems, we have used the simple LCAO forms; they have the correct $\mathbf{p}^{-4}$ asymptotic behaviour at large values of **p**. This particular asymptotic form is valid for the various hydrogen systems since for large values of **p**, the 1s function is more diffuse than any other atomic orbital. If one refers to the LCAO-MO formalism, one can anticipate that for large values of **p** the molecular orbital will decay like $\mathbf{p}^{-4}$.

It should be mentioned that the number of points on the integration grid should increase with the number of electrons, *i.e.* the number of occupied molecular orbitals or the number of atoms, for the same reasons that dictate the number of basis functions in a LCAO expansion. According to Navaza and Tsoucaris [9], this can be seen as a consequence of the Shannon theorem in information theory. In spite of its apparent complexity, the numerical technique should be less time-consuming than with the atomic basis expansion technique. The computing time scales like M^2 (M being the number points in the grid) in momentum space, while for files of LCAO integrals it grows like M^4 (M being the number of basis functions).

Let us now examine with some details the algorithm developed to evaluate the various terms of Eqs. (4) and (6) to (8). In general we have to compute quantities of the form,

$$I = \int_{R_1}^{R_2}\int_0^{\pi}\int_0^{2\pi} f(p,\theta,\varphi)\, p^2 \sin\theta \; dp d\theta d\varphi \ , \tag{12}$$

where the lower and upper limits of p are $R_1 = 0$ and $R_2 = R$ (the radius of the sphere) for a molecule of finite size. Integrating over the φ, θ and p variables, successively leads to

$$I = \int_{R_1}^{R_2} G(p)\, p^2 dp, \tag{13}$$

where G(p) is defined as follows,

$$G(p) = \int_0^{\pi} H(p,\theta) \sin\theta \; d\theta \ , \tag{14a}$$

$$H(p,\theta) = \int_0^{2\pi} f(p,\theta,\varphi)\; d\varphi \ . \tag{14b}$$

Numerically, integrals are approximated by weighted sums of integrand values, so that we can write :

$$I = \sum_i \sum_j \sum_k A_i B_j C_k f(p_i,\theta_j,\varphi_k) \ , \tag{15}$$

where the constants A_i, B_j, C_k are the weights assigned to each point in **p**. We use the Peirce-Mustard integration scheme [46] whose points are located at the intersections of the planes $\varphi = \varphi_k$, $\theta = \theta_j$ and the spheres $p = p_i$, as follows :

1) the φ_k's are given by :
   $\varphi_k = 2\pi k / (s+1)$; $C_k = 2\pi / (s+1)$ with $k = 1, 2, \ldots (s+1)$. Integration over φ can be performed immediately, giving a constant, because of the presence of a symmetry axis of infinite order in linear molecules.
2) the cosines of the θ_j's : $y_j = \cos\theta_j$ are the $2m + 2$ zeroes of the Legendre polynomials P_{2m+2} orthogonalized on the interval [-1,+1] and the B_j are the corresponding Gauss-Legendre weights. The existence of an inversion center [33,34] in probability distribution functions $\varphi^*(\mathbf{p})\varphi(\mathbf{p})$ makes it possible to limit integration over θ between 0 and $\pi/2$.
3) the p_i's and the A_i's are the zeroes of the Legendre polynomials P_n orthogonalized on the [-1,1] interval.

The accuracy of the integrals is determined by the values of s, m and n. If it be required that Eq.(12) has an accuracy of order s in $\sin\theta\sin\varphi$, $\sin\theta\cos\varphi$ and $\cos\theta$ and of order t in p, the following equalities have to be verified :

$s = 4m+3$ and $m = E[(t+4) / 2]$ where E[x] defines the integer part of real x.

2.3. RESULTS.

In our studies of H_3 molecule, we used a grid of 18 θ-angular points (NPA) and 37 radial points (NPR) inside a sphere of radius R = 5 au. The tails of the $\varphi(\mathbf{p})$ functions were taken into account by 5 outer radial points. About 4 SCF cycles were needed to obtain convergence of the wave functions within a 1% accuracy. At the present moment, the results obtained for the total energy E (-1.5906 hartree) and the equilibrium distance R (1.684 bohr) are inferior to those obtained by LCAO expansions, but our numerical technique still needs significant improvements.

The number of radial and angular points on the integration grid can be improved by several means. Namely, (a) by a simple increase of this number, (b) by another sampling of the integration range (*i.e.* the balance between the number of concentric spherical shells and the number of integration points per shell), (c) by an increase of the number of angular points for large values of p, (d) by a better fit of the molecular orbital tails by asymptotic forms. All these improvements are not necessarily straightforward;

for instance, for a given sphere of integration, an increase of the number of integration points does not automatically improve the numerical result [45].

Table I gives a survey of the molecules studied so far by numerical momentum space calculations. It includes recent results obtained through integration schemes different from ours, *i.e.* the $(\alpha, \theta, \varphi)$-grid for H_2^+ [40] and fast Fourier transformations for H_2 and H_3^+ [41]. The energy value indicated for the Li_2 molecule corresponds to a (p, θ, φ)-calculation where the inner electrons are described by frozen core atomic orbitals [45].

TABLE I.: Selected results of SCF-MO calculations in momentum space.

Bond length d (bohrs)	Total energy E (hartrees)	Integration scheme (momentum space)	Comparative data (position space)
H_2^+ d=2.0	-2.847656	$(\alpha, \theta, \varphi)$ [40]	-2.847656
H_2 d=1.3834	-1.1332	(p, θ, φ) [12]	(a)
H_2 d=1.4	-1.13301	FFT [41]	-1.13363 (b)
H_3^+ d=1.6405	-1.30016	FFT [41]	-1.29993 (c)
H_3 d=1.684	-1.5905	(p, θ, φ) [13]	(d)
Li_2 d=4.7915	-14.5340 (e)	(p, θ, φ) [45]	-14.8716 (f)

(a) E = -1.13295 for d = 1.4 bohr.
(b) Basis set expansion [47].
(c) Basis set expansion [48].
(d) Basis set expansion [29] : E = -1.5947 for d = 1.7 bohr.
(e) $(1s_1 \pm 1s_2)$ frozen cores.
(f) Numerical determination including variation of the core orbitals:
E = -14.8716 for d = 5.051 [49].

It is important to notice the major step accomplished with the treatment of H_3 which was the first numerical solution for a molecule with more than two atoms. Even though the way to routine applications of numerical calculations is still long, the ultimate advantages are very attractive and point to interest for further progress. Among other advantages, with respect to position space, it has to be noticed that : (a) numerical momentum wave functions for polyatomic molecules can be obtained without expansion in terms of basis sets, (b) electron-correlation studies can be included through configuration interaction or otherwise.

For H_3 , starting with Fourier transforms of simple Hückel MO's built from the 1s orbitals of the three hydrogens, Eqs.(11a) and (11b), the iterative treatment introduces polarization effects originating not from a conventional addition of basis orbitals with high angular quantum numbers, but from the iterative minimization process itself. This can be seen for instance by looking at the evolution (during the iteration steps) of the radial functions of H_3 already published [13].

In the case of H_2 molecule, computations have been performed not only at the SCF level, but also at the MC-SCF level, using one doubly excitated configuration from σ_g^+ to σ_u^-. The total energies E for H_2 are close to the corresponding results obtained in position space with a very large atomic basis [25]. The two-configuration SCF value is E = -1.1524 au for d= 1.4160 bohr (the LCAO value [25] is E = -1.1519 au for d = 1.4306 bohr). In a way parallel to ours, numerical MC-SCF calculations based on the same model, but including up to five diexcitations in position space, have been presented by Laaksonen *et al* [50], and both achievements in position and momentum spaces are genuine numerical examples of molecular MC-SCF wave functions.

Comparison of the accuracy obtained in the possible various methods is difficult to achieve since the present method leads directly to momentum space wave functions. The only way of appreciating its quality with other functions is by using Fourier transform position space wave functions. Even if position and momentum space wave functions are fully equivalent representations of the same information (given the dimensional reciprocity of **r**- and **p**-spaces in which the corresponding wave functions are related through a three-dimensional Fourier transform), we are always dealing with approximate wave functions because exact wavefunction determinations are not possible neither in position space nor in momentum space. Contributions to the most informative region in **p**-space come from not only the region well-described in position space but also from outside of this particular region and comparison is not easy.

In our application, using the previously defined integration grid, 4 SCF and 8 MC-SCF cycles were needed to obtain convergence of the wave functions within 1% accuracy. The convergence of the wave functions $\varphi(\mathbf{p})$ was estimated on the basis of the value of the quantity σ defined as,

$$\sigma = \Big[\sum_{i,j,k} A_i B_j C_k \left(\varphi^{(n+1)}(p_i,\theta_j,\varphi_k) - \varphi^{(n)}(p_i,\theta_j,\varphi_k) \right)^2 \Big]^{1/2} , \qquad (16)$$

where n denotes the n-th iteration step, (p_i,θ_j,φ_k) are the integration points and A_i, B_j,

C_k their related weights.

Self-consistency of the MC-SCF treatment for H_2 has been obtained by an iterative procedure on both the functions σ_g^+ and σ_u^- and the coefficients a_i: starting from the trial wavefunctions, the different form factors W are performed then the a_i's are calculated diagonalizing the 2x2 CI matrix in momentum space built from the trial functions and the corresponding energies.

Table II gives some rates of convergence for some orbitals of the H_2 and H_3 molecules at various levels of approximation.

Table II.: Rates of convergence for the different orbitals of H_2 (at SCF level), H_2 (at MC-SCF level) and H_3 (at SCF level).

Iteration number	H_2 SCF (a)	H_2 MC-SCF (b)	H_3 SCF (a)	
			φ_1	φ_2
1	3.28%	3.86%	5.50%	14.30%
2	1.09%	1.72%	1.52%	6.63%
3	0.33%	1.43%	1.00%	4.54%
4		4.26%		
5		4.23%		
6		3.63%		
7		0.73%		

Definition of the grid:

(a) NPA = 18, NPR = 37, R = 4.5 au, α = 1.23

(b) NPA = 16, NPR = 31, R = 5.5 au, α = 1.23

Note: the decrease by degrees observed for σ in the case of H_2 at MC-SCF level is connected with the fact that we used an iterative method where convergence on σ_g^+ and σ_u^- orbitals is achieved separately and the calculation is repeated till the requested accuracy.

The type of improvement obtained after one iteration can be perceived by other ways. A recent study on the hydrogen atom starting from Gaussian type orbitals [51] has shown that already after one iteration there is a qualitative change between the trial function and the iterated function. If the first iteration is performed analytically, in order to control the accuracy of the numerical steps, essential improvements on the nature of the orbitals are observed (better asymptotic decay, improved behavior near the origin ...)

The choice of a grid on which to solve Eqs.(4), (6), (7) and (8) determines to a large extent both the accuracy and efficiency of the momentum space Hartree-Fock algorithm. These conflicting requirements must be balanced for each case studied.

Before reporting some computation times, let us summarize below the needed steps to complete an iteration.

1. Initialization of the program.

step 1-1. Description of the grid: computation of the points and weights of integration Eq.(15).

step 1-2. Initialization of the orbitals: calculation of the trial one-electron functions Eqs.(11a) and (11b).

2. First part of an iteration.

step 2-1. Computation of the W terms Eq.(6c).

step 2-2. Computation of the orbital energies, the various integrals defined in the right hand side of Eqs. (7a) and (7b) are computed.

step 2-3. Continue unless the convergence threshold based on energy criterion is reached.

3. Second part of an iteration.

step 3-1. Computation of the $J_{n'}\varphi_n$ and $K_{n'}\varphi_n$ terms Eqs.(6a) and (6b).

step 3-2. Computation of the iterated functions Eqs.(8a) and (8b).

step 3-3. Normalization of the new functions.

step 3-4. Orthogonalization of the new orbitals (if necessary); a Schmidt orthogonalization has been used in the case of the $1\sigma_g$ and $2\sigma_g$ orbitals of Li_2.

step 3-5. Continue unless the convergence threshold based on σ criterion, Eq.(16), is reached.

The initialization of the program takes a constant time of computation for a given grid which is independent of the size of the molecule. Furthermore, if results of the description of the grid are stored, the initialization of the program for further calculations reduces to step 1-2 which does not consume too much time. Typical initialization times τ are:

- for a grid with NPA = 16, NPR = 31, τ = 55 CPUs

- for a grid with NPA = 18, NPR = 42, τ = 123 CPUs,

where CPUs means CPU second of a Cray I vector computer .

The iteration process itself, and namely the convolutions consume most of the time. The time of computation depends on two parameters: the number of orbitals to be determined and the grid density.

For a given number NO of one-electron orbitals, the number of W_{mn} Eq. (6c) as well as the number of $J_n.\varphi_n$ and $K_n.\varphi_n$ Eq.(6a) and (6b) computed in steps 2-1 and 3-1 is NO(NO+1)/2. Going from H_2 (NO=1) to H_3 (NO=2) and finally Li_2 (NO=3 with frozen core) increases the time for an iteration computation of a factor of 3. As a matter of fact, two iterations for H_2 last 2434 CPUs; for Li_2 (frozen core) or H_3, they last 6609 CPUs and 6645 CPUs, for a grid with NPA = 18 and NPR = 42.

The grid density ρ defined as being the number of integration points inside the sphere $\rho = (NPA+1)^2 x NPR x 8$ (since there are 8 octets per sphere) is the other determining parameter for the time spent in a calculation. The completion of an iteration of H_2 for a grid:

- with NPA = 16 and NPR = 31 ($\rho \approx$ 72000) takes 627 CPUs
- with NPA = 18 and NPR = 42 ($\rho \approx$ 121000) takes 1622 CPUs

Both times are proportional to M^2, as announced previously of, M being the number of points in the grid.

Finally, let us remind the corresponding accuracy. With the previous grids, within a sphere of radius 5.5 au, the energies found for the Helium atom are:

- for NPA = 16, NPR = 31, E= 2.8458 au
- for NPA = 18, NPR = 42, E= 2.8566 au.

where the last result stands within $5x10^{-3}$ au of the Hartree-Fock limit.

3. Hartree-Fock Equations in Momentum Space for $(\text{-H-})_\infty$.

First principle quantum mechanical calculations of the electronic structure of quasi-1D systems have almost exclusively been concerned with isolated periodic model chains.To help the reader unfamiliar with Bloch states concepts and approaches of the electronic structure of infinite chains, we first start with a brief account of the Restricted Hartree-Fock (RHF) equations in configuration space and then obtain their momentum space equivalent. As in section 2, we consider a simple case of model polymer, the infinite chain of hydrogen atoms, to illustrate the transformation of the classical

Hartree-Fock equations for regular polymers in momentum space. This simple model has been and is still used as a useful test case not only to analyse the formal aspects but also to apprehend the main computational problems encountered in the calculations of the electronic structure of polymers (see for example reference 52 and works quoted therein). The momentum space equations for more general cases can be found in reference 53.

3.1. THE RHF EQUATIONS IN CONFIGURATION SPACE.

The nuclear framework of atomic positions of an infinite model chain can be obtained through line-group symmetry operations, but without loss of generality only the translational symmetry is explicitly considered here [27,28].

The infinite chain of hydrogen atoms consists of a macroscopic number N ($\rightarrow \infty$) of unit cells of length a in the direction of periodicity corresponding to the z-axis in a Cartesian coordinate system. Each cell contains one hydrogen atom located at fixed positions $\mu a \mathbf{e}_z$ ($\mu = 0, \pm 1, \pm 2, ...$) relative to the origin of the coordinate system, $\mathbf{e}_z$ is a unit vector in the direction of periodicity. In the case of infinite systems, the cell electroneutrality must be imposed. The set of indices { $\mu, \mu', \mu'' = 0, \pm 1, \pm 2, ...$ } refer to the nodes of the 1D lattice.

In atomic units, the configuration space representation of the Hamiltonian of the hydrogen chain (nuclear motion, relativistic, magnetic, ... effects excluded) is,

$$H = \sum_{i=1}^{N} [-\frac{\nabla^2(\mathbf{r}_i)}{2} + V(\mathbf{r}_i)] + \sum_{1 \leq i < j \leq N} |\mathbf{r}_i - \mathbf{r}_j|^{-1} + U_N, \tag{17}$$

with $\mathbf{r}_i$ the position vector of electron i, $V(\mathbf{r}_i)$ the nuclear attraction operator,

$$V(\mathbf{r}_i) = - \sum_{\mu=-\infty}^{\infty} |\mathbf{r}_i - \mu a \mathbf{e}_z|^{-1}, \tag{18}$$

and U_N the nuclear repulsion energy,

$$U_N = \frac{1}{2} \sum_{\mu,\mu'=-\infty}^{\infty}{}'' \, |(\mu-\mu')a\mathbf{e}_z|^{-1} = \frac{N}{2} \sum_{\mu=-\infty}^{\infty}{}'' \, |\mu a \mathbf{e}_z|^{-1}. \tag{19}$$

In Eq.(19) the double prime indicates omission in the summations of the divergent terms, *i.e.* $\mu = \mu'$ on the left side and $\mu = 0$ on the right side. Unless otherwise indicated, all summations will be over the entire range for which the summed indices are defined.

The Hamiltonian is applied to single-determinant wave functions based on one-electron Bloch states $\varphi_n(k,\mathbf{r})$ which can generally be written as,

$$\varphi_n(k,\mathbf{r}) = e^{ikz}\, u_n(k,\mathbf{r}) \;, \tag{20}$$

where $u_n(k,\mathbf{r})$ is a function having the lattice periodicity, *i.e.* $u_n(k,\mathbf{r}-\mu a\mathbf{e}_z) = u_n(k,\mathbf{r})$; n = 1, 2, ... ∞ is the band index. Since our model system has only one occupied band (n=1), from now on the band index will be omitted in the expressions. The wave number k (k') is defined in the Brillouin zone (BZ) whose length is $2\pi/a$; $k \in [-\pi/a, \pi/a]$. The RHF Bloch states $\varphi(k,\mathbf{r})$ are doubly occupied up to the Fermi energy ε_F and orthonormal,

$$\int d\mathbf{r}\, \varphi^*(k,\mathbf{r})\, \varphi(k',\mathbf{r}) = \delta_{kk'} \,. \tag{21}$$

Applying the condition of minimum expectation value for the total energy E_N of the many-electron determinantal wave function subject to the above conditions yields the RHF equations, *i.e.*

$$F\, \varphi(k,\mathbf{r}_1) = \Big\{ -\frac{\nabla^2(\mathbf{r}_1)}{2} - \sum_{\mu} |\mathbf{r}_1-\mu a\mathbf{e}_z|^{-1}$$

$$+ \frac{a}{2\pi}\int_{BZ} dk'\, \theta(k')\, [\, 2\, J(k',\mathbf{r}_1) - K(k',\mathbf{r}_1)\,] \Big\}\, \varphi(k,\mathbf{r}_1) = \varepsilon(k)\, \varphi(k,\mathbf{r}_1) \,. \tag{22}$$

The detailed form of the operators $J(k',\mathbf{r}_1)$ and $K(k',\mathbf{r}_1)$, respectively describing the classical and exchange Coulombic interactions, is

$$J(k',\mathbf{r}_1)\, \varphi(k,\mathbf{r}_1) = \Big[\int d\mathbf{r}_2 \frac{\varphi^*(k',\mathbf{r}_2)\, \varphi(k',\mathbf{r}_2)}{|\mathbf{r}_1-\mathbf{r}_2|} \Big]\, \varphi(k,\mathbf{r}_1) \tag{23a}$$

$$K(k',\mathbf{r}_1)\, \varphi(k,\mathbf{r}_1) = \Big[\int d\mathbf{r}_2 \frac{\varphi^*(k',\mathbf{r}_2)\, \varphi(k,\mathbf{r}_2)}{|\mathbf{r}_1-\mathbf{r}_2|} \Big]\, \varphi(k',\mathbf{r}_1) \tag{23b}$$

$\varepsilon(k)$ is the occupied energy band and $\theta(k)$ is the occupation function defined as,

$$\theta(k) = \begin{cases} 1 & \text{if } \varepsilon(k) \le \varepsilon_F \\ 0 & \text{if } \varepsilon(k) > \varepsilon_F \end{cases} \tag{24}$$

In terms of the occupation function, the electroneutrality condition writes as :

$$2\left(\frac{a}{2\pi}\int_{BZ} dk\ \theta(k)\right) = 1\ . \tag{25}$$

Before proceeding with the deduction of the momentum space expression of the RHF equations for polymers, it is convenient to rewrite Eq.(22) in terms of its kinetic, T(k,**r**), Coulombic, C(k,**r**), and exchange, X(k,**r**), constituents,

$$F\,\varphi(k,\mathbf{r}_1) = T(k,\mathbf{r}_1) + C(k,\mathbf{r}_1) + X(k,\mathbf{r}_1) = \varepsilon(k)\,\varphi(k,\mathbf{r}_1)\ , \tag{26}$$

with

$$T(k,\mathbf{r}_1) = \left[-\frac{\nabla^2(\mathbf{r}_1)}{2}\right]\varphi(k,\mathbf{r}_1)\ , \tag{27}$$

$$C(k,\mathbf{r}_1) = \left[2\frac{a}{2\pi}\int_{BZ} dk'\ \theta(k')\int d\mathbf{r}_2\ \frac{\varphi^*(k',\mathbf{r}_2)\ \varphi(k',\mathbf{r}_2)}{|\mathbf{r}_1-\mathbf{r}_2|} - \sum_\mu |\mathbf{r}_1-\mu a e_z|^{-1}\right]\varphi(k,\mathbf{r}_1)\ , \tag{28}$$

and,

$$X(k,\mathbf{r}_1) = \left[-\frac{a}{2\pi}\int_{BZ} dk'\ \theta(k')\int d\mathbf{r}_2\ \frac{\varphi^*(k',\mathbf{r}_2)\ \varphi(k,\mathbf{r}_2)}{|\mathbf{r}_1-\mathbf{r}_2|}\right]\varphi(k',\mathbf{r}_1)\ . \tag{29}$$

3.2. THE RHF EQUATIONS IN MOMENTUM SPACE.

Using the properties of the Fourier transform , the momentum space representation of the RHF equation for infinite periodic chains is now obtained [53-55].

3.2.1. *Bloch orbitals in momentum space.* Applying the Fourier transform to the direct space Bloch states leads to the corresponding Bloch states in momentum space :

$$\phi(k,\mathbf{p}) = (2\pi)^{-3/2}\int d\mathbf{r}\ e^{-i\mathbf{p}.\mathbf{r}}\ \varphi(k,\mathbf{r}) = (2\pi)^{-3/2}\int d\mathbf{r}\ e^{-i\mathbf{p}.\mathbf{r}}\ e^{ikz}\ u(k,\mathbf{r})\ . \tag{30}$$

Owing to its translational invariance in direct space, u(k,**r**) can be expanded in Fourier series,

$$u(k,\mathbf{r}) = a^{-1}\sum_\mu u_\mu(k,\mathbf{r}_\perp)\ e^{i\mu\frac{2\pi}{a}z}\ \ ;\ \mathbf{r}_\perp = (x,y), \tag{31}$$

where $u_\mu(k,\mathbf{r}_\perp)$ is the μ-th coefficient of the above Fourier series, Eq.(31),

$$u_\mu(k,\mathbf{r}_\perp) = \int_0^a e^{-i\mu\frac{2\pi}{a}z}\, u(k,\mathbf{r})\, dz\,. \tag{32}$$

Insertion of Eq.(31) in Eq.(30) gives the momentum space expression for the Bloch states :

$$\phi(k,\mathbf{p}) = a^{-1}\int dz \sum_\mu e^{-i[p_z-(\mu\frac{2\pi}{a}+k)]z}\,(2\pi)^{-3/2}\int d\mathbf{r}_\perp\, e^{-i\mathbf{p}_\perp\cdot\mathbf{r}_\perp}\, u_\mu(k,\mathbf{r}_\perp)$$

$$= \frac{2\pi}{a}\sum_\mu \delta[\,p_z-(\mu\frac{2\pi}{a}+k)]\, u^T(k,\mathbf{p}_\mu)\,. \tag{33}$$

where $\mathbf{p}_\mu = (p_x,p_y,\mu(2\pi/a))$ and $u^T(k,\mathbf{p}_\mu)$ is given by :

$$u^T(k,\mathbf{p}_\mu) = (2\pi)^{-3/2}\int d\mathbf{r}_\perp\, e^{-i\mathbf{p}_\perp\cdot\mathbf{r}_\perp}\, u_\mu(k,\mathbf{r}_\perp)\,. \tag{34}$$

Notice that, while $\varphi(k,\mathbf{r})$ is a continuous function of x,y and z in configuration space, in momentum space its counterpart $\phi(k,\mathbf{p})$ is a continuous function of p_x and p_y only and a discrete function of p_z. The discrete part can be visualized as a set of peaks centered at $p_z = k + \mu(2\pi/a)$ with an amplitude equal to $u^T(k,\mathbf{p}_\mu)$.

3.2.2. *Fourier transform of orbital products.* Quantities denoted by $W(k',k,\mathbf{q})$ which we will use in the followings are related to the Fourier transform of products of pairs of Bloch orbitals; they play a central role in momentum space calculations. They are defined as,

$$W(k',k,\mathbf{q}) = \int d\mathbf{r}\, e^{i\mathbf{q}\cdot\mathbf{r}}\, \varphi^*(k',\mathbf{r})\,\varphi(k,\mathbf{r}) = \int d\mathbf{p}\,\phi^*(k',\mathbf{p})\,\phi(k,\mathbf{p}-\mathbf{q})$$

$$= \int d\mathbf{p} \sum_{\mu'} \frac{2\pi}{a}\delta\,[p_z-(\mu'\frac{2\pi}{a}+k')]\, u^{T*}_{\mu'}(k',\mathbf{p}_\perp)$$

$$\times \sum_\mu \frac{2\pi}{a}\delta\,[(p_z-q_z)-(\mu\frac{2\pi}{a}+k)]\, u^T_\mu(k,\mathbf{p}_\perp-\mathbf{q}_\perp)\,. \tag{35}$$

The above equation can be further simplified as,

$$W(k',k,\mathbf{q}) = \sum_\mu \frac{2\pi}{a}\delta\,[q_z-(\mu\frac{2\pi}{a}+k-k')]\, W(k',k,\mathbf{q}_\mu)\,, \tag{36}$$

where $W(k',k,\mathbf{q}_\mu)$ is :

$$W(k',k,q_\mu) = \frac{2\pi}{a}\sum_{\mu'}\int d\mathbf{p}_\perp\, u^{T*}_{\mu'}(k',\mathbf{p}_\perp)\, u_{\mu'-\mu}(k,\mathbf{p}_\perp - \mathbf{q}_\perp). \tag{37}$$

3.2.3. *The momentum space RHF equation.* The momentum space RHF equation is obtained by applying the Fourier transform to Eq.(26),

$$[\, F\,\varphi(k,\mathbf{r})\,]^T(\mathbf{p}) = \varepsilon(k)\,[\varphi(k,\mathbf{r})]^T(\mathbf{p}) = \varepsilon(k)\,\phi(k,\mathbf{p})\,. \tag{38}$$

The kinetic, Coulombic and exchange terms are treated separately :

(a) $T^T(k,\mathbf{p})$, the momentum space equivalent of $T(k,\mathbf{r})$ is easy to obtain and reads as :

$$T^T(k,\mathbf{p}) = [\, -\frac{\nabla^2(\mathbf{r})}{2}\varphi(k,\mathbf{r})\,]^T(\mathbf{p}) = \frac{\mathbf{p}^2}{2}\phi(k,\mathbf{p})$$

$$= \sum_\mu \frac{2\pi}{a}\delta\,[p_z - (\mu\frac{2\pi}{a} + k)]\,\mathbf{p}^2\, u^T_\mu(k,\mathbf{p}_\perp)\,. \tag{39}$$

(b) $C^T(k,\mathbf{p})$, the momentum space form of the electrostatic term $C(k,\mathbf{r})$, requires special attention. To ensure a convergent result, both the electron-electron and electron-nuclear Coulombic contributions must be combined together under the physical constraint that the partial sums represent neutral samples. To obtain $C^T(k,\mathbf{p})$, the momentum space expression for $C(k,\mathbf{r})$, it is convenient to insert the inverse Fourier transform of $|\mathbf{r}_1-\mathbf{r}_2|^{-1}$ and $|\mathbf{r}_1-\mu a\mathbf{e}_z|^{-1}$ in Eq.(28). Then, applying the Fourier transform to $C(k,\mathbf{r}_1)$ gives :

$$C^T(k,\mathbf{p}) = [\, C(k,\mathbf{r})]^T(\mathbf{p})$$

$$= (2\pi)^{-3/2}\int d\mathbf{r}_1\, e^{-i\mathbf{p}.\mathbf{r}_1}\,[\frac{a}{2\pi}\int_{BZ} dk'\,\theta(k')\int d\mathbf{r}_2\,\varphi^*(k',\mathbf{r}_2)\,\varphi(k',\mathbf{r}_2)$$

$$\times\frac{1}{2\pi^2}\int\frac{d\mathbf{q}}{\mathbf{q}^2}(\, e^{i\mathbf{q}.(\mathbf{r}_1-\mathbf{r}_2)} - \sum_\mu e^{i\mathbf{q}.(\mathbf{r}_1-\mu a\mathbf{e}_z)}\,)\,]\,\varphi(k,\mathbf{r}_1)$$

$$= \frac{1}{2\pi^2}\int\frac{d\mathbf{q}}{\mathbf{q}^2}[\,\frac{a}{2\pi}\int_{BZ} dk'\,\theta(k')\int d\mathbf{r}_2\,\varphi^*(k',\mathbf{r}_2)\,\varphi(k',\mathbf{r}_2)\, e^{-i\mathbf{q}.\mathbf{r}_2} - \sum_\mu e^{-iq_z\mu a}\,]$$

$$\times\,(2\pi)^{-3/2}\int d\mathbf{r}_1\, e^{-i(\mathbf{p}-\mathbf{q}).\mathbf{r}_1}\,\varphi(k,\mathbf{r}_1)$$

$$= \frac{1}{2\pi^2}\int \frac{d\mathbf{q}}{q^2}\left[\frac{a}{2\pi}\int dk'\,\theta(k')\,W(k',k',\mathbf{q}) - \sum_{\mu}\frac{2\pi}{a}\delta\left[q_z - \mu\frac{2\pi}{a}\right]\right]\phi(k,\mathbf{p}-\mathbf{q}) . \tag{40}$$

Using the expanded form of $W(k',k',\mathbf{q})$ yields,

$$C^T(k,\mathbf{p}) = \frac{1}{2\pi^2}\int \frac{d\mathbf{q}}{q^2}\sum_{\mu}\frac{2\pi}{a}\delta\left[q_z - \mu\frac{2\pi}{a}\right]$$

$$\times\left\{\frac{a}{2\pi}\int_{BZ} dk'\,\theta(k')\,W(k',k',\mathbf{q}_\mu) - 1\right\}\phi_\mu(k,\mathbf{p}-\mathbf{q}) , \tag{41}$$

and inserting a factor unity in terms of the electroneutrality condition, Eq.(25), leads to the final expression :

$$C^T(k,\mathbf{p}) = \frac{1}{2\pi^2}\sum_{\mu}\int \frac{dq_\perp}{q_\perp^2 + (\mu\frac{2\pi}{a})^2}\int_{BZ} dk'\,\theta(k')\left\{W(k',k',\mathbf{q}_\mu) - 1\right\}\phi(k,\mathbf{p}-\mathbf{q}) . \tag{42}$$

(c) $X^T(k,\mathbf{p})$, the momentum space expression for the exchange term $X(k,\mathbf{r})$ is obtained using similar steps as above for the electrostatic contribution $C^T(k,\mathbf{p})$. The final result for $X^T(k,\mathbf{p})$ is :

$$X^T(k,\mathbf{p}) = -\frac{1}{2\pi^2}\int \frac{d\mathbf{q}}{q^2}\frac{a}{2\pi}\int_{BZ} dk'\,\theta(k')\,W(k',k,\mathbf{q})\,\phi(k',\mathbf{p}-\mathbf{q})$$

$$= -\frac{1}{2\pi^2}\int_{BZ} dk'\int \frac{dq_\perp}{q_\perp^2 + (\mu\frac{2\pi}{a} + k-k')^2}\theta(k')\,W(k',k,\mathbf{q}_{\mu+k-k'})\,\phi(k',\mathbf{p}-\mathbf{q}_{\mu+k-k'}) . \tag{43}$$

3.2.4. *Total energy.* As in the case of the electrostatic contribution $C^T(k,\mathbf{p})$ to the Fock operator, the derivation of the expression for total energy E requires special attention in dealing with the lattice summations. These problems have been pointed out in details elsewhere [55,56], thus in this contribution we simply give the main equations leading to the final expression of E.

$$E = \frac{1}{2}\int_{BZ} dk\,\theta(k)\left[\varepsilon(k) + T(k)\right] + U + V \tag{44}$$

where T(k) is given by

$$T(k) = \int d\mathbf{p}\, \phi^*(k,\mathbf{p}) \left[\frac{p^2}{2}\right] \phi(k,\mathbf{p}) \tag{45}$$

and the combination U+V of the nuclear repulsion, U, and electron-nuclear attraction,V, terms is :

$$U + V = \int d\mathbf{q}_\perp \Big[\sum_\mu \frac{1}{\mathbf{q}_\perp^2 + (\mu\frac{2\pi}{a})^2} - \frac{\pi}{|\mathbf{q}_\perp|} - \frac{\int_{BZ} dk\, \theta(k)\, W(k,k,\mathbf{q}_\perp)}{\mathbf{q}_\perp^2}$$

$$- \sum_{\mu\,(\neq 0)} \frac{\int_{BZ} dk\, \theta(k)\, W(k,k,\mathbf{q}_\mu)}{\mathbf{q}_\perp^2 + (\mu\frac{2\pi}{a})^2} \Big]. \tag{46}$$

3.3. COMPUTATIONAL ASPECTS.

As seen above, applying the Fourier transform to the direct space form of the Bloch states leads to the corresponding expression in momentum space. It is worth stressing again that, while $\varphi(k,\mathbf{r})$ is a continuous function of x, y and z, its momentum transform $\phi(k,\mathbf{p})$ is continuous only in p_x and p_y and discrete in p_z. In momentum space, translational symmetry decouples p_z from the other two variables p_x and p_y. The discrete part can be represented as a set of equally spaced delta functions with weights $u^T(k,\mathbf{p})$ at $p_z = \mu\text{-}k$ (n = 0, ±1, ±2, ...). These functions as well as the descending quantities such as $W(k',k,\mathbf{p}_\mu)$ depend upon four variables : the wave number k, two continuous variables p_x and p_y and the lattice index μ. We found more convenient to work with polar coordinates for the variables $p_x$and p_y, *i.e.* $(p_x, p_y) \rightarrow (p,\theta)$. The ranges of p and θ, respectively, are $0 \le p \le \infty$ and $0 \le \theta \le 2\pi$. These features of infinite chains lead to specific choices for the numerical integration techniques. These are briefly indicated now.

3.3.1. *Integration over k.* In the application we have made on the infinite chain of hydrogen atoms [14,52], we have performed the numerical integrations over k in the Brillouin zone using the standard Gauss-Legendre method.

3.3.2. *Integration over* p *and* θ. To perform the corresponding numerical integrations, we have adapted to the two-dimensional case the procedure originally developed for molecules. The p-abscissas correspond to the nodes of the Gauss-Legendre quadrature in a finite interval. The θ-abscissas are evenly distributed on circles intersecting the radial axis at the p-nodes.

3.3.3.*The lattice summations.* As a consequence of the lattice orthogonality property, the integration over the variable $p_z = \mu\text{-}k$ reduces to an infinite summation of the functions defined at the points of a Dirac comb. The summation limits are, in principle, $-\infty$ and $+\infty$. In the case of the linear chain of hydrogen atoms [52], the terms of the lattice summations were of fast decay and could be summed directly. In general, however, one should be prepared to use acceleration techniques.

3.3.4. *Singular integrands.* The divergencies arising from the singularities at $\mathbf{q}_\mu = 0$ ($q_x = q_y = \mu = 0$) in some of the integrands, Eqs.(42) and (46), of the classical electrostatic contributions have been removed by combining terms equal in magnitude and of opposite signs. This is a well-known problem to which solutions have been proposed [54,55].

Much more cumbersome is the treatment of the singularities occuring in the exchange integrands which, contrary to the electrostatic situation, cannot be reduced on the basis of physical criteria such as the electroneutrality constraint which leads to the aforementioned combination of cancelling terms. The integrand in Eq.(43) is unbounded when $\mathbf{q}_{\mu+k-k'} \to 0$, however after integration the resulting contributions from the singular regions are finite. To solve this problem, we have used analytically fitted curves which we could then integrate explicitly [52].

The infinite chain of hydrogen atoms has been treated within a sphere of radius equal to 1 au for the radial coordinate. In that region, which corresponds to the inner region in the molecular cases discussed in section 2, 31 p-points and 6 θ-abscissas have been used. The number of k-points used in half the Brillouin zone was equal to 8.

3.4. NUMERICAL RESULTS.

So far the only application of this formalism to polymers has been on the numerical evaluation of the integrals entering the equations for the infinite chain of hydrogen atoms. The Bloch states were approximated as a linear combination of Gaussian functions to be compared with previously reported results based on the classical LCAO-CO approach. The Bloch orbital $\phi(k,p)$ writes as :

$$\phi(k,p) = N_k \sum_{\mu} \delta[p_z - (\mu\frac{2\pi}{a} - k)] \sum_{a=1}^{m} d_a (\frac{2\pi}{\alpha_a})^{3/4} \exp(-\beta_a p^2) \quad (47)$$

N_k is a normalization factor, d_a the coefficients of the Gaussian expansion and the β_a's are the Gaussian exponents in momentum space. The correspondence between the configuration space and momentum space exponents is straightforward and can be found in Ref. 53. In the illustrative case given below, we have selected the case m = 3 which corresponds to the usual STO-3G representation of the 1s atomic function. In Table III are compared the values of the Coulombic, C(k), exchange, X(k), and kinetic, T(k), energy contributions to the energy band ε(k) computed with a fully numerical procedure and the previously published results obtained with an algebraic approach [54].

TABLE III.: Selected values (in au) of the energy band ε(k) for $(\text{-H-})_\infty$ [53]. The Bloch are expressed as a linear combination of a single 1s (STO-3G) function . The results of the algebraic approach are given in italic.

k (in units of 2π/a)	C(k)	X(k)	T(k)	ε(k)
0.046	-0.4332	-0.6293	0.3570	-0.7056
	-0.4289	*-0.6365*	*0.3571*	*-0.7083*
0.131	-0.4731	-0.6009	0.4880	-0.5860
	-0.4655	*-0.6114*	*0.4871*	*-0.5926*
0.199	-0.5367	-0.5413	0.6939	-0.3841
	-0.5318	*-0.5531*	*0.6933*	*-0.3915*
0.240	-0.5907	-0.4643	0.8687	-0.1864
	-0.5860	*-0.4793*	*0.8684*	*-0.1856*

The main merit of this application has been the demonstration of the feasibility of fully numerical approaches for infinite polymers. Further works along this direction are needed before fully numerical momentum space band structure calculations of polymers become truly possible.

4. Conclusion.

Comparing with the situation in the middle of the century and noticing the rapid changes that have occured since 1981, we can be rather optimistic as to the future of numerical momentum space calculations of molecular structure.

In spite of its merits, the LCAO expansion technique, especially when large numbers of basis functions are involved, raises questions as to the actual convergence toward the exact solutions. At present, the momentum space approach looks as a promising direction to deal with polyatomic systems which constitute a difficult barrier to jump by numerical calculations in configuration space. The full numerical treatment of the H_3 molecule at the Hartree-Fock level has undisputably demonstrated the feasibility of such calculations in momentum space. However, the field is far from mature and approaches other than those already reported in these proceedings will be most welcome, and it can be anticipated that the increased awareness of the potential of fully numerical calculations in momentum space will boost such investigations.

It must also be pointed out that attempts to include correlation have already been reported, showing that it is quite possible to go beyond the one-particle schemes and carry out highly accurate calculations on polyatomic molecules.

Similarly, preliminary investigations on extended systems have been made which have shown that polymers are amenable to fully numerical treatments. This opens the prospect of going beyond sterile quarrels between solid state physicists privileging the use of localized basis functions and those preferring delocalized functions.

The quest for efficient algorithms toward fully numerical calculations should not deter research works from intermediate approaches which, at relatively low cost and with easier control on the numerics, could lead to substantial qualitative improvements on the existing LCAO approximate solutions.

Finally, an experimental counterpart of these theoretical developments can also be found in the promising results obtained by new techniques giving direct information on molecular electron momentum distribution, i.e., the (e, 2e) spectroscopy.

Appendix: Symmetry in momentum space

A consequence of Fourier transformation is that a position wavefunction $\psi(\mathbf{r})$ and a momentum wavefunction $\varphi(\mathbf{p})$ are different representations of the same Hilbert space, they possess the same eigenvalues for all operators and in particular for the symmetry operators of the molecular point group. Thus, $\psi(\mathbf{r})$ and $\varphi(\mathbf{p})$ belong to the same irreducible representation .

A point group operator T changes $\psi(\mathbf{r})$ into $\psi(\mathbf{r}') = \psi(T^{-1}\mathbf{r})$ with components of $\mathbf{r}'$ vector of the form:

$$r_k' = \sum_{j=1}^{3} t_{kj} r_j \quad ; \; j = 1, 2, 3 \tag{A1}$$

where t_{kj} denote the T^{-1} matrix elements. The Fourier transform of $\psi(\mathbf{r}')$ is given by:

$$[\psi(\mathbf{r}')]^T(\mathbf{p}) = (2\pi)^{-3/2} \int e^{-i\mathbf{p}.\mathbf{r}'} \psi(\mathbf{r}')\, d\mathbf{r}' \, . \tag{A2}$$

The volume element $d\mathbf{r}'$ is equal to $d\mathbf{r}$ since T is a unitary operator, and the scalar product $\mathbf{p}.\mathbf{r}'$ can be expressed as follows,

$$\mathbf{p}.\mathbf{r}' = \sum_k p_k \sum_j t_{kj} r_j = \sum_j r_j \sum_k p_k t_{kj} = \mathbf{r}.\,(T\mathbf{p})\, , \tag{A3}$$

where T is the transposed matrix of T^{-1}; therefore $[\psi(T^{-1}\mathbf{r})]^T(\mathbf{p})$ is given by:

$$[\,\psi(T^{-1}\mathbf{r})]^T(\mathbf{p}) = (2\pi)^{-3/2} \int e^{-i(T\mathbf{p}).\mathbf{r}} \psi(T^{-1}\mathbf{r})\, d\mathbf{r} \tag{A4}$$

So, if a point group operator T changes $\psi(\mathbf{r})$ into $\psi(T^{-1}\mathbf{r})$ then T^{-1} changes $\varphi(\mathbf{p})$ into $\varphi(T\mathbf{p})$. Consequently, if an orbital $\varphi(\mathbf{r})$ has a certain symmetry, say "A", under T, an orbital $\varphi(\mathbf{p})$ of symmetry "A" under T will correspond to it in momentum space.

Let us examine now the existing relations between **r**- and **p**- space electronic distributions. An electronic distribution is simply the square of the wavefunction; it writes:

$$\rho(\mathbf{r}) = \psi^*(\mathbf{r})\, \psi(\mathbf{r}) \quad \text{in position space} \tag{A5a}$$

$$\rho(\mathbf{p}) = \varphi^*(\mathbf{p})\, \varphi(\mathbf{p}) \quad \text{in momentum space} \tag{A5b}$$

Although no simple relation exists between these two quantities and, consequently, no immediate correspondence between both spaces, some symmetry relations remain which shall be considered now.

Eqs.(A5a) and (A5b) show that if the starting wave function is any real function, the momentum wave function may be complex -its complex conjugate $\varphi^*(\mathbf{p})$ being $\varphi(\mathbf{p})$ - but the momentum distribution is always real. Eq.(A5b) can be written in the following form:

$$\rho(\mathbf{p}) = [\mathrm{Re}\varphi(\mathbf{p})]^2 + [\mathrm{Im}\varphi(\mathbf{p})]^2 = [\mathrm{Re}\varphi^*(\mathbf{p})]^2 + [\mathrm{Im}\varphi^*(\mathbf{p})]^2 \quad \text{(A6)}$$

For an electron momentum in the opposite direction, one has:

$$\rho(-\mathbf{p}) = [\mathrm{Re}\varphi(-\mathbf{p})]^2 + [\mathrm{Im}\varphi(-\mathbf{p})]^2 = \rho(\mathbf{p}) \quad \text{(A7)}$$

Hence, the momentum distribution will contain an inversion center.This inversion symmetry is necessary to ensure that the space function $\psi(\mathbf{r})$ corresponding to a given momentum function $\varphi(\mathbf{p})$ has no net translational motion (otherwise the molecule would fall apart) [33].

The qualitative similarity in both representations implies that atoms have s-, p- and d-type atomic orbitals which have the same appearence in both spaces. But things are not always so simple: the additional inversion center at the p-space origin might lead to ambiguities in interpreting experimental results (e.g. (e,2e) measurements directly related to momentum densities [7]): ethylene (D_{2h} in r-space) and formaldehyde (C_{2v} in r-space) have both D_{2h} symmetry in p-space [57]; carbonyl sulfide ($C_{\infty v}$ in r-space) and carbon dioxide or carbon disulfide ($D_{\infty h}$ in r-space) have both $D_{\infty h}$ symmetry in p-space [58].Therefore they lead to similar spectroscopic patterns.

Acknowledgements.

Two of us (M.D. and J.D.) acknowledge with appreciation the support of this work under the agreements (1988) for scientific cooperation between the Commissariat Général aux Relations Internationales de la Communauté Française de Belgique and the CNRS-France. The authors thank the Ecole Polytechnique (Palaiseau-France) for giving access to computing facilities on the Cray I.

References.

1. E.A. Hylleraas, Z. Physik, **74**, 216 (1932)
2. H.A. Bethe, E.E. Salpeter, *Quantum Mechanics of One and Two-Electron Atoms* (Springer-Verlag, Berlin, 1957), pp. 36-47
3. R. McWeeny, C.A. Coulson, Proc. Phys. Soc., **A62**, 509 (1949)
4. R. McWeeny, Proc. Phys. Soc., **A62**, 519 (1949)
5. N.V. Svartholm, *The Binding Energies of the Lightest Atomic Nuclei*, Thesis, Lund (1945)
6. W.E. Duncanson, C.A. Coulson, Proc. Phys. Soc., **A57**, 190 (1945)
7. C.E. Brion, Int. J. Quantum Chem., **29**,1397 (1986)
8. E.N. Lassettre, J. Chem. Phys., **58**, 1991 (1973)
9. J. Navaza, G. Tsoucaris, Phys. Rev.**A24**, 683 (1981)

10. C. Carter, Proc. Roy. Soc., **A235**,321 (1956)
11. See, e.g., E.A. McCullough Jr., J. Chem. Phys., **62**, 3991 (1975), L. Laaksonen, P. Pyykkö, D. Sundholm, Chem. Phys. Lett., **96**,1 (1983) and subsequent papers
12. M. Defranceschi, M. Suard, G. Berthier, C. R. Acad. Sci. Paris, **299**, Serie II, 9 (1984)
13. M. Defranceschi, M. Suard, G. Berthier, C. R. Acad. Sci. Paris, **296**, Serie II, 1301(1983) and Int. J. Quantum Chem. **25**, 863 (1984)
14. M. Defranceschi, J. Delhalle, C. R. Acad. Sci. Paris, **301**, Serie II, 1405 (1985)
15. M. Defranceschi, Chem. Phys. to be published
16. M. Defranceschi, A. Lahmam-Bennani, J. Electr. Spectr. Rel. Phenom. (in press)
17. C.C.J. Roothaan, Rev. Mod. Phys., **23**, 69 (1951)
18. G.G. Hall, Proc. Roy. Soc., **A205**, 541 (1951)
19. G. Berthier, C. R. Acad. Sci. Paris, **238**, 91 (1954); J. Chim. Phys., **51**, 363 (1954) and **52**, 141 (1955)
20. J.A. Pople, R.K. Nesbet, J. Chem. Phys., **22**, 571 (1954)
21. C.C.J. Roothan, Rev. Mod. Phys., **32**, 179 (1960)
22. J. Hinze, J. Chem. Phys., **59**, 6424 (1973)
23. A. Veillard, Theoret. Chim. Acta, **4**, 22 (1967)
24. D.R. Hartree, W. Hartree, B. Swirles, Phil. Trans. Roy. Soc., **A238**, 229 (1939)
25. G. Das, A.C. Wahl, J. Chem. Phys., **44**, 87 (1970)
26. A. Veillard, E. Clementi, Theoret. Chim. Acta, **7**, 133 (1967)
27. G. Del Re, J. Ladik, G. Biczo, Phys. Rev., **155**, 977 (1967)
28. J.M. André, L. Gouverneur, G. Leroy, Int. J. Quant. Chem., **1**, 427 (1967)
29. B. Liu, J. Chem. Phys., **58**, 1925 (1973)
30. P.J. Reynolds, R.N. Barnett, W.A. Lester Jr., Int. J. Quant. Chem. Symp., **18**, 709 (1984)
31. G. Berthier, M. Defranceschi, J. Delhalle, *Self-Consistency Field: Theory and Applications*, Edited by R. Carbo and M. Klobukowski (Elsevier, The Nederlands, 1988), Studies in Physical and Theoretical Chemistry
32. J. Higuchi, J. Chem. Phys., **26**, 151 (1957)
33. P. Kaijser, V.H. Smith Jr., *Quantum Chemistry Methods and Structure. A tribute to P.O. Löwdin,* Eds. J.L. Calais, O. Goscinski, J. Linderberg and Y. Öhrn (Plenum Press, New York, 1976), p. 417
34. M. Defranceschi, G. Berthier, submitted for publication
35. V.A. Fock, Z. Phys., **98**, 145 (1935)
36. H.J. Monkhorst, B. Jeziorski, J. Chem. Phys., **71**, 5268 (1979)
37. B.K. Novosadov, Opt. Spectrosc. **41**, 490 (1976)
38. T. Shibuya, C.E. Wulfman, Proc. Roy. Soc., **A286**, 376 (1965)
39. B.K. Novosadov, A.I. Pogonin, Theoret. Chim. Acta (Berl), **60**, 495 (1982)
40. W. Rodriguez, Y. Ishikawa, Chem. Phys. Lett., **146**, 515 (1987) and Int. J. Quant. Chem.Symp. (1988) in press
41. S.A. Alexander, H.J. Monkhorst, Int. J. Quantum Chem., **32**, 361 (1987).
42. S.A. Alexander, R.L. Coldwell, H.J. Monkhorst, to be published
43. M. Defranceschi, M. Suard, to be published
44. J. Delhalle, M. Defranceschi, Int. J. Quantum Chem. Symp. **21**, 425 (1987)
45. M. Defranceschi, unpublished results
46. W.H. Peirce, Math. Tab. Aid. Comp. **11**, 244 (1957); D. Mustard, Math. Comp., **18**, 578 (1964)
47. W. Kolos, C.C.J. Roothan, Rev. mod. Phys. **32**, 219 (1960)

48. M.E. Schwartz, L.J. Schaad, J. Chem. Phys., **47**, 5325 (1967)
49. D. Sundholm, P. Pyykkö, L. Laaksonen, Mol. Phys., **56**, 1411 (1985)
50. L. Laaksonen, D. Sundholm, P. Pyykkö, Chem. Phys. Lett., **105**, 573 (1984)
51. J. Delhalle, J.G. Fripiat, M. Defranceschi, Annales Soc. Scient. Bruxelles, **101**, Fasc. I, 9 (1987)
52. J. Delhalle, M.H. Delvaux, J.G. Fripiat, J.M. André, J.L. Calais, J. Chem. Phys., **88**,3141 (1988)
53. M. Defranceschi, J. Delhalle, Phys. Rev. **B34**, 5862 (1986)
54. J. Delhalle, F.E. Harris, Phys. Rev., **B31**, 6755 (1985)
55. F.E. Harris, J. Chem. Phys., **56**, 4422 (1972)
56. F.E. Harris, *Theoretical Chemistry, Advances and Perspectives*, edited by D. Henderson and H. Eyring (Academic Press, New York, 1975), Vol 1, pp. 147-218
57. D.C.Rawlings, E.R. Davidson, J. Phys. Chem.**89**, 969(1985)
58. K.T. Leung, C.E. Brion, Chem. Phys., **96**, 241(1985)

Should Numerical Orbital Calculations be done with Basis Sets?

S.A. Alexander and H.J. Monkhorst
Quantum Theory Project
Departments of Physics and Chemistry
University of Florida
Gainesville, Fl 32611

Abstract. The two most common methods of solving the Hartree-Fock equations, basis sets and numerical orbitals, are in many ways complementary. Today basis set calculations are relatively fast but for most molecular orbitals their accuracy is rather limited. In contrast, numerical orbital calculations are still relatively slow but they have been able to achieve highly accurate results in a diverse assortment of systems. We outline how the best properties of these two approaches could be combined.

1. Introduction

Numerical orbital calculations must, by necessity, be performed on a grid. For atoms the Hartree-Fock equations reduce to a problem involving only one variable and excellent results can be obtained with a small number of grid points [1]. For diatomic systems elliptic coordinates can be used to recast the Hartree-Fock equations as a two variable problem. Although a larger number of points are needed to obtain accurate results [2,3] the amount is easily manageable on most computers. For general polyatomic systems, however, one is faced with a set of equations containing three independent variables and no 'best' choice of grid. In addition, the position space representation of these equations contain a number of singularities. This last

M. Defranceschi and J. Delhalle (eds.), Numerical Determination of the Electronic Structure of Atoms, Diatomic and Polyatomic Molecules, 239–244.

difficulty can be solved by transforming everything into momentum space where the usual integrodifferential form of the Hartree-Fock equations is changed into a series of convolution integrals [4] which contain only one, removeable singularity at the origin. Recently a number of calculations have been performed in this representation but no particular grid has proven overwhelmingly successful, and the two most common choices, spherical [4-7] and Cartesian [8,9] grids, each exhibited a number of disadvantages. Perhaps the most serious problem is the inability of either grid to efficiently cover a large volume of space. In ref. 8 we showed that a large number of equally spaced points on a Cartesian grid could be used to accurately compute some properties of simple two-electron systems and that the truncation error, the contribution from outside the grid, was the major source of error. In principle, this error can be reduced by simply using larger grids but, as Delhalle and Defranceschi have shown [10], the volume needed to obtain even modest accuracies can sometimes be enormous. Such large regions of space require proportional increases in the number of grid points though, and so this idea quickly becomes computationally prohibitive. Clearly some other solution to this problem is needed.

2. A mixed numerical/analytic formalism

We propose to reduce the computational requirements of accurate numerical orbital calculations by writing each Hartree-Fock orbital as a sum of an analytic part, ϕ^a, and a numerical part, ϕ^n

$$\phi_i \, (p) = \phi_i^{\,a} \, (p) + \phi_i^{\,n} \, (p) \tag{1}$$

This ansatz can be implemented in two very different ways. The first approach is to use the analytic part of this function to cover the region outside the grid. Since the asymptotic behavior of a Hartree-Fock orbital is known [11,12],

$$\phi_i(\mathbf{p}) \underset{\mathbf{p}\to\infty}{\simeq} \sum_{a=1}^{\#\,atoms} Z_a \frac{e^{i\,\mathbf{p}.\mathbf{R}_a}}{(p^2 - 2\,\varepsilon_i)\,p^2} \phi_i(\mathbf{R}_a) + O(\frac{1}{\mathbf{p}^6}) \tag{2}$$

this combination will certainly reduce the truncation error. One could perform such a calculation either by first computing the fully numerical result as before and then fitting the analytic part on at the end (using perhaps a least-square type procedure) or by using equation 1 in a full self-consistent calculation with some perscription for matching the numerical part and the analytic part. The disadvantage of either method is that equation 2, because of its form, has to be integrated numerically from the edge of the grid out to infinity. Such calculations scale like M^2 where M is the total number of points. Furthermore, any numerical integration technique, e.g. a Gauss-Laguerre quadrature, could just as easily be used to perform a purely numerical calculation. Finally this combination would destroy any possibility of using the Fast Fourier Transform to evaluate the convolutions, a technique which scales like $N \log_2 N$ where N is the total number of grid points.

The alternative is to let the analytic part range over both the grid and region outside the grid. In this approach the analytic part acts very much like a control deviate [13]. This method is most commonly used in Monte Carlo integrations and involves subtracting a simple function from the integrand to be computed

$$\int f(x)\,dx = \int [f(x)-g(x)]\,dx + \int g(x)\,dx \tag{3}$$

If the function g(x) is a good approximation to the integrand and can be integrated analytically, then the numerical evaluation of f(x)-g(x) should be more accurate than the evaluation of f(x) alone. In the case of a numerical Hartree-Fock calculation, such a scheme should enable us to reduce the size of the grid and/or the number of points needed to achieve a given accuracy. This technique would also reduce the truncation error since the analytic part provides some information about the behavior of the orbital at large momenta. In addition, this method allows us to use Fast Fourier Transforms to evaluate convolutions. Such integrals are simply separated into an analytic part and a numerical part e.g.

$$\int \phi^*_i(q)\ \phi_j(q+p)\ d^3q = [\ \phi^{*a}_i(r)\ \phi^a_j(r)\ ]^T(p) + \\ [\ \phi^{*a}_i(r)\ \phi^n_j(r) + \phi^{*n}_i(r)\ \phi^a_j(r) + \phi^{*n}_i(r)\ \phi^n_j(r)\ ]^T(p) \tag{4}$$

The one requirement of this approach is that all the integrals involving only the analytic part have algebraic forms. Thus in equation 4 the first Fourier transform is a known expression and the Fourier transform involving the mixture of numerical and analytic parts is solved as before using the Fast Fourier Transform. Although this restriction on the choice of the analytic part may appear severe, a number of interesting possibilities have not been excluded. One choice is an expansion of Gaussian-type orbitals from a small Hartree-Fock calculation. Others include one of the pseudopotential expressions developed to treat core orbitals. One possible problem with this procedure is that it requires the analytic part to be evaluated at each grid point. If a large number of grid points are needed, these calculations could become a major computational expense.

3. Conclusions

We have outlined a method of combining the best features of basis sets and numerical orbitals. In this procedure the numerical part is used to correct for the inadequacies of the analytic part and the analytic part is used to approximate the overall shape of each Hartree-Fock orbital so that in principle, a smaller grid size and/or fewer numerical points are needed to accurately represent these functions. Because the analytic part also supplies information about the behavior of the orbital at large momenta, the truncation error should be reduced. In addition, this procedure can be applied with equal benefit to either of the two major algorithms now in use, those which scale like N $\log_2$ N or M^2. While it is still too early to say just how effective this idea of mixing analytical and numerical calculations will be for the Hartree-Fock problem, similar applications in other areas seem encouraging [14].

Acknowledgements

We wish to thank Dr. R.L. Coldwell for many valuable discussions with the authors. This work was supported by a PACER fellowship from Control Data.

References

[1] C. Froese Fischer, The Hartree-Fock Method for Atoms (Wiley, New York,1977) and references therein.

[2] E.A. McCullough Jr., Comp. Phys. Rep. 4,265 (1986) and references therein.

[3] L. Laaksonen, P. Pyykko and D. Sundholm, Comp. Phys. Rep. 4,315 (1986) and references therein.

[4] J. Navaza and G. Tsoucaris, Phys. Rev. A 24,683 (1981).

[5] M. Defranceschi, M. Suard and G. Berthier, Comptes Rendus 299,9 (1984).

[6] M. Defranceschi, M. Suard and G. Berthier, Int. J. Quantum Chem. 25,863 (1984).

[7] G. Berthier, M. Defranceschi, J. Navaza, M. Suard and G. Tsoucaris, Theochem 120,343 (1985).

[8] S.A. Alexander and H.J. Monkhorst, Int. J. Quantum Chem. 32,361 (1987).

[9] S.A. Alexander, R.L. Coldwell and H.J. Monkhorst, to appear J. Comp. Phys.

[10] J. Delhalle and M. Defranceschi, Int. J. Quantum Chem. S21,425

(1987).

[11] E.N. Lassettre, J. Chem. Phys. 43,4479 (1965); 57,4357 (1972); 70,3468 (1979); 72,2374 (1980); 74,1225 (1981); 83,1709 (1983); Comments At. and Mol. Phys. 16,35 (1985); this proceedings.

[12] C.L. Davis, H.J.A. Jensen and H.J. Monkhorst, J. Chem. Phys. 80,840 (1984).

[13] J.M. Hammersley and D.C. Handscomb, Monte Carlo Methods (Wiley, New York, 1964).

[14] See for example P.J. Feibelman, J. Chem. Phys. 81,5864 (1984).

QUANTUM CHEMISTRY BY RANDOM WALK: HIGH ACCURACY FOR LARGE MOLECULES

JAMES B. ANDERSON
Department of Chemistry
The Pennsylvania State University
University Park, Pennsylvania 16802

ABSTRACT. In random walk calculations for the systems methane and F-H-H more than 96 percent of the correlation energy has been recovered when node structures are taken from SCF calculations. The energies obtained are significantly lower than those of the lowest-energy variational calculations and lie only about 5.0 and 9.0 kcal/mole, respectively, above the experimentally determined values for these systems. It appears that similar accuracies may be obtained in calculations for molecular systems of many more electrons without greatly increasing the computation effort required.

1. Introduction

The chemical reaction F + HH ---> HF + H provides an excellent example of a recurring problem in chemical kinetics and molecular dynamics, as well as in physical chemistry in general: the difficulty in obtaining accurate potential energy surfaces. For this reaction a large number of experiments of many types have yielded a wealth of detailed experimental data. Most of the experimental observations have been explained qualtitatively, but a number of questions remain. The answers to these and the complete theoretical description of the reaction await a very accurate potential energy surface (possibly surfaces) for kinetics and dynamics calculations of several kinds. The F-H-H system is just one of hundreds of systems which might be completely understood if accurate surfaces could be computed.

In the past year (1,2) we have found it possible to determine electronic energies for small molecules such as methane, hydrogen fluoride, and the reaction intermediate F-H-H with high accuracies by using the random walk method. The energies obtained for these systems are 30 to 50 kcal/mol lower than those from the most accurate variational calculations to date. The errors due to inexact node locations are less than 10 kcal/mol and the 'statistical' uncertainties are less than 2 kcal/mol. Nevertheless, there remain a number of opportunities for further improvement of the random walk method which are under investigation in several laboratories and many of these will be successful.

M. Defranceschi and J. Delhalle (eds.), Numerical Determination of the Electronic Structure of Atoms, Diatomic and Polyatomic Molecules, 245–250.

The methane and F-H-H systems contain second-row atoms and totals of ten and eleven electrons. In considering applications to larger systems, those with a greater number of electrons, it is important to distinguish between systems containing heavier atoms and systems containing merely more atoms of the second row. The latter includes most of the chemical species important in the atmosphere (oxygen and nitrogen molecules, water, ozone), in biological systems (the 'organics' containing carbon, hydrogen, and oxygen), and in industrial processes. It is for these systems that it is now possible to consider all-electron quantum calculations by random walk methods.

In fixed-node random walk calculations the nodes are usually specified by a trial or 'guide' wavefunction expressed in the form of a Slater determinant or a series of Slater determinants. For the case of methane the determinant is a 10x10 determinant, but in practice this reduces to the product of two determinants, each one of dimension 5x5. The calculation effort required for the random walk calculation is largely the effort associated with the evaluation of these determinants and their derivatives with respect to electron positions. For determinants of modest size the effort is roughly proportional to m factorial, where m is the dimension of the determinant. The dimension m is about equal to N/2 where N is the number of electrons in the system. Thus, one might expect that the effort for any system would vary as (N/2) factorial which increases very rapidly with the number of electrons. We show below that the effort for extended systems of second-row atoms is not as great as this suggests and that accurate results for such systems might be obtained without an excessive computation effort.

2. Recent Results for Methane and F-H-H

Simulation of the Schrodinger equation by a random walk process, an idea attributed to Fermi(3), was first investigated in the late 1940's, but it was not until the 1970's that faster computers and refined techniques made it practical for applications in determining the energies of atomic and molecular systems. The first applications to polyatomic ions(4) and molecules(5), to systems of more than one fermion(5), and to excited states(5) made use of what is now called the fixed-node constraint and gave energies as precise as 0.001 au (0.6 kcal/mol). With use of importance sampling as introduced by Grimm and Storer(6) for a related method the accuracy was improved and the number and variety of applications were greatly increased(7,8). Although several methods of releasing nodes have been suggested and have been successfully applied to a few systems, the use of fixed nodes appears at present to be the only practical way of treating systems of ten or more electrons. We have found for methane, hydrogen fluoride, and F-H-H that nodes taken from SCF calculations give remarkably accurate results. Since details have been reported previously(1,2) only a short summary is given here.

Results for these systems are given in Tables I and II and in Figure 1. Although calculations were made for F, HF, and F-H-H using the same basis set for the SCF node structure of each, we list only the result for F + HH which may be compared

TABLE I. Energies for methane.

Method	Energy (au)	Correlation Energy (percent)	Reference
Hartree-Fock			
Minimal SZ STO's	-40.1282		9
DZ STO's	-40.1728		1
Extended basis	-40.2142		10
Estimated limit	-40.219	0	11
Variational			
Extended-basis CI	-40.4584	81	M 10
Random Walk			
DZ STO's	-40.506 (0.002)	97	GA 1
Experimental			
(Nonrelativistic)	-40.514 (0.002)	100	11,12

TABLE II. Energies for separated F + HH .

Method	Energy (au)	Correlation Energy (percent)	Reference
Hartree-Fock			
Estimated limit	-100.5349	0	13
Variational			
SRCI	-100.5425	2	BPOS 14
SRCI	-100.5620	7	BOPS 15
MRCI	-100.7126	48	USL9 16
SRCI	-100.7975	71	USL8 16
MP4	-100.7892	69	FBS 13
MRCI	-100.7875	68	SSBT 17
FCI	-100.7101	47	BT 18
MRCI	-100.8221	77	BWLTJ 19
Random Walk			
SCF nodes	-100.891 (0.002)	96	GA 2
Experimental			
(Nonrelativistic)	-100.906 (0.002)	100	20

directly with the experimentally determined energy. The total electronic energy of methane is about -40 au and that for F + HH is about -100 au. Since these correspond to approximately 25,000 and 63,000 kcal/mol, attaining accuracies of better than 1 kcal/mol is not a trivial pursuit. As indicated in the tables, the correlation energy recovered for methane is 97 percent and that for F + HH is 96 percent. The calculated energies for both are well below those from the best available variational calculations. Since there is little room left for error between the calculated energies and the experimental energies, one may expect relative values of the energies for various structures of the two systems to be accurate. In subtracting the energy for methane with one bond length from that for methane with another bond length one may expect a partial cancellation of the error due to inexact node structure for each of the species. For F + HH vs. HF + H the calculated energy difference matches that from experiment within 2.0 kcal/mol.

3. Simplifications for Larger Systems

If the successes with methane and the F-H-H systems can be improved somewhat and can be repeated for extended systems of the same or similar atoms, the results for these larger systems will have immediate applications in the prediction of chemical reaction paths, reaction rates, equilibrium structures, and thermodynamic properties. Fortunately, the problem of greater numbers of electrons is not as difficult as it might seem for these systems.

In an all-electron random walk calculation the typical trial wavefunction is, as noted above, normally written as a product of two determinants multiplying additional Jastrow terms, but it may as easily be written as the product of two series of permutations multiplying the additional terms. For an extended system of atoms as in the molecules propane and butane most of the terms in the series are negligible in comparison to the leading terms for typical electron distributions. In most cases the magnitude of a term is easily estimated and it is unnecessary to complete a full calculation of its value to determine if it may be neglected. If calculation of the negligible terms is eliminated the computation effort is much more manageable.

Two simple examples serve to illustrate the reduction in effort which may be obtained. Consider first a system of two helium atoms, the helium dimer with two nuclei A and B, electrons 1 and 2 of one spin, and electrons 3 and 4 of the other. An excellent trial function is the product A13xB24 of those for two helium atoms (identified as A13 for electrons 1 and 3 with nucleus A and B24 for electrons 2 and 4 with nucleus B). With the permutations required for antisymmetry this wavefunction becomes + A13xB24 - A23xB14 - A14xB23 + A24xB13. At the equilibrium internuclear distance of 5.6 au the relative values of these terms for a typical electron distribution are 1.0E-00, 1.0E-07, 1.0E-07, 1.0E-14. The last may be neglected.

For the helium trimer with nuclei A,B,C and electrons 1,2,3 of one spin and 4,5,6 of the other, a similar treatment results in thirty-six terms of the type A14xB25xC36. If the arrangement of the nuclei is linear with a separation of 5.6 au, there is one term

Figure 1. Comparisons of calculated and measured total electronic energies for methane and for F + HH. Data from Tables I and II. Referred to zero for the total electronic energy.

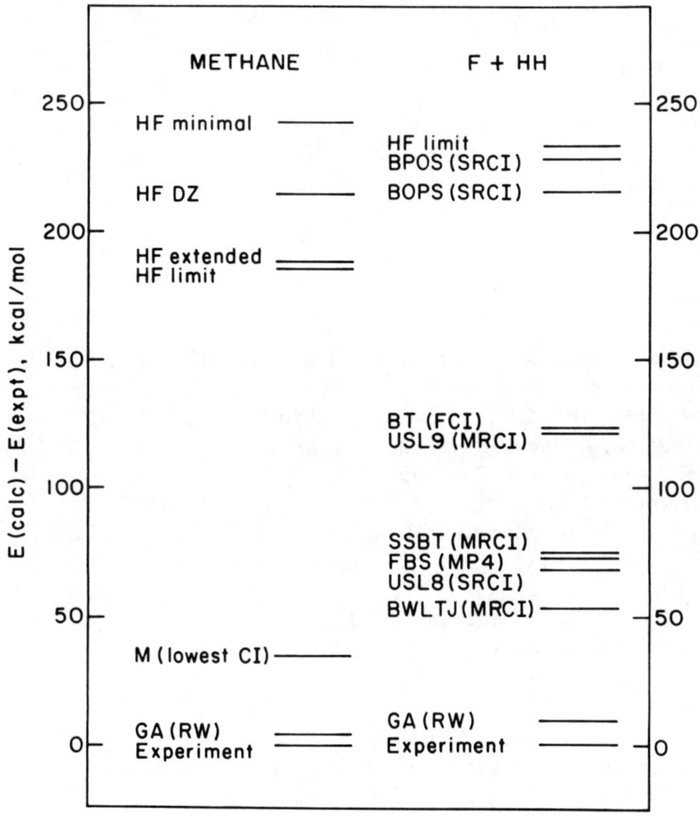

with a value 1.0E-00, four with 1.0E-07, ten with 1.0E-14, twelve with 1.0E-21, and nine with 1.0E-28. Thirty-one of the thirty-six may be neglected.

Other things being equal, the helium dimer requires three times the effort for a single helium atom and the linear trimer requires only five times the effort for the single atom. For the series methane, ethane, propane, and onward similar effects are observed and similar conclusions may be reached. Thus, extended systems of atoms of the same nuclear charge need not be immune to treatment by a complete all-electron random walk calculation.

4. Acknowledgements

Support of this work by the National Science Foundation under Grant CHE-8714613, the American Chemical Society - Petroleum Research Fund under Grant 18854-AC6, and the John von Neumann Computation Center under Grant NAC-826 is gratefully acknowledged.

5. References

(1) D. R. Garmer and J. B. Anderson, J. Chem. Phys. **86**, 4025 (1987).
(2) D. R. Garmer and J. B. Anderson, J. Chem. Phys. **86**, 7237 (1987).
(3) N. Metropolis and S. Ulam, J. Am. Stat. Assoc. **44**, 335 (1949).
(4) J. B. Anderson, J. Chem. Phys. **63**, 1499 (1975).
(5) J. B. Anderson, J. Chem. Phys. **65**, 4121 (1976).
(6) R. C. Grimm and R. G. Storer, J. Comp. Phys. **7**, 134 (1971).
(7) For example: F. Mentch and J. B. Anderson, J. Chem. Phys. **74**, 6307 (1981).
(8) For example: P. J. Reynolds, D. M. Ceperley, B. J. Alder, and W. A. Lester, Jr., J. Chem. Phys. **77**, 5593 (1982).
(9) R. M. Pitzer, J. Chem. Phys. **46**, 4871 (1967).
(10) W. Meyer, J. Chem. Phys. **58**, 1017 (1973).
(11) J. A. Pople and J. S. Binkley, Mol. Phys. **29**, 599 (1975).
(12) A. Veillard and E. Clementi, J. Chem. Phys. **49**, 2415 (1968).
(13) M. J. Frisch, J. S. Binkley, and H. F. Schaefer, J. Chem. Phys. **81**, 1882 (1985).
(14) C. F. Bender, P. K. Pearson, S. V. O'Neill, and H. F. Schaefer, J. Chem. Phys. **56**, 4626 (1972).
(15) C. F. Bender, S. V. O'Neill, P. K. Pearson, and H. F. Schaefer, Science **176**, 1412 (1972).
(16) S. R. Ungemach, H. F. Schaefer, and B. Liu, Discuss. Faraday Soc. **62**, 330 (1977).
(17) D. W. Schwenke, R. Steckler, F. B. Brown, and D. G. Truhlar, Chem. Phys. Lett. **121**, 475 (1985). See also J. Chem. Phys. **84**, 5706 (1986).
(18) C. W. Bauschlicher and P. R. Taylor, J. Chem. Phys. **86**, 858 (1987).
(19) C. W. Bauschlicher, S. P. Walch, S. R. Langhoff, P. R. Taylor, and R. L. Jaffe, preprint (1988).
(20) F. Sasaki and M. Yoshimine, Phys. Rev. **A 9**, 17 (1974).

PROLATE SPHEROIDAL WAVEFUNCTIONS

M. AUBERT-FRECON and G. HADINGER
Laboratoire de Spectrométrie Ionique et Moléculaire
Université Lyon I, Campus de la Doua, Bât. 205
F 69622 Villeurbanne Cedex, France

ABSTRACT. Exact Born-Oppenheimer solutions (Diatomic Orbitals) are derived in prolate-spheroidal coordinates for the one-electron two-center problem. A simple eigenvalues determination process is proposed. Useful relations between the energy and the separation constant eigenvalues are derived using the computer algebraic manipulation language REDUCE.

Such (Shielded) Diatomic Orbitals are seen to be interesting basis functions for the description of diatomic systems such as alkali dimers (when combined with Pluvinage's method) and their positive ions as well as multiply charged diatomic ions involved in molecular theory of ion-atom low-energy collisions.

1. Introduction

The problem of an electron moving in the field of two fixed center of nuclear charges Z_A and Z_B respectively represents the simplest diatomic system for which exact Born-Oppenheimer solutions (that we have named [1] Diatomic Orbitals D.O.) may be derived in prolate spheroidal coordinates. Because of their intrinsic diatomic feature such functions can provide interesting basis sets for the description of more complicated diatomic systems such as multiply charged diatomic ions involved in ion-atom low-energy collisions.

2. Determination of D.O. Eigenvalues

As is well known the Born-Oppenheimer Schrödinger equation for the one-electron two-center problem is separable in prolate spheroidal coordinates ($\mu \in (-1,+1)$, $\lambda \in (1,\infty)$, $\varphi \in (0,2\pi)$) and the wavefunctions may be factorized :

$$\psi(\lambda, \mu, \varphi) = M(\mu, \varphi)\, \Lambda(\lambda)$$

M. Defranceschi and J. Delhalle (eds.), Numerical Determination of the Electronic Structure of Atoms, Diatomic and Polyatomic Molecules, 251–256.

with

$$\left[\frac{d}{d\mu}(1-\mu^2)\frac{d}{d\mu}-\frac{m^2}{1-\mu^2}-R(Z_A-Z_B)\mu+p^2\mu^2-\mathcal{A}\right]M(\mu,\varphi)=0 \qquad (1)$$

and

$$\left[\frac{d}{d\lambda}(\lambda^2-1)\frac{d}{d\lambda}-\frac{m^2}{\lambda^2-1}+R(Z_A-Z_B)\lambda-p^2\lambda^2+\mathcal{A}\right]\Lambda(\lambda)=0 \qquad (2)$$

where R is the internuclear distance, Z_A and Z_B are the nuclear charges, m is the eigenvalue of the z component of the electronic angular momentum L_z, $p^2=-ER^2/2$ and the separation constant $\mathcal{A}$ is the eigenvalue of the following invariant operator $\mathcal{Q}$ [2].

$$\mathcal{Q}=-L^2-\frac{R^2}{4}\left\{\Delta-\frac{\partial^2}{\partial z^2}\right\}-R(Z_A\cos\theta_A-Z_B\cos\theta_B)-p^2.$$

D.O. are the eigenfunctions of L_z, of the B.O. hamiltonian and of $\mathcal{Q}$ and are labelled in the united atom approximation $\psi_{n\ell m}$. Except the work of Bailey [3] in which the B.O. eigenvalues $\mathcal{A}$ and p of one-electron diatomic systems were determined from a method based on the numerical integration of some initial value problems for Eqs.(1) and (2), most of the highly accurate determination methods [4–17] (one of which was proposed by us some ten years ago [1]) are based on expansions of the solution $M(\mu,\varphi)$ and $\Lambda(\lambda)$ in terms of various analytical basis functions

$$M(\mu,\varphi)=\sum_\ell f_\ell F(\mu,\varphi)\ , \qquad \Lambda(\lambda)=\sum_k c_k X_k(\lambda)$$

The pair of coupled differential equations (1) and (2) is then replaced by recurrence formulas for the coefficients f and c respectively and the problem of finding the eigenvalues p and $\mathcal{A}$ reduces to the problem of finding the common roots of either two infinite continued fractions or two Hill determinants.

We proceed here in a somewhat different way by adapting a simple method which was proposed recently by Killingbeck [18] to calculate the energy levels for the perturbed one-dimensional oscillator. Provided that three-term recurrence relations may be derived for the f and c coefficients Killingbeck's method is applicable. Three-term recurrence relations are obtained by choosing the Baber and Hasse [19] expansion

$$F_\ell(\mu,\varphi) = e^{-p\mu}\, P_\ell^m(\mu)\, e^{im\varphi} \tag{3}$$

and the Hylleraas [20] expansion

$$X_k(\lambda) = (\lambda^2-1)^{m/2}\; e^{-p(\lambda-1)}\; L^m_{m+k}(2p(\lambda-1)) \tag{4}$$

or the Jaffé [21] one

$$X_k(\lambda) = (\lambda^2-1)^{m/2}\; (\lambda+1)^{R(Z_A+Z_B)/2p-m-1}\; e^{-p\lambda} \left(\frac{\lambda-1}{\lambda+1}\right)^k \tag{5}$$

The associated Legendre functions $P_\ell{}^m$ and the Laguerre polynomials [Eqs(3–4)] used here correspond to the definitions of Ref. [6].

With these choices, the recurrence relations for the expansions coefficients f and c may both be expressed in the following simplified forms after some straightforward manipulations :

$$g_{\ell+1} + \alpha_\ell\, g_\ell + \beta_\ell\, g_{\ell-1} = 0 \tag{6}$$

$$d_{k+1} + \gamma_k\, d_k + \delta_k\, d_{k-1} = 0 \tag{7}$$

with

$$\alpha_\ell = \mathcal{A} - p^2 + \ell(\ell+1)$$

$$\beta_\ell = (\ell^2-m^2)\,(R^2(Z_A-Z_B)^2 - 4p^2\ell^2)\; /\; (4\ell^2-1) \tag{8}$$

$$f_\ell = g_\ell \prod_{n=m}^{\ell-1} \frac{2n+3}{(n+m+1)(2p(n+1)+R(Z_A-Z_B))}$$

and

$$\gamma_k = \mathcal{A} - p^2 + R(Z_A+Z_B) + (m+1)\,(R(Z_A+Z_B)/2p-1-2p) - - 2k(k+m+1+2p-R(Z_A+Z_B)/2p)$$

$$\delta_k = k(k+m)\,(k-R(Z_A+Z_B)/2p)\,(k+m+1-R(Z_A+Z_B)/2p) \tag{9}$$

$$c_k = d_k \;/ \prod_{t=0}^{k-1} (t+m+1)(t+1-R(Z_A+Z_B)/2p)$$

for the Hylleraas expansion, and

$$C_k = d_k \,/\, \prod_{t=0}^{k-1}(t+m+1)(t+1)$$

for the Jaffé expansion.

By differentiating (6) and (7) with respect to $\mathcal{A}$ and p, recurrence relations are also obtained for the quantities $g'_{\mathcal{A}}$ ($\equiv \partial g_\ell/\partial\mathcal{A}$), g'_p, $d'_{\mathcal{A}}$ and d'_p.

The calculation starts with $g_{\ell<m}=0$, $g_m=1$ and $d_{k<0}=0$, $d_0=1$ and for a set of initial values (p_0, $\mathcal{A}_0$) provides at each step the values of g_ℓ, d_k and their various derivatives. Then the corrections $\Delta\mathcal{A}$ $[=(d_k . g'_p - g_\ell . d'_p)/(g'_{\mathcal{A}} . d'_p - d'_{\mathcal{A}} . g'_p)]$ and Δp are computed. As ℓ and k increase the values of $\Delta\mathcal{A}$ and Δp stabilize fairly well and when this is achieved the run is repeated with a set of corrected values ($\mathcal{A}_i+\Delta\mathcal{A}_i$, $p_i+\Delta p_i$) until the absolute value of the corrections are both less than a given limit. This process first proposed by Kerner [22] is very efficient.

Approximations have been derived up to the third order for the eigenvalue $\mathcal{A}$ from both recurrence relations (6)($\mathcal{A}_\mu{}^*$) and (7)($\mathcal{A}_\lambda{}^*$) by using the computer algebraic manipulation language REDUCE. The quality of these approximations has been systematically tested by comparison with the exact value $\mathcal{A}$ (for the corresponding exact value of p) for the ions $H_2{}^+$ and HeH^{2+} in the range $0<R\ll 30$ a.u. for numerous excited states $\{n\ell m\}$. Briefly stated the range of R for which $\mathcal{A}_\mu{}^*$ is a very good approximation for $\mathcal{A}$ is enlarged (from 0 towards larger values of R) when one goes up through the excited states while this range is reduced for $\mathcal{A}_\lambda{}^*$ (excluding the smallest values of R).

3. D.O. as a Basis Set

3.1. ONE(-ACTIVE)-ELECTRON DIATOMIC SYSTEMS

As expected D.O. have been seen to be very efficient for the theoretical investigation of the electron transfer between hydrogen atoms and fully stripped ions [23] (i.e. C^{6+}, N^{7+}, O^{8+}...). This problem has received increasing attention both experimentally and theoretically during the recent years.

The positive ions of homo and heteronuclear alkali dimers and alkali hydrides as well as quasi molecules of interest in multiply charged ion-atom collisions (i.e. CH^{4+}, NH^{5+}, OH^{6+}...) may be considered as one-active-electron diatomic systems, where the overall interaction effects between the external electron and the atomic cores are represented by a potential model (such as that proposed by Klapisch [24]). The monoelectronic model Schrödinger equation may then be solved in a variational way using D.O. as a basis set. Very satisfactory results have been obtained in this way with compact wavefunctions for numerous states of LiH^+ [25], Li_2^+ [26] and NHe^{6+} [27].

3.2. MANY ELECTRON DIATOMIC SYSTEMS

3.2.1. *Shielded Diatomic Orbitals (S.D.O.).* Following earlier work by Teller [12] S.D.O. have been defined [28–29] as exact solutions in prolate spheroidal coordinates of the Schrödinger equation for one electron moving in the field of two fixed centre plus an additional parametric cylindrical potential representing in some way for a given electron the interactions due to the other electrons of the system. In the one-configuration approach S.D.O. greatly improve the corresponding results obtained with D.O. producing results close to Hartree-Fock reference values (generally derived in the L.C.A.O. approximation).

D.O. and S.D.O. have also been used successfully within the configuration interaction scheme.

3.2.2. *Pluvinage Type Correlated Wavefunctions for Two-Active-Electron Diatomic Systems*. Following an earlier work of Pluvinage [30] for two-active-electron atoms, we have proposed as basis functions for two-active-electron diatomic systems, products of exact solutions of the corresponding one-active-electron diatomic system by exact solutions of the differential equation representing the correlated movement of the two electrons in interaction by the term $1/r_{12}$ and in the absence of the two atomic cores. The monoelectronic part of these functions is described in terms of D.O. while the explicit dependence of the interelectronic distance r_{12} is taken into account through Coulombic functions.

In the one (explicitly correlated) configuration approach, such wavefunctions have provided very interesting results for the ground and various excited states of H_2 [31–32] and He_2^{2+} [33–34] for instance.

References

1. M. Aubert, N. Bessis and G. Bessis, Phys. Rev. **A10**, 51 (1974).
2. K. Helfrich, Theor. Chim. Acta **21**, 381 (1971).
3. P.B. Bailey, J. Chem. Phys. **69**, 1676 (1971).

4. D.R. Bates, K. Ledsham, and A.L. Stewart, Philos. Trans. R. Soc. (London), Ser. **A246**, 215 (1953).
5. D.R. Bates and T.R. Carson, Proc. R. Soc. (London), Ser. **A234**, 207 (1956).
6. R.A. Buckingham, in *Quantum Theory I. Elements*, edited by D.R. Bates (Academic, New York, 1961).
7. H. Wind, J. Chem. Phys. **42**, 2371 (1965).
8. J.M. Peek, J. Chem. Phys. **43**, 3004 (1965).
9. L.Y. Wilson and G.A. Gallup, J. Chem. Phys. **45**, 586 (1966).
10. G. Hunter and H.O. Pritchard, J. Chem. Phys. **46**, 2146 (1967).
11. H. Hartmann and K. Helfrich, Theor. Chim. Acta **10**, 406 (1968).
12. E. Teller and H.L. Sahlin, in *Physical Chemistry : An Advanced Treatise* (Academic, New York, 1970), vol. V.
13. K. Helfrich and H. Hartmann, Theor. Chim. Acta **16**, 263 (1970).
14. S. Khel, K. Helfrich, and H. Hartmann, Theor. Chim. Acta **21**, 44 (1971).
15. J.M. Peek and E.N. Lassettre, J. Chem. Phys. **38**, 2382 (1963) ; **38**, 2395 (1963).
16. M.M. Madsen and J.M. Peek, At. Data **2**, 171 (1971).
17. J.D. Powers, Phil Trans. R. Soc. (London), Ser. **A274**, 663 (1973).
18. J. Killingbeck, J. Phys. A : Math. Gen. **18**, L-1025 (1985).
19. W.G. Baber and H.R. Hasse, Proc. Cambridge Philos. Soc. **31**, 564 (1935).
20. E.A. Hylleraas, Z. Phys. **71**, 739 (1931).
21. G. Jaffe, Z. Phys. **87**, 535 (1934).
22. E.H. Kerner, Phys. Rev. **83**, 71 (1951).
23. A. Salin, Phys. Rev. **A34**, 4436 (1986) and references therein.
24. M. Klapisch, Thesis, Université de Paris-Sud, Orsay, France (1969).
25. A. Alikacem and M. Aubert-Frécon, J. Mol. Spectrosc. **111**, 20 (1985).
26. A. Alikacem and M. Aubert-Frécon, Phys. Rev. **A31**, 2096 (1985).
27. A. Alikacem, Thesis, Université Lyon I, France (1986).
28. M. Aubert, N. Bessis and G. Bessis, Phys. Rev. **A10**, 61 (1974) ; Phys. Rev. **A12**, 2298 (1975).
29. M. Aubert-Frécon, Phys. Rev. **A15**, 1344 (1977).
30. P. Pluvinage, Ann. de Physique **5**, 145 (1950) ; J. de Physique et le Radium **12**, 789 (1951).
31. M. Aubert-Frécon and C. LeSech, J. Chem. Phys. **74**, 2931 (1981) ; J. Phys. B : At. Mol. Phys. **15**, L-135 (1982).
32. M. Aubert-Frécon, C. LeSech and P. Pluvinage, J. Mol. Spectrosc. **105**, 276 (1984).
33. C. LeSech, A.M. Jorus and M. Aubert-Frécon, J. Chem. Phys. **75**, 2932 (1981).
34. M. Aubert-Frécon, P. Ceyzeriat, A.M. Jorus and C. LeSech, J. Phys. B : At. Mol. Phys. **17**, L-7 (1984).

ERROR ESTIMATE IN VARIATIONAL CALCULATIONS OF EIGENVALUES AND EIGENVECTORS

G. Fonte
Department of Physics Catania University and INFN
Corso Italia 57, I 95129 Catania, Italy
G. Schiffrer
Department of Physics Ferrara University and INFN
via Paradiso 12, I 44100 Ferrara, Italy
D. Stanzial
Cemoter CNR Institute, via Canalbianco 28, I 44044 Cassana (Ferrara), Italy.

ABSTRACT. We show that rigorous error estimates for variational approximations of eigenvalues and eigenfunctions can be given for the problem of hydrogen atom in strong magnetic fields using Temple's lower bounds and Kato's inequalities. These calculations can be made by a method already tested for anharmonic oscillators: use of a basis making the Hamiltonian matrix of Jacobi type; basis optimization by a new criterion; use of a direct minimization algorithm instead of the usual Rayleigh-Ritz diagonalization. We show that this method gives already good upper bounds, even if we do not fully exploit the basis optimization. More complete calculations will follow.

1. Introduction.

A rigorous error estimate is of fundamental importance in quantum mechanical variational calculations of eigenvalues and eigenvectors for a meaningful comparison of theory and experimental data. The most common method of error estimate of eigenvalues is that of computing a lower and an upper bound to each eigenvalue. Methods for calculation of lower bounds have been developed since the earliest days of quantum mechanics (see ref. [1] for a general review). Kato has shown long ago [2] that Temple's formula

$$E_N - \varepsilon_N^2/(\beta - E_N) \le E_0 \tag{1}$$

for any non-degenerate eigenvalue E_0 of the Hamiltonian H is the best one if only the quantities E_N, ε_N, α and β are

M. Defranceschi and J. Delhalle (eds.), Numerical Determination of the Electronic Structure of Atoms, Diatomic and Polyatomic Molecules, 257–261.

known. Here, $E_N = \langle\Psi_N|H\Psi_N\rangle$, Ψ_N being the normalized approximation to the exact eigenfunction in an N-dimensional subspace, $\varepsilon_N = \|(H-E_N)\Psi_N\|$ and (α,β) is a spectral interval, containing the only eigenvalue E_o, such that $\varepsilon_N^2 < (E_N-\alpha)(\beta-E_N)$. Putting $F_N^o = 1 - |\langle\Psi_N|\Psi_o\rangle|^2$, $(H\Psi_o=E_o\Psi_o)$, the following inequalities are proved in ref. [2]

$$F_N^o \le \varepsilon_N^2/[(E_N-\alpha)^2+\varepsilon_N^2] \ , \text{ if } E_N+\varepsilon_N^2/(E_N-\alpha) \le (\alpha+\beta)/2 \ , \qquad (2)$$
$$F_N^o \le \varepsilon_N^2/[(E_N-\beta)^2+\varepsilon_N^2] \ , \text{ if } E_N-\varepsilon_N^2/(\beta-E_N) \ge (\alpha+\beta)/2 \ .$$

F_N^o is a convenient measure of the eigenfunction error, since it is a continuous function of Ψ_N, vanishing if and only if $\Psi_N=e^{i\vartheta}\Psi_o$ (ϑ real); it can be related to $\|\Psi_N-\Psi_o\|$ if one multiplyes Ψ_N by a phase factor, such that $\langle\Psi_N|\Psi_o\rangle$ becomes real. In this case we have namely

$$F_N^o = \|\Psi_N-\Psi_o\|^2 - \|\Psi_N-\Psi_o\|^4/4 \ .$$

Thus, the error estimate of both eigenvalue and eigenfunction is reduced to the calculation of the quantity ε_N, involving the exact (not truncated) operator H. This difficulty can be overcome if H is represented by a Jacobi matrix (all elements outside a strip centred on the diagonal vanish) by a proper basis choice. In this work, we consider two cases in which this can be easily obtained. One is the case of anharmonic oscillators, in which two of us (see ref. [3]) obtained ε_N values among the smallest in the literature. The other is the more complicated case of the hydrogen atom in strong magnetic fields (quadratic Zeeman effect), where we shall try to follow, as long as possible, the method of ref. [3].

This method has two distinctive features:

a) optimization of the basis parameter according to a new criterion;

b) use of an algorithm [4] for the direct minimization of the energy functional instead of the more common method of the Rayleigh-Ritz diagonalization.

The criterion a) has turned out to be extraordinarily effective for anharmonic oscillators. The method b) allows us to obtain very accurate upper bounds by means of an iterative sequence of operations, easily implementable on a computer (matrix inversion is not required). Furthermore, it has a rather satisfactory numerical stability and is to be preferred to the classical method of Rayleigh-Ritz if only the first few levels are required.

2. Anharmonic oscillators.

In ref. [3] we have considered anharmonic oscillators with Hamiltonian $H = p^2+\mu x^2+\nu x^{2q}$, where $\mu=0,\pm1$, $q=2,3,..,6$ and $\nu\in [10^{-6},10^{6}]$. The basis is that of the harmonic oscillator of given parity. The eigenvalue and eigenfunction of the ground state and the first excited state have been calculated. The results are very good: in the best cases we have obtained ε_N of the order of 10^{-22} and in the worst cases of the order of 10^{-7}. These results have been obtained in subspaces of dimension less than 60 (with one exception) and are due essentially to our method [3] for basis optimization, consisting in minimizing with respect to the basis parameter the norm of the $N\times N$ matrix of the *effective interaction*

$$W_{PP}(E) = H_{PQ}(EI_{QQ} - H_{QQ})^{-1}H_{QP}. \tag{3}$$

The idea is that if the norm of (3) is minimal, the approximate eigenvalue problem in N-dimensional subspace

$$(H_{PP} - EI_{PP})\Psi_P = 0 \tag{4}$$

turns out to be the nearest to the exact one, which differs from eq.(4) by a term containg just the matrix (3).

3. The hydrogen atom in strong magnetic fields.

This problem has received in recent years increasing attention, for its importance in astrophysics of dense objects and in solid state physics. The problem has been investigated both theoretically and numerically. Several of the lowest eigenvalues and eigenfunctions are well known (see ref. [5]). Nevertheless, neither rigorous lower bounds for eigenvalues, nor error estimates for the eigenfunctions have been reported, as far as we know. We want to show here how this can be done.

It is well known that the spectrum of the hydrogen atom in a strong uniform magnetic field $\mathbf{B} = (0,0,B)$ is accurately described in large B intervals by the Schrödinger equation

$$[-\Delta/2-1/r+\gamma^2(x^2+y^2)/2]\Phi = E\Phi \tag{5}$$

(atomic units, linear term in B omitted, $\gamma = B/2c$, infinite nuclear mass). Changing to paraboloidal coordinates (u,v,ϕ)

$$x = uv\cos\phi,$$
$$y = uv\sin\phi,$$
$$z = (u^2-v^2)/2, \quad u,v \ge 0, \quad 0 \le \phi < 2\pi,$$

and putting $\Phi(u,v,\phi)=\Psi(u,v)\exp(im\phi)$, where m is the projection of the angular momentum along the z-axis, one can easily show that eq.(5) transforms to the following generalized eigenvalue problem in $L^2(\mathbb{R}^+\times\mathbb{R}^+, du\,dv)$

$$A\Psi = EB\Psi \tag{6}$$

where

$$A\Psi = \{-\Psi_{uu}+[(m^2-1/4)/u^2+\omega_1 u^2]\Psi\}+\{-\Psi_{vv}+[(m^2-1/4)/v^2+\omega_1 v^2]\Psi\}+$$
$$+[-4-\omega_1 u^2-\omega_2 v^2+\gamma(u^2+v^2)u^2v^2]\Psi,$$
$$B = 2(u^2+v^2) .$$

Here the two operators in {} are radial parts of plane harmonic oscillators and [] is the coupling term. The operators A and B are symmetric and the functional

$$E(\Psi) = \langle\Psi|A\Psi\rangle/\langle\Psi|B\Psi\rangle \tag{7}$$

is bounded from below. Then, a well known theorem [6] states the equivalence of problem (6) and the minimum problem for (7). We can, therefore, directly minimize (7). Using a basis made of symmetrized or antisymmetrized products (depending on the parity of the required state) of eigenfunctions of the two oscillators above, the calculation of the matrix elements of A and B is very simple and the corresponding matrices are of Jacobi type. This feature is important for the computation of ε_N and, therefore, for the error estimate through formulae (1),(2), which, being only valid for a self-adjoint problem of the usual type ($B = I$), can nevertheless be extended to the case of a general B replacing E_N by $\tilde{E}_N=\langle\tilde{\Psi}_N|A\tilde{\Psi}_N\rangle$ ($\tilde{\Psi}_N$ is the approximate eigenfunction of (6) in an N-dimensional subspace, normalized as $\langle\tilde{\Psi}_N|B\tilde{\Psi}_N\rangle=1$) and ε_N by

$$\tilde{\varepsilon}_N = \|B^{-1/2}(A - \tilde{E}_N B)\tilde{\Psi}_N\|.$$

The new formula corresponding to (2) is obtained by replacing E_N and ε_N as above, and F^O_N by

$$\tilde{F}^O_N = 1 - \langle\tilde{\Psi}_N|B\tilde{\Psi}_O\rangle^2, \tag{8}$$

where $\tilde{\Psi}_O$ is the exact eigenfunction of (6), corresponding to the eigenvalue E_O and normalized as $\tilde{\Psi}_N$. The quantity (8) has similar properties as the corresponding (2), so that it is again a convenient measure of the eigenfunction error. However, the relation between $\tilde{F}^O_N$ and $\|\tilde{\Psi}_N - \tilde{\Psi}_O\|$ is now the following (the space is real)

$$\tilde{F}^O_N = \|B^{1/2}(\tilde{\Psi}_N-\tilde{\Psi}_O)\|^2-\|B^{1/2}(\tilde{\Psi}_N-\tilde{\Psi}_O)\|^4/4 .$$

The basis parameters ω_1 and ω_2 can be optimized by the same method used for the oscillators: one easily finds that now one should minimize the norm of the matrix $\widetilde{W}_{PP}(E)$, obtained from (3) replacing the operator H by $B^{-1}A$. Now, since $\tilde{\varepsilon}_N$ can be written as

$$\tilde{\varepsilon}_N = \langle (A-\widetilde{E}_N B)\widetilde{\Psi}_N | B^{-1}(A-\widetilde{E}_N B)\widetilde{\Psi}_N \rangle^{1/2},$$

this problem differs from that of the oscillators by the presence of the operator B and its inverse. Anyway, this is not a serious computational difficulty, due to the Jacobian structure of B and to the symmetry of its inverse: in this case only a finite-dimensional submatrix of B^{-1} appears in the calculations. The problem seems more serious for the minimization of the norm of $\widetilde{W}_{PP}(E)$, since it involves the whole operator B^{-1}. In any case, we expect, on the basis of the oscillators calculations, that even a rough minimization of $\widetilde{W}_{PP}$ could be successful. This expectation is further supported by the fact our method provides already good upper bounds to the eigenvalues, even with an empirical basis optimization, as shown by the following results.

With the notation $E=E_0$, $\omega=\omega_1=\omega_2$, we find:

γ	ω	E(present work)	E(ref.[5])
1	13.3137	-0.04442779	-0.044428
2	16.5681	1.43840399	1.438404
7	55.327	10.07759	10.07757
10	106.117	15.56926	15.56920

We see that at this stage our results compare well with those of ref. [5]

References.

[1] M.A. Abdel-Arouf: *Phys. Rep.* **84**, 163 (1982).

[2] T. Kato: *Proc. Phys. Soc. Japan* **4**, 334 (1949).

[3] G. Schiffrer and D. Stanzial: *Nuovo Cim.* **B90**, 24 (1985).

[4] B. Pchénitchny et Y. Daniline: *Méthodes numériques dans les problèmes d'extrémum*, Ed. Mir, Moscou (1977), p 127.

[5] W. Rösner, G. Wunner, H. Herold and H. Ruder: *J. Phys.* **B 17**, 29 (1984).

[6] S.G. Mikhlin: *Variational Methods in Mathematical Physics*, Pergamon Press, London (1964).

A MOMENTUM SPACE APPROACH TO IMPROVE *ab initio* HARTREE-FOCK RESULTS BASED ON THE LCAO-GTF APPROXIMATION.

J. G. FRIPIAT+, J. DELHALLE+ and M. DEFRANCESCHI*

+*Laboratoire de Chimie Théorique Appliquée, Facultés Universitaires Notre-Dame de la Paix, 61, rue de Bruxelles, B-5000 Namur , Belgium*

**Laboratoire de Chimie, Ecole Normale Supérieure, 1, rue Maurice Arnoux, F-92120 Montrouge (France).*

ABSTRACT. Almost all routine calculations are made within the framework of the LCAO approximation where the atomic functions are Gaussian functions (GTF's) which have inherent deficiencies. A preliminary study toward a systematic use of the iteration-variation method expressed in momentum space to improve molecular properties calculated at the Hartree-Fock level by standard quantum chemistry methods is presented. The results obtained for the hydrogen atom, whose exact solutions are known, show an important qualitative improvement when the trial wave function is a simple Gaussian . The properties considered are the electronic energy, the Compton profile J(p) at p =0 and $<p^4>$, all of which are very sensitive to the quality of the approximate solution.

1. Introduction.

Deviations of LCAO *ab initio* calculations from the Hartree-Fock (HF) limit are caused by truncations of the one-particle basis set. The error resulting from this truncation is difficult to estimate, and its incidence largely depends on the nature of the properties [1]. According to the work by Davis and co-workers, chances for remedying these shortcomings from a purely analytical point of view look rather gloomy [2-4]. Solving numerically the HF equations seems to be the ultimate answer to this problem. Unfortunately, this approach which is now routine in the case of atoms [5] and nearly so for diatomic molecules [6,7] is not yet possible for polyatomic molecules. One reason is that nowaydays numerical evaluations of the atomic and molecular wave functions are

M. Defranceschi and J. Delhalle (eds.), Numerical Determination of the Electronic Structure of Atoms, Diatomic and Polyatomic Molecules, 263–268.

performed in direct (or position) space. In this representation it is apparently not possible to achieve enough separability among the integration variables to efficiently deal with polyatomic cases. Recently, it was pointed out that this restriction can be removed in momentum space [8,9], and attempts to solve numerically the HF equations in momentum space have been reported for atoms, diatomic and polyatomic molecules, and polymers [9-14]. In spite of the fact that the momentum space approach is still in infancy and further progress is needed before it reaches the level of routine applicability, interesting possibilities based on the general features of the approach are conceivable.

In this paper we would like to propose a simple route to improve the HF-Roothaan results when the atomic functions are Gaussian functions [15]. Gaussian functions (GTF's) have inherent deficiencies near the nuclear centers (cusp problem) and exhibit a wrong asymptotic decay at large distances from the nuclear centers. These problems, which affect the calculated molecular properties, can in principle be reduced by increasing the number of GTF's in the basis sets, but at the expense of prohibitive computer time and integral storage.Though it is still an open question, iterative computations [16-18] can provide improved approximations to the wave functions, even after one iteration. A preliminary study [19] indicates that in the case of 1s Gaussian functions it seems possible to cast the first iterate in terms of reasonably closed mathematical expressions involving the Dawson function. This opens the prospect of applying a one iteration step in momentum space to MO-LCAO-GTF's obtained by standard programs to build into these orbitals a more appropriate information regarding the physical nature of the system. In this contribution we assess the type improvement achieved after one iteration on a trial function, approximating the exact 1s ground state of the hydrogen atom respectively by one (1G) and three (3G) Gaussian functions [20]. The concluding section summarizes some of the possible implications of this approach for more general situations.

2. One Iteration on a Trial Wave Function Approximated by GTF's.

The momentum space representation of the Schrödinger equation for the H atom is obtained by applying the Fourier transformation,

$$f^T(\mathbf{p}) = [\, f(\mathbf{r})]^T(\mathbf{p}) = (2\pi)^{-3/2} \int d\mathbf{r}\, e^{-i\mathbf{p}.\mathbf{r}}\, f(\mathbf{r}) \qquad (1)$$

to the corresponding position space expression. In atomic units, the position space Schrödinger equation, a partial differential equation, for the hydrogen atom is,

$$H\,\psi_i(\mathbf{r}) = \Big[-\frac{\nabla^2(\mathbf{r})}{2} + V(\mathbf{r}) \Big]\, \psi_i(\mathbf{r}) = \varepsilon_i\, \psi_i(\mathbf{r})\,, \tag{2}$$

where $V(\mathbf{r}) = -\,|\mathbf{r}|^{-1}$. In momentum space, the equivalent of Eq.(2) reads

$$[H\,\psi_i(\mathbf{r})\,]^T(\mathbf{p}) = \varepsilon_i\,[\,\psi_i(\mathbf{r})\,]^T(\mathbf{p}) = \varepsilon_i\,\phi_i(\mathbf{p}), \tag{3}$$

its detailed expression reveals the integral nature of the equation :

$$\frac{\mathbf{p}^2}{2}\phi_i(\mathbf{p}) - \frac{1}{2\pi^2}\int \frac{d\mathbf{q}}{\mathbf{q}^2}\,\phi_i(\mathbf{p}-\mathbf{q}) = \varepsilon_i\,\phi_i(\mathbf{p})\,. \tag{4}$$

If the exact bound state orbitals $\phi_i(\mathbf{p})$, i.e. the solutions to Eq.(3) such that $\varepsilon_i < 0$, were not known, it would still be possible, starting with an initial guess, to use Eq.(4) as an iterative procedure to obtain approximations to the $\phi_i(\mathbf{p})$'s. The iterative scheme is directly obtained after rewriting Eq.(4) as

$$\phi_i(\mathbf{p}) = [\,\frac{\mathbf{p}^2}{2} - \varepsilon_i\,]^{-1}\,\frac{1}{2\pi^2}\int \frac{d\mathbf{q}}{\mathbf{q}^2}\,\phi_i(\mathbf{p}-\mathbf{q})\,. \tag{5}$$

The condition $\varepsilon_i < 0$ prevents the denominator from diverging. Eq.(5) is nothing but an integral equation formulated in terms of the bound-state free-particle Green's operator $[\mathbf{p}^2/2-\varepsilon]^{-1}$, it defines a variation-iteration method from which systematic improvement of the initial trial function can be obtained. Having in mind ground-state problems, we will use the subscript "iG" (i = 1 or 3) to indicate the number of Gaussian functions in the trial function, then the iterative expression takes the form,

$$\phi_{iG}^{(n+1)}(\mathbf{p}) = [\,\frac{\mathbf{p}^2}{2} - \varepsilon^{(n)}]^{-1}\,\frac{1}{2\pi^2}\int \frac{d\mathbf{q}}{\mathbf{q}^2}\,\phi_{iG}^{(n)}(\mathbf{p}-\mathbf{q}), \tag{6}$$

where n denotes the n-th iteration. To keep the paper short we give the explicit result in the case a trial wave function approximated by one GTF, $(2\pi\alpha)^{-3/4}\exp(-\mathbf{p}^2/4\alpha)$:

$$\phi_{1G}^{(1)}(\mathbf{p}) = 4\,K^{-1}\,(\frac{2\alpha}{\pi^5})^{1/4}\,\frac{\mathrm{daw}(\sqrt{\mathbf{p}^2/4\alpha}\,)}{[\mathbf{p}^2 + \mathbf{p}_0^2]\,|\mathbf{p}|}, \tag{7}$$

In Eq.(7), daw(z) is the so-called Dawson function [21-23] and K^{-1} is the normalized constant obtained numerically by a Gauss-Legendre integration [19] of :

$$\int d\mathbf{p}\,|\,\phi_{1G}^{(1)}(\mathbf{p})\,|^2 = K^2. \tag{8}$$

The type of improvement obtained after one iteration can already be perceived by analyzing the behaviour of $\phi_{1G}^{(1)}(\mathbf{p})$ for $|\mathbf{p}|$ tending to ∞. At large values of z , daw(z) $\approx$ $(2z)^{-1}$ which, when inserted in Eq.(7), corresponds to a $|\mathbf{p}|^{-4}$ decay typical of the exact solution $1s^T(\mathbf{p})$ when $|\mathbf{p}| \rightarrow \infty$:

$$1s^T(\mathbf{p}) = \frac{(8\zeta^5)^{1/2}}{\pi [\mathbf{p}^2 + \zeta^2]^2} \quad ; \quad \zeta = 1 . \tag{9}$$

The improvement is further assessed by comparing (Table I) the orbital energy ε, Compton profile $J(\mathbf{p})$ at $\mathbf{p} = 0$, i.e. $J(0)$ or $<|\mathbf{p}|^{-1}/2>$, and $<|\mathbf{p}|^4>$ calculated with the intial function $\phi_{1G}^{(0)}(\mathbf{p})$, the first iterate $\phi_{1G}^{(1)}(\mathbf{p})$ and the exact solution $1s^T(\mathbf{p})$. These properties provide sensitive measures of the quality of the wave function with specific emphasis on different portions of the range in which $\mathbf{p}$ is defined [24].

TABLE I.: Value f(0) of the functions $\phi_{iG}^{(n)}(\mathbf{p})$ and $1s^T(\mathbf{p})$ at $\mathbf{p} = 0$, ε the orbital energy and its kinetic ε_{kin} and nuclear attraction ε_{nuc} parts, the virial ratio, $<|\mathbf{p}|^{-1}/2>$, $<|\mathbf{p}|^4>$ and the mean square deviation : $<| \phi_{iG}^{(n)}(\mathbf{p}) - 1s^T(\mathbf{p})|^2>$.

	$\phi_{1G}^{(0)}(\mathbf{p})$	$\phi_{1G}^{(1)}(\mathbf{p})$	$\phi_{3G}^{(0)}(\mathbf{p})$	$\phi_{3G}^{(1)}(\mathbf{p})$	$1s^T(\mathbf{p})$
f(0)	0.649519	0.881834	0.815290	0.876949	0.900316
ε	-0.424413	-0.487088	-0.496976	-0.499913	-0.5
ε_{kin}	0.424413	0.402145	0.496982	0.495890	0.5
ε_{nuc}	-0.848826	-0.889233	-0.993958	-0.995803	-1.0
virial	-0.5	-0.452238	-0.500003	-0.497980	-0.5
$<|\mathbf{p}|^{-1}/2>$	0.75	0.865001	0.824323	0.843121	0.848826
$<|\mathbf{p}|^4>$	1.200844	2.086410	3.836949	4.625970	5.0
m.s.dev.	0.043641	0.006976	0.001247	0.000075	0.0

Both for $\phi_{1G}^{(0)}(\mathbf{p})$ and $\phi_{3G}^{(0)}(\mathbf{p})$, one iteration step leads to considerable improvements in all properties considered. The most difficult being $<|\mathbf{p}|^4>$ because it samples the tail of the wave function. Notice however that already with one iteration on $\phi_{3G}^{(0)}(\mathbf{p})$ the value of $<|\mathbf{p}|^4>$ becomes reasonably close to the exact value. Also to be noticed is the improvement of $<|\mathbf{p}|^{-1}/2>$ after one iteration. This property is important to watch out because it is specific of the wave function at low values of $\mathbf{p}$. It is mainly in this range, for molecular systems, that the bonding characteristics are embodied.

3. Concluding Remarks.

The use of the Green's integral operator ensures that a considerable amount of information regarding the physical nature of the system has been built into the wave function by means of the associated Green's function. Already after one iteration very significant changes are forced in the approximate wavefunction.

One advantage of starting with trial functions represented by Gaussian functions lies in the fact that the resulting first iterates express in terms of Dawson functions. Here, contrary to purely numerical schemes, analytical information is still available and can be used to control the accuracy of the numerical steps involved in the normalization procedure and computations of molecular properties such as electric responses (induced dipole moment, polarisability, etc.). Indeed, we can assume that, if the starting trial orbitals $\phi_i^{(0)}(\mathbf{p})$ obtained with standard softwares are already of good quality, one step of iteration could result in very essential improvements on the nature of the orbitals (better asymptotic decay, improved behaviour near the origin, etc.). This preliminary study on the hydrogen atom provides the basis for a simpler and possibly more efficient strategy than extending basis sets to remove the adverse incidences of one-particle basis set truncation.

Acknowledgements.

Two of us (M.D., J. D.) acknowledge with appreciation the support of this work under the agreements (1986) for scientific cooperation between the Commissariat Général aux Relations Internationales de la Communauté Française de Belgique and the CNRS-France. The authors thank the Belgian National Fund for Scientific Research, IBM-Belgium and the Facultés Universitaires Notre-Dame de la Paix for the use of the Namur Scientific Computing Facilities.

References.

1. E.R. Davidson, D. Feller, Chem. Rev., **86**, 681 (1986).
2. C.L. Davis, H.J.A. Jensen, H. Monkhorst, J. Chem. Phys., **80**, 840 (1984).
3. C.L. Davis, H. Monkhorst, J. Chem. Phys., **81**, 5845 (1984).
4. C.L. Davis, H. Monkhorst, Chem. Phys. Lett., **111**, 562 (1984).
5. Ch. Froese Fischer, *The Hartree-Fock Method for Atoms. A Numerical Approach*, (Wiley, New York, 1977).
6. E.A. McCullough, Comput. Phys. Rep., **4**, 313 (1986).
7. L. Laaksonen, P. Pyykkö, D. Sundholm, Comp. Phys. Rep., **4**, 265 (1986).

8. B.K. Novosadov, J. Mol. Struct., **54**, 269 (1979).
9. J. Navaza, G. Tsoucaris, Phys. Rev.**A24**, 683 (1981).
10. M. Defranceschi, M. Suard, G. Berthier, Int. J. Quantum Chem. **25**, 863 (1984).
11. M. Defranceschi, J. Delhalle, Phys. Rev. **B34**, 5862 (1986).
12. S.A. Alexander, H.J. Monkhorst, Int. J. Quantum Chem., **32**, 361 (1987).
13. S.A. Alexander, R.L. Coldwell, H.J. Monkhorst, J. Comp. Phys., in press.
14. W. Rodriguez, Y. Ishikawa, Chem. Phys. Lett., in press
15. S.F. Boys, Proc. Roy. Soc.(London), **200**, 542 (1950).
16. N.V. Svartholm, Ph.D. thesis, *The Binding Energies of Light Nuclei*, Stockholm-Lund, 1945.
17. R. McWeeny and C.A. Coulson, Proc. Phys. Soc. London, Sect.A **62**, 509 (1949).
18. H.J. Monkhorst, K.J. Szalewicz, J. Chem. Phys., **75**, 5782 (1981).
19. J. Delhalle, J.G. Fripiat, M. Defranceschi, Annales Soc. Scient. Bruxelles, **101**, Fasc. I, 9 (1987).
20. S. Huzinaga, J. Chem. Phys., **42**, 1293 (1965).
21. M. Abramowitz, I.A. Stegun, *Handbook of Mathematical Functions*, (Dover, New York, 1968)
22. N.N. Lebedev, *Special Functions and their Applications*, (Dover, New York, 1972), pp. 19-21.
23. J. Spanier, K.B. Oldham, *An Atlas of Functions*, (Hemisphere Publishing Co., Washington, 1987), chapter 42.
24. J. Delhalle, M. Defranceschi, Int. J. Quantum Chem. Symp. **21**, 425 (1987).

ISOTOPE SHIFT MCHF CALCULATIONS IN STRONTIUM

M. Godefroid* and N. Vaeck†
Laboratoire de Chimie Physique Moléculaire, CP 160
U.L.B. , av. F.D. Roosevelt 50
B-1050 Brussels, Belgium

Jørgen E. Hansen
Zeeman Laboratory, University of Amsterdam
Plantage Muidergracht, 4
NL-1018 TV Amsterdam, The Netherlands

ABSTRACT. Extensive multiconfiguration Hartree-Fock calculations have been performed in the 5snp $^1P^o$, 5snd ^1D and 5snf $^1F^o$ series in neutral strontium. The <u>ab initio</u> wavefunctions are used for evaluating the effects of the perturbers $5p^2$, $4d^2$ and 5p4d on the field and specific mass shifts of the low-lying levels of these series.

1. INTRODUCTION

The Rydberg levels of alkaline-earth atoms can be strongly perturbed by doubly-excited states[1]. The strong electrostatic interaction between a Rydberg series and a doubly-excited state leads often to some ambiguities in level designations[2] and the admixture of the perturber(s) may induce drastic changes in the electronic properties of the perturbed levels. The theoretical study of the correlation effects responsible of these perturbations, is very delicate and constitutes a real challenge for excited states. The use of <u>numerical</u> multiconfiguration Hartree-Fock wavefunctions (MCHF), taking the configuration mixing effects into account, has been shown to be successful in many cases[3,4]. We apply this method for evaluating isotope shifts of strongly perturbed singlet terms in neutral strontium (Z=38).

---

* Research Associate of the Belgian National Fund for Scientific Research (FNRS)

† IRSIA fellowship

M. Defranceschi and J. Delhalle (eds.), Numerical Determination of the Electronic Structure of Atoms, Diatomic and Polyatomic Molecules, 269–274.

2. MCHF CALCULATIONS

The numerical multiconfiguration Hartree-Fock approach is described by Froese Fischer[5,6]. The atomic wavefunction $\Psi^{MCHF}(\Gamma LS)$ is a superposition of orthonormal configuration state functions (CSF) $\Phi(\gamma_i LS)$:

$$\Psi^{MCHF}(\Gamma LS) = \sum_i c_i \Phi(\gamma_i LS) \quad (1)$$

built on the one-electron orbital basis $\{P_{n\ell}(r)Y_{\ell m}(\theta,\phi)\}$. The final radial functions $\{P_{n\ell}(r)\}$ are solutions of the non-linear MCHF differential equations arising when requiring the total energy $E=\langle\Psi^{MCHF}|H|\Psi^{MCHF}\rangle$ to be stationary with respect to variations in the radial functions. The MCHF equations are coupled with the configuration interaction secular equations $\mathbb{H}\mathbb{C} = E\mathbb{C}$, the radial functions depending on the mixing coefficients $\{c_i\}$ and vice-versa.

We treat the strontium atom as a "two-electron" problem, using the reduced forms of the valence-pair-correlation function[7]. Multiconfiguration wavefunctions have been calculated for the ground level $5s^2\ {}^1S$, for the five lowest ${}^1P^o$ terms, the three lowest ${}^1D$ terms, the five lowest ${}^1F^o$ terms and for the lowest ${}^1G$ level. The MCHF expansions and eigenvector compositions are described by Vaeck *et al*[8] (hereafter noted VGH).

3. ISOTOPE SHIFTS

Our theoretical lifetime values calculated[8] for the ${}^1P^o$ and ${}^1F^o$ series reproduce very well the observed trends[9]. The deviation from the expected simple scaling law $\tau \div (n^*)^3$ in the ${}^1P^o$ series can be explained through the interference effects due to the perturbers[8]. Similarly to the lifetime trends, the optical isotope shifts (IS) can probe finely the configuration mixing effects[10]. In this section, we investigate the action of the perturbers on the transition isotope shifts.

The strontium is of particular interest due to the effect of the N=50 neutron shell closure at ${}^{88}Sr$ and to the increasing nuclear deformation of neutron deficient Sr isotopes, giving rise to negative isotope shifts[11]. We test the *ab initio* calculation techniques of optical IS using our MCHF numerical wavefunctions described in VGH[8]. We will only consider the isotope pair ${}^{86,88}Sr$ for which the IS of some low-lying states have been investigated extensively by laser-atomic-beam spectroscopy[10,12,13,14].

3.1. Theory

The IS of an atomic energy level arises from a mass shift (MS) and a field shift (FS) contributions. The MS takes into account the finite nuclear mass and is usually separated into the normal mass shift (NMS) and the specific mass shift (SMS) contributions[15]. The FS arises from the finite nuclear volume, i.e. from the departure of real nuclei from the pure Coulombic potential of a nuclear point charge. The residual isotope shift (RIS) is defined as the difference between the total isotope shift and the normal mass shift.

In a first-order perturbation theory, the SMS and FS are given by the expectation values of the SMS and FS operators[15,16]:

$$H^{SMS} = M^{-1} \sum_{i<j} \underline{p}_i \cdot \underline{p}_j \qquad (2)$$

$$H^{FS} = C \sum_{i} \delta(\underline{r}_i) \qquad (3)$$

where $\underline{r}_i$ and $\underline{p}_i$ are respectively the vectors position and momentum of electron i, M is the nuclear mass and C contains the dependence on the nuclear quantities. The level IS is the variation, from one isotope A to another A' (A'>A), of the level shift. The transition IS $\delta\nu_i^{AA'}$ between the two isotopes observed in a spectral line of frequency ν_i can then be written as the sum of the three effects

$$\delta\nu_i^{AA'} = \delta\nu_{i,NMS}^{AA'} + \delta\nu_{i,SMS}^{AA'} + \delta\nu_{i,FS}^{AA'} \qquad (4)$$

The last two terms define the residual transition isotope shift (RIS) and can be written as :

$$\delta\nu_{i,SMS}^{AA'} = K_i(A'-A)/AA' \qquad (5)$$

$$\delta\nu_{i,FS}^{AA'} = E_i f(Z)\,\delta\langle r^2\rangle^{AA'} \qquad (6)$$

where K_i contains the difference $k(\Gamma_2L_2S_2) - k(\Gamma_1L_1S_1)$ of the two level-SMS involved in the transition i; the level specific mass shift $k(\Gamma LS)$ can be evaluated[17] using the gradient form of the operator (2) and the MCHF wavefunction $\Psi^{MCHF}(\Gamma LS)$ (1) :

$$k(\Gamma LS) = \sum_{pq} c_p c_q k_{pq} \qquad (7a)$$

$$k_{pq} = \langle\Phi(\gamma_p LS)| \sum_{i<j} \underline{p}_i \cdot \underline{p}_j |\Phi(\gamma_q LS)\rangle \qquad (7b)$$

The factor f(Z) in (6) accounts for the finite charge distribution of the nucleus; $\delta\langle r^2\rangle^{AA'}$ is the change in the mean-square nuclear charge radius between the two isotopes. The electronic factor E_i is directly related to the change of the electronic charge density at the nucleus[11]

$$E_i = \pi a_o^3 \Delta|\Psi(0)|_i^2/Z \qquad (8)$$

The field shifts are calculated from (6) and (8) by summing the weighted one-electron ns-densities at the nucleus :

$$|\phi_{ns}(0)|^2 = A_0(ns)/4\pi \qquad (9a)$$

$$A_0(ns) = \lim(r\to 0)\,[P_{n\ell}(r)/r^{\ell+1}] \qquad (9b)$$

corresponding to the MCHF expansion (1). In our calculations, the nuclear parameter $f(Z)\delta\langle r^2\rangle^{86,88}$ is included in the "calibrating" factor obtained by Aspect et al[10] from a scaling of the FS-Hartree-Fock result for 5s→5p in Sr^+ to the experimental value.

3.2. The $^1P^o$ series

We compare in Table I the theoretical and experimental transition isotope shifts for the $5s^2\ ^1S \rightarrow n^1P^o$ transitions.

Table I : Transition isotope shifts (in MHz) for $5s^2\ ^1S \rightarrow n^1P^o$

transition[a]	$\delta\nu_{FS}^{86,88}$		$\delta\nu_{SMS}^{86,88}$		$\delta\nu_{RIS}^{86,88}$	
	MCHF[b]	obs.[c]	MCHF[b]	obs.[c]	MCHF[b]	obs.[c]
$5s^2\ ^1S \rightarrow 5s5p\ ^1P^o$	46	42.1(5.5)	-56	-11.0(5.5)	-10	31.1(3)
$\rightarrow$ 5s6p	41	45.6(3.8)	- 6	33.6(3.7)	35	79.1(7)
$\rightarrow$ 5s7p	38	-	0	-	38	-
$\rightarrow$ 4d5p	38	-	6	-	44	-
$\rightarrow$ 5s8p	36	-	- 3	-	33	-

[a]adopting level notation of Ref.1 ; [b]this work ; [c]Ref.14

The separation of experimental IS in MS and FS was performed using a King-plot procedure with muonic x-ray data[14].

The theoretical and observed field shifts agree quite well. The s-electron density decreases from $5s^2\ ^1S$ to n^1P^o but this effect is smaller for high n. Indeed, the density at the nucleus for the n^1P^o wavefunctions increases with n, in correlation with the 5s-orbital contraction and the 4d5p-character reduction.

The MCHF specific mass shift results disagree with the experimental values. The sign of $\delta\nu_{SMS}^{86,88}$ is even wrong for 5s6p. However, the theoretical values reproduce the observed increase of the level-SMS from 5s5p to 5s6p. This trend has already been explained by Eliel *et al*[18] through the admixture of the 5p4d configuration. For illustrating the difficulty of evaluating *ab initio* the transition SMS in that case, we report in Table II the two largest (diagonal) contributions to $\delta\nu_{SMS}^{86,88}$ which can be rewritten as :

$$\delta\nu_{SMS}^{86,88}(\mathrm{MHz}) = 3610246.1\frac{(88-86)}{88.86}\left[\sum_i c_i^2\left[k_{ii}-k(5s^2\ ^1S)\right]+2\sum_{i<j} c_i c_j k_{ij}\right] \quad (10)$$

Table II : Main diagonal contributions $c_i^2\left[k_{ii}-k(5s^2\ ^1S)\right]$ to $\delta\nu_{SMS}^{86,88}$ (MHz) for the transition $5s^2\ ^1S \rightarrow n^1P^o$ (see eq.10)

CSF level:	$1\ ^1P^o$	$2\ ^1P^o$	$3\ ^1P^o$	$4\ ^1P^o$	$5\ ^1P^o$
$5snp_2$	-68	-63	-74	-82	-79
$5p_14d_1$	26	61	82	96	81

The negative interference of the two diagonal terms $c_i^2\left[k_{ii}-k(5s^2\ ^1S)\right]$ associated with the CSF $5snp_2$ and $5p_14d_1$ makes the resulting transition SMS highly sensitive to small change in the MCHF wavefunctions. The 5p4d-character is decreasing with n but the corresponding diagonal value increases faster. In a two-configuration model approximation, the weight of the perturber should change from 13% to 20% in order to get the positive transition SMS +33.6 MHz for $5s6p\ ^1P^o$.

3.3. The ^1D series

The FS, SMS and RMS contributions to the frequency isotope shifts are reported in Table III for the transitions $5s^2\ ^1S \rightarrow n^1D$. Our residual mass shift MCHF results agree very nicely with the observed values for the two lowest ^1D levels. The FS and SMS have not been separated experimentally.

Table III : Transition isotope shifts (in MHz) for $5s^2\ ^1S \rightarrow n^1D$

transition	$\delta\nu^{86,88}_{FS}$ theory		$\delta\nu^{86,88}_{SMS}$ theory		$\delta\nu^{86,88}_{RIS}$ theory		obs.
$5s^2\ ^1S \rightarrow 5s4d\ ^1D$	70[a]	43[b]	208[a]	153[b]	278[a]	196[b]	271[c]
$\rightarrow 5s5d$	51[a]	-	88[a]	-	139[a]	-	151(9)[d]
$\rightarrow 5p^2$	84[a]	101[b]	100[a]	170[b]	184[a]	271[b]	339[b]
		92[e]		372[e]		464[e]	

[a]this work ; [b]Ref.10 ; [c]Ref.12 ; [d]Ref.13 ; [e]Ref.19

We can compare our results with the theoretical values of Aspect *et al*[10] for the lowest and third terms. Their MCHF expansions were much more limited than ours[8]; moreover, the core- and the 5s-orbitals have been systematically frozen during the MCHF procedure in their calculations. The resulting difference in the eigenvector composition alters considerably the RIS calculation. They showed in their paper that the so-called $5p^2$ state (the third ^1D) should contain large admixture of the $4d^2$ configuration in order to obtain the large positive SMS. We find the same effect : the transition SMS for $5s^2\ ^1S \rightarrow 3^1D$ arises mainly from the diagonal contributions to (10) -33, -71 and 218 MHz associated with the CSF $5s5d_1$, $5p_1^2$ and $4d_2^2$ respectively, and is then highly sensitive to the composition of the wavefunction.

In view of this, the model-potential calculations of Chidichimo[19] based on the limited two-configuration ($5p^2 + 4d^2$) approximation for the 3^1D state are unreliable.

4. CONCLUSION

In 1976, the *ab initio* evaluation of mass shifts was considered, though extremely difficult, as a possible solution to the problem of the separation of mass- and field-shift quantities for extracting the nuclear properties[18]. In the last decade, only a few calculations have been carried out to calculate the SMS in two-electron systems[13,17,20]. The results of the present contribution illustrate the power of the *numerical* multiconfiguration Hartree-Fock method for investigating strong perturbation effects on the atomic structure.

5. REFERENCES

[1] W.R.S. Garton and K. Codling, J. Phys. B 1, 106 (1968)
[2] C. Froese Fischer and M. Godefroid, Phys. Scripta 25, 394 (1982)
[3] C. Froese Fischer and M. Godefroid, Nucl. Instrum. Meths. 202, 307 (1982)
[4] C. Froese Fischer and J.E. Hansen, J. Phys. B 18, 4030 (1985)
[5] C. Froese Fischer, The Hartree-Fock method for atoms (Wiley, New York, 1977)
[6] C. Froese Fischer, Comput. Phys. Rep. 3, 275 (1986)
[7] C. Froese Fischer, J. Comput. Phys. 13, 508 (1973)
[8] N. Vaeck, M. Godefroid and J.E. Hansen, submitted for publication in Phys. Rev. A (1988)
[9] G. Jönsson, C. Levinson, A. Persson and C.-G. Wahlström, Z. Phys. A 316, 225 (1984)
[10] A. Aspect, J. Bauche, A.L.A. Fonseca, P. Grangier and G. Roger, J. Phys. B 17, 1761 (1984)
[11] P. Aufmuth, K. Heilig and A. Steudel, Atomic Data Nucl. Data Tables 37, 455 (1987)
[12] P. Grundevik, M. Gustavsson, I. Lindgren, G. Olsson, T. Olsson and A. Rosén, Z. Phys. A 311, 143 (1983)
[13] C.-J. Lorenzen, K. Niemax and L.R. Pendrill, Phys. Rev. A 28, 2051 (1983)
[14] D. Bender, H. Brand and V. Pfeufer, Z. Phys. A 318, 291 (1984)
[15] J. Bauche and R.-J. Champeau, Adv. At. Mol. Phys. 12, 39 (1976)
[16] P. Aufmuth, J. Phys. B 15, 3127 (1982)
[17] C. Froese Fischer and L. Smentek-Mielczarek, J. Phys. B 16, 3479 (1983)
[18] E.R. Eliel, W. Hogervorst, T. Olsson and L.R. Pendrill, Z. Phys. A 311, 1 (1983)
[19] M.C. Chidichimo, Z. Phys. A 321, 549 (1985)
[20] E. Lindroth, A.-M. Mårtensson-Pendrill, Z. Phys. A 316, 265 (1984)

FINITE ELEMENT METHOD FOR THE ACCURATE SOLUTION OF DIATOMIC MOLECULES

D. Heinemann, B. Fricke, D. Kolb
Physics Department
University of Kassel
Heinrich Plett Str. 40
D-3500 Kassel
Germany

ABSTRACT. We present the Finite-Element-Method (FEM) in its application to quantum mechanical problems solving for diatomic molecules. Results for Hartree-Fock calculations of H_2 and Hartree-Fock-Slater calculations of molecules like N_2 and CO have been obtained. The accuracy achieved with less then 5000 grid points for the total energies of these systems is 10^{-8} a.u. , which is demonstrated for N_2.

1. INTRODUCTION

The Finite Element Method (FEM) is a well known technique for solving static and dynamical problems in engineering science. Examples can be found in the textbooks of Ref. 1-3. The idea of this method is to divide the space into a number of 2 or 3 dimensional domains and describe the properties of interest on each of these elements separately connected via boundary conditions. In engineering science one uses low order polynomials to specify the interesting quantities on each element.

We applied the FEM to quantum mechanical problems where the accuracy required is very high. To achieve this accuracy we were forced to use high order polynomials on each element and, in order to minimize the total number of points, to use a small number of elements.

2. THE HARTREE-FOCK-SLATER EQUATIONS

The Hartree-Fock equations for a diatomic molecule read :

$$H_{HF}(\underline{r})\ \varphi_i(\underline{r}) = \varepsilon_i\ \varphi_i(\underline{r}) \qquad ; \ i = 1,2,\ldots,N \quad . \qquad (1)$$

M. Defranceschi and J. Delhalle (eds.), Numerical Determination of the Electronic Structure of Atoms, Diatomic and Polyatomic Molecules, 275–282.

With

$$H_{HF}(\underline{r}) = h(\underline{r}) + V^{c}(\underline{r}) + V^{x}(\underline{r}) \qquad (2)$$

$$h(\underline{r}) = -\frac{1}{2}\nabla^{2} - \frac{Z_1}{|\underline{r}_1 - \underline{r}|} - \frac{Z_2}{|\underline{r}_1 - \underline{r}|} \qquad (3)$$

$$V^{c}(\underline{r}) = \sum_{i=1}^{N} \int \varphi_i^{*}(\underline{r}') \frac{1}{\underline{r} - \underline{r}'} \varphi_i(\underline{r}')\, d\underline{r}'$$

and

$$V^{x}(\underline{r})\ \varphi_j(\underline{r}) = \sum_{i=1}^{N} \int \varphi_i^{*}(\underline{r}') \frac{1}{|\underline{r} - \underline{r}'|} \varphi_j(\underline{r}')\, d\underline{r}'\ \varphi_i(\underline{r}) .$$

The interelectronic Coulomb potential $V^{c}(\underline{r})$ satisfies the Poisson equation

$$\nabla^{2}\ V^{c}(\underline{r}) = -\ 4\pi\ \rho(\underline{r}) \qquad (4)$$

where $\rho(\underline{r})$ is the total electronic density.
In the Hartree-Fock-Slater equations the nonlocal exchange potential $V^{x}(\underline{r})$ of the Hartree-Fock equations is replaced by the local exchange potential

$$V^{x}(\underline{r}) = -\ 3/2\ \alpha\ (3/\pi\ \rho(\underline{r}))^{1/3} \qquad (5)$$

where α is a constant. We use this simplified form of a local exchange with $\alpha = 0.7$.
The total molecular energy then is

$$E_{tot} = \sum_{i=1}^{N} \varepsilon_i - 1/2 \int \rho(\underline{r})\ V^{c}(\underline{r})\ d\underline{r}\ - \qquad (6)$$

$$-\ 1/4 \int \rho(\underline{r})\ V^{x}(\underline{r})\ d\underline{r} \qquad .$$

The differential equations to be solved are the Schrödinger type equation derived from Equ. (1) with potentials kept fixed, and the Poisson equation (4) relating the potential V^{c} to the electronic density

3. THE COORDINATE SYSTEM

To solve the Hartree-Fock-Slater differential equations for a diatomic molecule numerically we chose the prolate spheroidal coordinates

$$
\begin{aligned}
x_1 &= R/2 \sinh(s) \sin(t) \cos(\varphi) \\
x_2 &= R/2 \sinh(s) \sin(t) \sin(\varphi) \qquad (7) \\
x_3 &= R/2 \cosh(s) \cos(t)
\end{aligned}
$$

which where also used by Laaksonen et al.[4] and Becke[5] for their num erical approach. Within these coordinates the unknown one electron wave functions are:

$$\varphi_i(s,t,\ ) = \varphi_i(s,t) \exp(i\ m\ \varphi) \qquad ; \qquad (8)$$

m denotes the projection of the angular momentum onto the internuclear axis.
Using this ansatz the Schrödinger type equation (1) can be rewritten as a two dimensional differential equation

$$
\begin{aligned}
&- \frac{1}{2} \frac{\partial}{\partial s}\left(K_1(s,t) \frac{\partial \varphi_i}{\partial s} \right) - \frac{1}{2} \frac{\partial}{\partial t}\left(K_2(s,t) \frac{\partial \varphi_i}{\partial t} \right) - \\
&- 1/2\ m^2\ K_3(s,t)\ \varphi_i(s,t) + K_4(s,t)\ V(s,t)\ \varphi_i(s,t) \qquad (9) \\
&= \varepsilon_i\ K_4(s,t)\ \varphi_i(s,t)
\end{aligned}
$$

Similary we get for the Poisson equation (4)

$$
\begin{aligned}
&- \frac{1}{2} \frac{\partial}{\partial s}\left(K_1(s,t) \frac{\partial V^c}{\partial s} \right) - \frac{1}{2} \frac{\partial}{\partial t}\left(K_2(s,t) \frac{\partial V^c}{\partial t} \right) \qquad (10) \\
&= -4\pi\ K_4(s,t)\ \rho(s,t)
\end{aligned}
$$

with

$$
\begin{aligned}
K_1(s,t) &= K_2(s,t) = R/2 \sinh(s) \sin(t) \\
K_3(s,t) &= R/2\ (\sinh^2(s) + \sin^2(t))/(\sinh(s) \sin(t)) \qquad (11) \\
K_4(s,t) &= R^3/2\ (\sinh^2(s) + \sin^2(t)) \sinh(s) \sin(t) \quad .
\end{aligned}
$$

4. THE FINITE ELEMENT METHOD[1-3]

The FEM orginally was developed in engineering science in order to calculate static and dynamic stresses of complicated constructions. In order to apply the FEM to the solution of the Hartree-Fock-Slater equations, one best starts from the equivalent variational principle

$$I_1 = \iint \{ -\frac{1}{4} K_1 \left(\frac{\partial \varphi_i}{\partial s}\right)^2 - \frac{1}{4} K_2 \left(\frac{\partial \varphi_i}{\partial t}\right)^2 - \frac{1}{4} m^2 K_3 (\varphi_i)^2 - \frac{1}{2} (V - \varepsilon_i) K_4 (\varphi_i)^2 \} \, ds \, dt \tag{12}$$

for the Schrödinger type equation (9) and

$$I_2 = \iint \{ -\frac{1}{2} K_1 \left(\frac{\partial V^c}{\partial s}\right)^2 - \frac{1}{4} K_2 \left(\frac{\partial V^c}{\partial t}\right)^2 + 4\pi K_4(s,t) \rho(s,t) V^c \} \, ds \, dt \tag{13}$$

for the Poisson equation (10).
We subdivide the two dimensional space by the use of triangles. On each element e a number of points, the nodal points, are chosen. If $u^e(s,t)$ is the approximation of the unknown function on the element with the number e, this function is written as a linear combination of the nodal values

$$u^e(s,t) = \sum_{i=1}^{p} u_i^e \, N_i^e(s,t) \tag{14}$$

were u_i^e is the i-th nodal value of the e-th element. The function $N_i^e(s,t)$ is the i-th shape function. The sum runs over all p nodal values of the element.
Because of the subdivision of the region into N_s elements the functionals I_1 equ. (12) and I_2 equ. (13) can be written as a sum of element integrals

$$I_1 = \sum_{e=1}^{N_s} I_1^e \qquad \text{and} \qquad I_2 = \sum_{e=1}^{N_s} I_2^e \tag{15}$$

The substitution of the trial function (14) into the element integrals (15) leads to

$$I_1^e = (\underline{u}^e)^t \cdot \underline{\underline{H}}^e \cdot \underline{u}^e - (\underline{u}^e)^t \cdot \underline{\underline{S}}^e \cdot \underline{u}^e \tag{16}$$

with the matrix elements

$$H_{ij} = -\frac{1}{4} \iint \{ K_1 \frac{\partial N_i^e}{\partial s} \frac{\partial N_j^e}{\partial s} + K_2 \frac{\partial N_i^e}{\partial t} \frac{\partial N_j^e}{\partial t} + (K_3 m^2 + 2 K_4 V) N_i^e N_j^e \} \, ds \, dt \tag{17}$$

and

$$S_{ij} = \iint \frac{1}{2} K_4 \, N_i^e \, N_j^e \; ds \; dt \tag{18}$$

and for the Poisson equation similary

$$I_2^e = (\underline{u}^e)^t \cdot \underline{\underline{D}}^e \cdot \underline{u}^e - (\underline{u}^e)^t \cdot \underline{d}^e \tag{19}$$

with the matrix elements

$$D_{ij} = \frac{1}{2} \iint \{ K_1 \frac{\partial N_i^e}{\partial s} \frac{\partial N_j^e}{\partial s} + K_2 \frac{\partial N_i^e}{\partial t} \frac{\partial N_j^e}{\partial t} \} \; ds \; dt \tag{20}$$

and the vector elements

$$d_i = 4\pi \iint K_4 \, \rho(s,t) \, N_i^e \; ds \; dt \quad . \tag{21}$$

The vector $\underline{u}^e$ is the element nodal vector of the e-th element.
Adding the contributions of all elements leads to the expressions :

$$I_1 = \underline{u}^t \cdot \underline{\underline{H}} \cdot \underline{u} - \underline{u}^t \cdot \underline{\underline{S}} \cdot \underline{u} \tag{22}$$

and

$$I_2 = \underline{u}^t \cdot \underline{\underline{D}} \cdot \underline{u} - \underline{u}^t \cdot \underline{d} \tag{23}$$

where $\underline{u}^t$ denotes the global nodal vector. Minimizing I_1 and I_2 with respect to the nodal values leads to the matrix eigenvalue problem

$$\underline{\underline{H}} \cdot \underline{u} = \varepsilon \cdot \underline{\underline{S}} \cdot \underline{u} \tag{24}$$

for the Schrödinger type equation and to the matrix equation

$$\underline{\underline{D}} \cdot \underline{u} = \underline{d} \tag{25}$$

for the Poisson equation.
The calculation of the matrix elements (17, 18, 20, 21) is done numerically with a conical product Gauss integration rule[6].
To account for the asymptotic decrease of the potential V^c we introduce a function g(s,t)

$$V^c = g(s,t) \cdot f^v(s,t) \tag{26}$$

and solve for the function $f^v(s,t)$ by the FEM.

5. RESULTS

For the calculations the points were distributed equidistant in both coordinates for simplicity. Such an equidistant point distribution is not at all optimal and it will be shown that point distributions which are physically more adequate will increase the accuracy.
In Tab. I we compare the convergence of the results with increasing grid size for both 5-th and for 6-th order polynomials for the system N_2. These results are expected to be accurate within 10^{-8} a.u. .

Table I:

N_2, internuclear distance R = 2.07 a.u.

	Laaksonen[4]	this work, 5-th order polynomials	
Points	5989	3136	4356
E_T	-108.346622	-108.3466076	-108.3466090
$(1\sigma_g)$	-13.981070	-13.98106828	-13.98106840
$(1\sigma_u)$	-13.979661	-13.97965838	-13.97965850
$(2\sigma_g)$	-1.007215	-1.00721472	-1.00721471
$(2\sigma_u)$	-0.460725	-0.46072505	-0.46072505
$(1\pi_u)$	-0.404235	-0.40423462	-0.40423462
$(3\sigma_g)$	-0.350058	0.35005852	-0.35005852

this work, 6-th order polynomials

Points	2401	3025	4489
E_T	-108.3466090	-108.34660925	-108.34660934
$(1\sigma_g)$	-13.9810686	-13.98106844	-13.98106844
$(1\sigma_u)$	-13.9796587	-13.97965854	-13.97965854
$(2\sigma_g)$	-1.00721474	-1.00721471	-1.00721471
$(2\sigma_u)$	-0.46072506	-0.46072505	-0.46072505
$(1\pi_u)$	-0.40423461	-0.40423462	-0.40423462
$(3\sigma_g)$	-0.35005852	0.35005852	-0.35005852

Total energy and energy eigenvalues of the system N_2 for different grid sizes and different order of the polynomials on the elements. For the largest grids the last figure is uncertain. All values are given in a.u. .

In Tab. II a step towards an optimized point distribution is documented. In this table we compare the results for the system N_2 for a 31*31 points grid of 6-th order. In mesh A we use an equidistant distribution of the points whereas in mesh B the s-coordinates (Equ. 7) of the vertices of the triangles are distributed logarithmically improving the accuracy by about a factor of 15.

Table II:

N_2, internuclear distance R = 2.07 a.u.
6-th order polynomials

	Mesh A	Mesh B
Points	961	961
E_T	-108.34645	-108.34659
$(1\sigma_g)$	-13.98104	-13.981066
$(1\sigma_u)$	-13.97963	-13.979656
$(2\sigma_g)$	-1.0072143	-1.0072150
$(2\sigma_u)$	-0.4607255	-0.4607253
$(1\pi_u)$	-0.4042346	-0.4042347
$(3\sigma_g)$	-0.3500577	0.3500587

Total energy and energy eigenvalues of the system N_2 for two different point distributions. Mesh A refers to a equidistant grid with all elements of equal size whereas mesh B refers to a logarithmically point distribution in the s-coordinate. All values are given in a.u. .

The results presented are by 2 orders of magnitude more accurate then the results achieved by Laaksonen et al.[4] with the finite difference method, although we used a much smaller number of points.
A Hartree-Fock approach to diatomic molecules will be one of our next goals. The first attempt towards an optimized element distribution shows a very nice increase in accuracy. With better adapted element distributions one may tackle the 3-dimensional problems within reasonable computer times.

Acknowledgement

This work was partially supported by the Deutsche Forschungsgemeinschaft (DFG) and the Gesellschaft für Schwerionenforschung Darmstadt (GSI).

References

1. G. Strang, G. Fix, An Analysis of the Finite Element Method, Prentice-Hall, Englewood Cliffs (1973)

2. D.H. Norrie, G. de Vries, An Introduction to Finite Element Analysis, Academic Press, New York (1978)

3. H.R. Schwarz, Methode der finiten Elemente, Teubner, Stuttgart (1980)

4. L. Laaksonen, D. Sundholm and P. Pyykkö, Comp. Phys. Rep. 4, 313 (1986)

5. A.D. Becke, J. Chem. Phys. 76, 6037 (1982)

6. A.H. Stroud, Approximate calculation of multiple integrals, Prentice-Hall, Englewood Cliffs (1971)

MOMENTUM SPACE COORDINATE TRANSFORMATIONS AND THEIR USE IN NUMERICAL ORBITAL CALCULATIONS.

Y. ISHIKAWA
Department of Chemistry, University of Puerto Rico
Rio Pedras, Puerto Rico 00931 USA

S.A. ALEXANDER
Department of Physics, University of Florida
Gainesville, Florida 32611

ABSTRACT. The truncation of momentum space is a major source of error in the calculation of accurate numerical orbitals for polyatomic systems. We show that coordinate transformations can eliminate this problem. Transformations of both polar and Cartesian coordinates are discussed.

1. Introduction.

For atoms and diatomic molecules, the fully numerical solution of the Hartree-Fock (HF) equations has produced very accurate results[1-6]. The success of these methods is due in large part to the use of appropriate coordinate systems. These choices have analytically removed the Coulomb singularities from the HF equations and made the evaluation of this operator much more efficient. For a general polyatomic molecule, however, no coordinate system in position space can eliminate all of the Coulomb singularities. In 1981 Navaza and Tsoucaris showed that the momentum space formulation of the HF equations has only one, removable singularity at the origin [7]. A number of different algorithms have been developed to evaluate these equations using both polar [7-15] and Cartesian [16-17] coordinates systems. In this paper we show that coordinate transformations of momentum space variables can eliminate several computational problems which currently limit the accuracy of these calculations.

M. Defranceschi and J. Delhalle (eds.), Numerical Determination of the Electronic Structure of Atoms, Diatomic and Polyatomic Molecules, 283–288.

2. Coordinate transformations

In all of the momentum space algorithms which have been developed to date, there are two major impediments to the calculation of accurate numerical orbitals. The first is the error caused by the truncation of momentum space in the numerical integrations [14,17,18,19]. Because the wavefunction decreases rather slowly as a function of p, like p^{-4}, this error will dominate all the others unless a large volume of space is sampled. The second impediment is caused by singularities which are located on or near the imaginary axis. That each orbital should have such singularities should come as no surprise. Liouville's theorem states that a complex function which has no singularities is a constant. Thus, any analytical function which deviates from a constant value must have at least one singularity in the finite complex plane [20,21]. Although these singularities are less of a computational problem than those found on the real axis, under the proper circumstances they can drastically affect the convergence of numerical integrations [22-24].

Both these problems can be easily solved using coordinate transformations. One example of such a transformation was recently examined in Refs. 18 and 19. In these papers, Rodriguez and Ishikawa evaluated the momentum space HF equations for the Helium atom and the H_2^+ ion using a Gauss- Legendre quadrature and the polar coordinates $\mathbf{p} = (p, \theta, \phi)$ where $p = [0,\infty]$, $\theta = [0,\pi]$ and $\phi = [0,2\pi]$. Since the variable p has an infinite range, this calculation cannot be done with a Gauss-Legendre quadrature. Earlier calculations avoided this problem by integrating only up to some value of p [7-15]. Rodriguez and Ishikawa showed, however, that the transformation $p = \tan(\alpha/2)$ was much more accurate. Not only does this approach eliminate truncation error but because the variable α has a finite range, $0 \leq \alpha \leq \pi$, a Gauss- Legendre quadrature is quite appropriate. In the model problem used by Delhalle and Defranceschi [14], the 1S orbital of the He atom was approximated by

$$\Phi(\mathbf{p}) = (8\zeta^5)^{1/2}\,\pi^{-1}\,(p^2 + \zeta^2)^{-2} \qquad (1)$$

where ζ=1.6875. Delhalle and Defranceschi found that both a large number of grid points and a large volume were needed to obtain accurate results when the integration over p was performed only up to some maximum value. In contrast, a 20 point Gauss-Legendre integration in the transformed coordinates was able to compute the kinetic energy of this system to eleven significant digit accuracy [25]. Such a rapid convergence also suggests that this coordinate transform is not significantly affected by the singularities in Eq. 1 at $\pm i\zeta$.

When this coordinate transformation was applied to the H_2^+ ion, the convergence was found to be much less rapid than in the Helium calculations. In particular, a relatively large number of gridpoints in both the α and θ variables were needed in order to compute the nuclear attraction term

$$\sum_m Z_m \int d^3q\, \Phi(\mathbf{q})\, e^{-i(\mathbf{p}-\mathbf{q}).\mathbf{R}_m} / |\mathbf{p}-\mathbf{q}|^2 \qquad (2)$$

to a high accuracy. One reason is because of the singularity at $\mathbf{p}=\mathbf{q}$. Although this singularity was numerically removed by an addition and substraction procedure, the Kantorovitch technique [26], a more sophisticated approach might be able to improve the convergence. Another reason for the large number of gridpoints is the presence of the scattering factor. In momentum space this function contains all of the information about the geometry of the molecule. Not only can it oscillate quite rapidly but in a Hartree - Fock calculation the scattering factor also induces oscillations in each

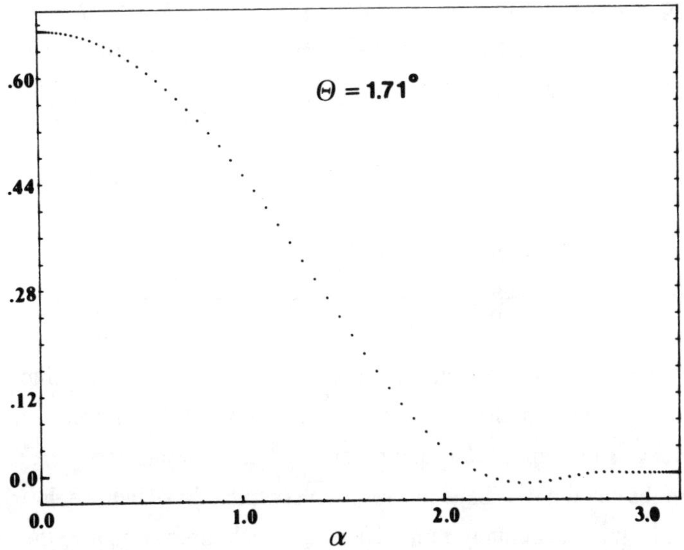

Fig. 1. *Amplitude of the H_2^+ wave function in transformed polar coordinates.*

momentum space orbital. In any type of numerical integration these oscillations will have a slow convergence. Figure 1 is a plot of the numerical orbital for H_2^+ as a function of the transformed coordinates. Although only a single value of θ is shown, the oscillations occur in both α and θ. In their calculations on H_2^+, Rodriguez and Ishikawa found that many more gridpoints over the angle θ were needed to achieve a high accuracy than over the variable α [19]. We believe this to be a feature of polar coordinates in general. As the number of atoms in a molecule increases, we expect this problem to become much worse.

In order to soften the difficulties associated with the nuclear attraction term, we are currently experimenting with a coordinate transformation of Cartesian variables. This expression has the form $p_x=\tan(\alpha_x)$, $p_y=\tan(\alpha_y)$ and $p_z=\tan(\alpha_z)$ where $-\pi/2 \leq \alpha_x, \alpha_y, \alpha_z \leq \pi/2$. To study the effectiveness of this transform, we first examined the kinetic energy of the Helium for the orbital in Eq. 1. Below we compare these results with those obtained using a trapezoidal integration on an evenly spaced grid. Like the calculations in Refs. 16 and 17, this grid covers the finite cube [-P,P].

Cube	$N=64^3$	$N=128^3$	$N=256^3$	$N=512^3$
P=10	1.410758	1.410768	1.410771	1.410771
P=20	1.422044	1.422116	1.422116	1.422116
P=40	1.332083	1.423540	1.423611	1.423611
P=80	0.328201	1.332265	1.423730	1.423801
	$N=20^3$	$N=40^3$	$N=60^3$	$N=80^3$
Transformed	1.424477	1.423842	1.423829	1.423828

These numbers show that this new coordinate transformation, like the earlier one, converges quickly to the exact energy i.e. 1.423828 Hartrees. In constrast, the calculations using trapezoidal integration need more points and quite a large size cube. Although the calculations have not yet been completed, we believe that the singularity in Eq. 2 can be more accurately removed using a technique similar to the one described in

Ref. 17. We also expect to be able to more evenly sample momentum space with this Cartesian transform than with the polar version.

3. Conclusions

Coordinate transformations provide an alternative to the direct use of the momentum space variables in numerical orbital calculations. These new coordinates cover the entire space and thus eliminate truncation error. Several model calculations have been performed and the results are encouraging. Future work will examine the effectiveness of this idea in more detail.

Acknowledgements

We authors wish to thank Drs. E. Lassettre, J. Delhalle and D. Kolb for helpful discussions. This research has been supported by the U.S. National Science Foundation, the Army Research Office and by Control Data with a PACER fellowship.

References

[1] C. Froese Fisher, The Hartree-Fock Method for Atoms (Wiley, New York, 1977) and references therein.

[2] E.A. McCullough Jr., Comp.Phys.Rep. 4, 265(1986) and references therein.

[3] L. Laaksonen, P. Pyykkö and D. Sundholm, Comp.Phys.Rep. 4, 315(1986) and references therein.

[4] A.D. Becke, J.Chem.Phys. 76, 6037(1983)

[5] P. Pyykkö, this proceedings

[6] L. Adamowicz, this proceedings

[7] J. Navaza and G. Tsoucaris, Phys.Rev. A 24, 683(1981)

[8] M. Defranceschi, M. Suard and G. Berthier, Comptes Rendus 299, 9(1984)

[9] M. Defranceschi, M. Suard and G. Berthier, Int. J. Quantum Chem. 25, 863(1984)

[10] M. Defranceschi, M. Suard and G. Berthier, Comptes Rendus Acad. Sci. serie II 299, 9(1984)

[11] G. Berthier, M. Defranceschi, J. Navaza, M.Suard and G. Tsoucaris, Theochem

120, 343(1985)

[12] M. Defranceschi and J. Delhalle, Comptes Rendus Acad. Sci. serie II 301, 1405(1985)

[13] M. Defranceschi and J. Delhalle, Phys. Rev. B 34, 5862(1986)

[14] J. Delhalle and M. Defranceschi, Int. J. Quantum Chem S21, 425(1987)

[15] G. Berthier, M. Defranceschi and J. Delhalle, this proceedings

[16] S.A. Alexander and H.J. Monkhorst, Int. J. Quantum Chem. 32, 361(1987)

[17] S.A. Alexander, R.L. Coldwell and H.J. Monkhorst, to appear in J. Comp. Phys.

[18] W. Rodriguez and Y. Ishikawa, to appear in Chem. Phys. Lett.

[19] W. Rodriguez and Y. Ishikawa, to appear in Int. J. Quantum Chem. Symp. 22

[20] E.N. Lassettre, J. Chem.Phys. 83, 1709(1985)

[21] E.N. Lassettre, this proceedings

[22] E. Eisner, CACM 10, 239(1969)

[23] D. Kolb, private communications

[24] P.J. Davis and P. Rabinowicz, " Methods of Numerical Integration", (Academic Press, 1975), pages 141-142

[25] Y. Ishikawa, unpublished results

[26] L.V. Kantorovitch, Mat. Sb 41, 235(1934)

MØLLER PLESSET CALCULATIONS WITH EXPLICITLY CORRELATED WAVE FUNCTIONS

Werner Kutzelnigg and Wim Klopper
Lehrstuhl für Theoretische Chemie
Ruhr-Universität Bochum
4630 Bochum

ABSTRACT: Møller-Plesset calculations to second order have been carried out for various closed-shell atoms and molecules with a new functional, including r_{12}-dependent pair correlation functions, which takes care of the correlation cusp. The calculated second-order pair energies are accurate to within a few millihartree in comparison with the estimated exact values. In particular, second-order energies of 425.1, 819.0, 360.3 and 265.9 mE_h have been obtained for the ground states of Mg, Ca, H_2O and CH_4 respectively.

1. Introduction

One of the big bottlenecks of traditional ab-initio quantum chemistry is the slow convergence of the CI (configuration interaction) expansion. The reasons for this slow convergence are now essentially understood [1-4]. Fast convergence is only possible if the basis used for the expansion is able to describe the singularities of the wave function correctly. Test calculations on He like ions in ground and excited states have shown that the ansatz [5]

$$\psi(1,2) = \frac{1}{2} r_{12}\phi(1,2) + \chi(1,2) \qquad (1)$$

with ϕ the unperturbed wave function and χ written in a CI or partial wave expansion, is very powerful. While the energy increments for singlet states (in 2nd order perturbation theory) converge like $\sim (l+\frac{1}{2})^{-4}$ for the conventional partial wave expansion (of ψ) - or like $\sim (l+\frac{1}{2})^{-6}$ for triplet states -, the convergence of the energy increments corresponding to the partial wave expansion of χ goes $\sim (l+\frac{1}{2})^{-8}$. A similar fast convergence (though not exactly $\sim (l+\frac{1}{2})^{-8}$) with the same ansatz (1) has been observed in variational calculations for He-like ions [5]. With l-values up to 5, microhartree accuracy is achieved, while with the traditional partial wave expansion one has to go up to $l = 50$ for the same accuracy [6]. The slowly convergent contributions in the conventional CI come mainly from terms that can easily be evaluated in closed form. If one expands the expectation value for the He ground state

$$\frac{<\phi|(1+\frac{1}{2}r_{12})H(1+\frac{1}{2}r_{12})|\phi>}{<\phi|(1+\frac{1}{2}r_{12})^2|\phi>} \qquad (2)$$

in its partial wave expansion one obtains

$$-0.071(l+\frac{1}{2})^{-4} - 0.029(l+\frac{1}{2})^{-5} + 0[(l+\frac{1}{2})^{-6}] \qquad (3)$$

M. Defranceschi and J. Delhalle (eds.), Numerical Determination of the Electronic Structure of Atoms, Diatomic and Polyatomic Molecules, 289–293.

while the corresponding expansion of the variational energy of the He ground state is [6]

$$E = -0.074(l + \frac{1}{2})^{-4} - 0.031(l + \frac{1}{2})^{-5} + 0[(l + \frac{1}{2})^{-6}] \tag{4}$$

2. Møller-Plesset-calculations to 2nd order

The Hylleraas functional for the second-order Møller-Plesset correlation energy decouples into a sum of pair contributions

$$\begin{aligned} f(u_{ij}) = 2Re < [ij]|r_{12}^{-1}|u_{ij}(1,2) > \\ + < u_{ij}(1,2)|F(1) + F(2) - \varepsilon_i - \varepsilon_j|u_{ij}(1,2) > \end{aligned} \tag{5}$$

where $[i,j]$ is the unperturbed pair function and u_{ij} is the 1st order pair function, F is the Fock operator and ε_i an orbital energy. We make the ansatz

$$u_{ij}(1,2) = \frac{1}{2}\{1 - P(1)\}\{1 - P(2)\}r_{12}[ij] - \omega_{ij}(1,2) \tag{6}$$

where P is the projector to the occupied Hartree-Fock states, c_{ij} is a variational parameter and ω_{ij} is expanded in terms of orbital products. In practice singlet and triplet pairs rather than simple spin-orbital product pairs are taken. In order to evaluate the pair functions and the pair correlation energies some assumptions and approximations are made that are described in ref. 7 and that all become exact in the limit of a complete basis. The errors due to these approximations decrease much faster with increasing basis size than those of a conventional calculation . Applications to the ground states of Ne, HF and H_2O [7] and Ar [8] have been published. In the following tables so far unpublished results for Mg, Ca, H_2O and CH_4 are collected. For Mg our results are compared with the extrapolated supposedly exact value of Jankowski et al. [10]. The agreement is within 3 millihartree ($<$ 1% of the correlation energy), a similar agreement has been found for Ne [7] or Ar [8], such that we can expect that our result for Ca is similarly close to the exact 2nd order energy. This should also be the case for H_2O (the present results have been obtained with a larger and better optimized basis than in ref. 7) and CH_4. For H_2O the agreement between our calculated values and the rather tricky extrapolations of Petersson [12] et al. is striking.

The method has also been implemented in a localized representation, 3rd order calculations are in progress and a coupled cluster approach on the same line is planned.

References

1. Schwartz C (1962) Phys. Rev. 126:1015
2. Kato T (1957) Commun. Pure. Appl. Math. 10:151
3. Hill RN (1985) J. Chem. Phys. 38:1173
4. Morgan JD, in this volume
5. Kutzelnigg W (1985) Theoret. Chim. Acta 68:445
6. Carrol DP, Silverstone HJ, Metzger RM (1979) J. Chem. Phys 71:4142
7. Klopper W, Kutzelnigg W (1986) Chem. Phys. Lett. 134:17
8. Kutzelnigg W, J. Mol. Struct. (Theochem) in press
9. Müller W, Flesch J, Meyer W (1984) J. Chem. Phys. 80:3297
10. Jankowski K, Malinowski P, Polasik M (1979) J. Phys. B12:3157
11. Müller W, Flesch J, Meyer W (1984) J. Chem. Phys. 80:3297
12. Petersson GA, Braunstein M (1985) J. Chem. Phys. 83:5129

Table 1

Second order pair energies of the Mg atom in mE_h with signs reversed[a)]

Pair ij	e_{ij}[b)]			f_{ij}[c)]			e_{JMP}[e)]	c_{ij}[d)]
	singlet	triplet	sum	singlet	triplet	sum		s/t
$1s^2$	34.9	-	34.9	39.4	-	39.4	39.7	.84
$2s^2$	9.6	-	9.6	11.4	-	11.4	11.6	.92
$2s2p$	40.3	21.6	61.9	55.7	23.4	79.1	79.3	.87/.39
$2p^2$	111.0	81.4	192.3	135.0	84.2	219.2	218.6	.81/.34
$3s^2$	22.6	-	22.6	25.7	-	25.7	27.7	.90
$1s2s$	3.9	1.4	5.3	4.3	1.5	5.8	6.0	.69/.26
$1s2p$	7.7	15.9	23.5	9.6	16.4	26.1	26.3	.63/.27
$1s3s$	.16	.08	.23	.17	.09	.26	.27	.24/.15
$2s3s$	.75	.67	1.4	.90	.77	1.7	1.7	.56/.32
$2p3s$	4.9	9.8	14.7	5.8	10.6	16.4	16.6	.42/.26
total			366.4			425.1	427.9	

a) Uncontracted 14s/11p basis set taken from ref. 9
Extended by 5d (38.4, 9.6, 2.4, .6, .15) + 1f (6.4).
b) Conventional second order pair energy in this basis. *c*) Calculated pair energy.
d) Variational parameter. *e*) Extrapolated pair energy of ref. 10

Table 2

Second order pair energies of the CH_4 molecule in mE_h with signs reversed[a)]

Pair ij	e_{ij}[b)]			f_{ij}[c)]			c_{ij}[d)]
	singlet	triplet	sum	singlet	triplet	sum	s/t
$1a_1^2$	31.6	-	31.6	41.2	-	41.2	.69
$1a_1 2a_1$	2.0	1.1	3.1	2.3	1.2	3.6	.27/.12
$1a_1 t_2$	1.5	5.3	6.8	2.5	6.0	8.5	.24/.13
$2a_1^2$	8.7	-	8.7	10.2	-	10.2	.57
t_2^2	82.2	38.3	120.6	97.2	40.1	137.4	.50/.19
$2a_1 t_2$	38.6	14.0	52.6	50.1	15.0	65.0	.64/.17
total			223.4			265.9	

a) Tedrahedral geometry, $R_{C-H} = 2.067\ a_0$.
C: 5111111/211111 contracted 11s/7p Huzinaga + 2d (.35, 1.4).
H: 3111 contracted Huzinaga + 1p (.65).
b) Conventional second order pair energy in this basis. b) c) d) as on Table 1

Table 3

Second order pair energies of the Ca atom in mE_h with signs reversed[a)]

Pair ij	e_{ij} [b)]			f_{ij} [c)]			c_{ij} [d)]
	singlet	triplet	sum	singlet	triplet	sum	s/t
$1s^2$	31.4	-	31.4	37.2	-	37.2	.79
$2s^2$	8.1	-	8.1	11.2	-	11.2	.83
$2s2p$	22.0	15.9	37.9	50.0	20.4	70.4	.84/.33
$2p^2$	93.6	69.5	163.1	142.8	77.6	220.3	.80/.34
$3s^2$	8.2	-	8.2	11.3	-	11.3	.91
$3s3p$	21.9	9.1	30.9	49.4	13.4	62.8	.81/.37
$3p^2$	87.7	68.8	156.5	140.1	76.9	217.1	.78/.35
$4s^2$	19.2	-	19.2	21.4	-	21.4	.64
$1s2s$	4.6	1.1	5.6	5.1	1.4	6.5	.54/.25
$1s2p$	5.3	17.0	22.2	14.0	20.8	34.7	.60/.31
$1s3s$	.56	.18	.74	.57	.18	.75	.22/.0[e)]
$1s3p$	.46	1.6	2.0	1.0	1.8	2.8	.38/.16
$1s4s$	.03	.01	.04	.03	.01	.04	.0/.0[e)]
$2s3s$	1.9	1.1	2.9	2.3	1.3	3.6	.68/.33
$2p3s$	5.9	7.7	13.6	7.1	8.8	15.9	.40/.19
$2s3p$	3.6	5.8	9.4	4.4	6.6	11.0	.37/.22
$2p3p$	23.7	31.5	55.2	26.7	34.1	60.8	.28/.17
$2s4s$	.11	.08	.19	.11	.08	.19	.0/.0[e)]
$2p4s$	.50	.84	1.3	.50	.90	1.4	.04/.06
$3s4s$	.94	.67	1.6	.99	.68	1.7	.19/.05
$3p4s$	8.3	17.3	25.6	9.4	18.5	27.9	.26/.17
total			595.7			819.0	

a) Uncontracted 16s/13p basis set taken from ref. 11
Extended by 5d (25.6, 6.4, 1.6, .4, .1) – 1f (4.0).
b) Conventional second order pair energy in this basis.
c) Calculated pair energy.
d) Variational parameter.
e) No r_{12}-dependent corrections included.

Table 4

Second order pair energies of the H_2O molecule in mE_h with signs reversed[a)]

Pair ij	e_{ij} [b)]			f_{ij} [c)]			e_{PB} [d)]	c_{ij} [e)]
	singlet	triplet	sum	singlet	triplet	sum		s/t
$1a_1^2$	31.5	-	31.5	40.9	-	40.9	41.0	.70
$1a_1 2a_1$	2.4	1.3	3.7	2.9	1.4	4.3	5.0	.33/.13
$1a_1 1b_2$	.34	1.8	2.1	1.3	2.5	3.8	3.9	.32/.18
$1a_1 3a_1$	.66	2.2	2.9	1.9	3.2	5.1	5.3	.35/.20
$1a_1 1b_1$	.46	2.6	3.1	2.0	3.8	5.8	6.0	.36/.21
$2a_1^2$	10.3	-	10.3	13.1	-	13.1	12.7	.70
$1b_2^2$	21.0	-	21.0	25.6	-	25.6	25.3	.52
$3a_1^2$	20.5	-	20.5	25.5	-	25.5	25.3	.53
$1b_1^2$	20.7	-	20.7	26.4	-	26.4	25.6	.52
$2a_1 1b_2$	14.3	7.2	21.5	21.9	8.1	29.9	29.4	.68/.21
$2a_1 3a_1$	11.7	7.3	19.0	17.7	8.5	26.2	27.4	.64/.23
$2a_1 1b_1$	12.1	7.8	19.9	18.8	9.5	28.3	28.6	.64/.24
$1b_2 3a_1$	14.2	21.6	35.8	18.0	23.3	41.3	42.7	.51/.22
$1b_2 1b_1$	10.9	23.1	33.9	15.1	25.1	40.2	40.0	.52/.23
$3a_1 1b_1$	12.5	23.9	36.4	17.9	26.1	43.9	43.1	.56/.23
total			282.3			360.3	361.1	

a) $R_{O-H} = 1.8088\ a_0, \alpha = 104.5°$.
O: 5111111/211111 contracted 11s/7p Huzinaga + 2d (.35, 1.4),
H: 3111 contracted Huzinaga + 1p (.65).

b) Conventional second order pair energy in this basis.

c) Calculated pair energy.

d) Extrapolated pair energy of ref. 12

e) Variational parameter.

ON THE COULOMB STURMIAN BASIS

Alfred Maquet, Philippe Martin and Valérie Véniard
Laboratoire de Chimie Physique, Université Pierre et Marie Curie
11, Rue Pierre et Marie Curie, F-75231 Paris Cedex 05. France.

Abstract. The Coulomb Sturmian basis is briefly introduced and some recent applications are discussed.

The Coulomb Sturmian basis

Although Coulomb, Sturmian-like, basis have been used since the early days of Quantum Mechanics,(1),(2),(3) it is only in the sixties that a formal theory has been explicited, in particular by Rotenberg (4). The main characteristics of the Coulomb Sturmian basis are most easily displayed by reference to the usual (physical) hydrogenic basis. Indeed, the Sturmian basis can be constructed by solving a Sturmian eigenvalue problem, which is in some sense "conjugate" to the usual Schrödinger eigenenergy problem. Within the Sturmian approach, one considers the energy (E<0) of the system as a fixed parameter and one solves the eigenvalue equation for the coupling parameter of the potential :

Schrödinger Equation	Sturmian Equation
$\left(\frac{p^2}{2} - \frac{Z}{r}\right)\psi = E\psi$	$\left(\frac{p^2}{2} - E\right)\psi = \frac{\alpha}{r}\psi$ [1]
Z = charge of the nucleus (fixed parameter)	E<0 is considered as a fixed parameter
Eigenvalue problem for the energy E (Z fixed)	Eigenvalue problem for the coupling constant α (E<0, fixed)
Spectrum : discrete + continuous	Spectrum : discrete
a) E<0 ; $E_n = -Z^2/(2n^2)$	$\alpha_n = n\,x$; $x = (-2E)^{1/2}$ [2]
$\psi_{nlm}(\mathbf{r}) = Y_{lm}(\theta,\phi)\,R_{nl}(Zr/n)$	$\psi_{nlm}(\mathbf{r}) = Y_{lm}(\theta,\phi)\,S_{nl}(xr)$
b) E>0 ; $n/Z \rightarrow -i/k$	(if one fixes $\alpha_n = Z$ one recovers
$E_k = k^2/2$	$E_n = -Z^2/(2n^2)$)

M. Defranceschi and J. Delhalle (eds.), Numerical Determination of the Electronic Structure of Atoms, Diatomic and Polyatomic Molecules, 295–299.

The Coulomb Sturmian functions S_{nl} have the same formal structure than the usual hydrogenic functions R_{nl} for discrete states :

$$S_{nl}(xr) = N_{nl}\, e^{-xr}\, r^l\, L_n^{2l+1}(2xr) \quad , \qquad [3]$$

they, however differ notably of the hydrogenic ones in several respects : (3)-(5)

i) they are not orthogonal and one has, more precisely : (3)-(5)

$$<S_{n'l'} \mid S_{nl}> = \delta_{l,l'}\, x^{-1}\, [\, n\, \delta_{n,n'} + A_{n'}\, \delta_{n',n-1} + A_n\, \delta_{n',n+1}\,] \qquad [4]$$

where $A_n = -\,[(n-1)\,(n+l+1)]^{1/2}/2$

ii) they are orthogonal with respect to the weight $1/r$, namely, one has :

$$<S_{n'l'} \mid 1/r \mid S_{nl}> = \delta_{n,n'}\, \delta_{l,l'} \qquad [5]$$

which implies a normalization slightly different from the usual one for hydrogenic functions. An important feature of the Sturmians is that they are energy dependent through the parameter $x = (-2E)^{1/2}$. We will come back to this property in the next sections.

The Sturmian ideas have been used in many areas of Physics, including particle physics(6), nuclear physics (8), the three-body problem (4), collision theory (7),(9) and as a technique to solve the Schrödinger equation(5),(6),(10). However, given the importance of the Coulomb interaction, it is in the context of atomic and molecular physics that the Coulomb Sturmian basis has been most often used : a non-exhaustive list of applications includes the multicenter Coulomb problem (11)-(14), the Helium atom (3),(15),(16) (and more generally atomic structure problems (17)) and also the response of atomic Hydrogen in strong external fields. These latter topics will be briefly considered below.

The Helium atom revisited

In order to illustrate the interest of the Sturmian approach in a well defined context we discuss here the Helium problem. The Schrödinger equation $(H - E)\,\psi = 0$ reads explicitly (in atomic units) :

$$\left(\frac{p_1^2}{2} - \frac{2}{r_1} + \frac{p_2^2}{2} - \frac{2}{r_2} + \frac{1}{r_{12}} - E\right) \psi(\mathbf{r}_1, \mathbf{r}_2) = 0 \qquad [6]$$

In order to fully exploit the properties (and in particular its energy dependence) of the Sturmian basis it is convenient to reexpress the total energy of the system as

$$E = -\,(x_1^2 + x_2^2)\,/\,2 \qquad [7]$$

where x_i will be the Sturmian parameter (Eq.[2]) attributed to the electron labelled i. The matrix elements of (H - E) in the two-particle Coulomb Sturmian basis with parameter x_1 and x_2 are easily determined and the Helium bound state energies are then obtained by merely finding the roots in x_1 and x_2 of the secular equation : (5),(10),(16)

$$\det [H - E] = 0 \qquad [8]$$

the energies being obtained from Eq[7]. As the Sturmian set is complete, one expects that a standard truncation procedure will permit to smoothly and efficiently converge towards the correct eigenvalues as one includes more and more configurations (5),(10)(16).

As a simple illustration of the specificity of the method we have considered the simplest ground state case and more precisely the so-called radial limit obtained when only including s-state configurations (18). One has then $x_1=x_2=x=(-2E)^{1/2}$, the value of x being obtained for each configuration by solving the algebraic equation [8]. See the table below. It should be stressed that no variational procedure is involved here, and although x looks like the usual variational parameter η contained in the basis functions, it retains a precise meaning related to its definition in terms of the energy of the system. The extension to excited states, which involves some variational treatment, has been briefly discussed in ref.(16).

Configurations	Energy $\eta = 1.6875$ (18)	Energy $\eta = 2$ (18)	Sturmian parameter x	Energy
$<1s>^2$	-2.847 656	-2.750 000	1.687 500	-2.847 656
$<1s>^2,\ldots,<2s>^2$	-2.860 408	-2.850 225	1.696 281	-2.860 430
$<1s>^2,\ldots,<3s>^2$	-2.877 266	-2.878116	1.696 274	-2.877 346
$<1s>^2,\ldots,<4s>^2$	-2.878 368	-2.878 597	1.696 580	-2.878 382
$<1s>^2,\ldots,<5s>^2$	-2.878 742	-2.878 868	1.696 687	-2.878 748
$<1s>^2, \ldots,<6s>^2$	-2.878 871	-2.878 939	1.696 724	-2.878 873

Table I . Comparison of the Sturmian approach with the variational calculation by Shull and Löwdin, Ref.(18), who used a Laguerre L^{2l+2} basis. The first two columns correspond to the results of Ref.(18) with the indicated variational parameter η.The sturmian parameter x, which formally corresponds to η in the argument of the basis functions, is linked to the energy via Eq.[2].

Hydrogen atom in a strong external field

The Coulomb Sturmian functions provide a natural basis to study the properties of atomic Hydrogen submitted to a strong external electromagnetic field (19).For example, following a suggestion by Edmonds [20], they have been used with particular success in the study of Rydberg atoms in a magnetic field [21]. In this perspective, symmetry-adapted Sturmian basis have been recently introduced, allowing to accurately predict the onset of quantum chaos in such systems [22].

An atom submitted to a strong laser field can experience multiphoton transitions. In Hydrogen, the perturbative description of such processes (description valid in a wide range of laser intensities) requires the evaluation of high order time dependent perturbative amplitudes, expressed in terms of the Coulomb Green's function, $G(E) = (E-H_c)^{-1}$ where H_c is the Coulomb hamiltonian. G(E) can be conveniently expressed as an expansion over the Sturmian basis (for the energy E) [23] which allows to compute with great accuracy high order multiphoton amplitudes [24]. We have very recently used such a Sturmian representation of G(E) within the framework of a study of electron H-atom collisions in the presence of a laser field. We are now able to compute, within any desired accuracy, second-order transition matrix elements of the general form : [25],[26]

$$T(\mathbf{e},\mathbf{q}) = \langle n',l',m'| e^{i\mathbf{q}.\mathbf{r}}\, G(E)\, \mathbf{r}.\mathbf{e}\, |1s\rangle \ , \qquad [9]$$

the final state belonging either to the bound or continuous spectrum [25],[26]. More details on the physical context in which such amplitudes do contribute will be provided at the Conference.

Acknowledgments. The Laboratoire de Chimie Physique de l'Université Pierre et Marie Curie, is an Unité Associée au CNRS, UA 176.

References

(1) P. S. Epstein, Proc. Nat. Acad. Sci. 12, 630 (1926).

(2) B. Podolsky, Proc. Nat. Acad. Sci. 14, 253 (1928).

(3) E. A. Hylleraas, Z. Phys. 48, 469 (1928).

(4) M. Rotenberg, Ann. Phys. (NY) 19, 262 (1962); in *Advances in Atomic and Molecular Physics*, edited by D. R. Bates and I. Estermann (Academic Press, New York, 1970).

(5) J. P. Gazeau, J. Phys. A 13, 3605 (1980).

(6) S. Weinberg, Phys. Rev. 131, 440 (1962).

(7) G. H. Rawitscher and G. Delic, J. Math. Phys. 27, 816 (1986) and references therein.

(8) J. Bang and F. A. Gareev, Phys. Scripta 18, 297 (1978) and references therein.

(9) E. J. Heller and H. A. Yamani, Phys. Rev. A 9, 1201 (1974); R. Shakeshaft, Phys. Rev. A 18, 1930 (1978) and references therein; see also : T. G. Winter, Phys. Rev. A 25,

697 (1982); ibid 33, 3842 (1986); ibid 35, 3799 (1987); P. C. Ojha, J. Math. Phys. 28, 392 (1987).
(10) J. P. Gazeau and A. Maquet, Phys. Rev. A 20, 727 (1979).
(11) B. K. Novosadov, Opt. Spectrosc. 41, 490, (1976).
(12) H. J. Monkhorst and B. Jeriorski, J. Chem. Phys.71, 5268 (1979).
(13) C. Duchon, M. C. Dumont-Lepage and J. P. Gazeau, J. Phys. A 15, 1227 (1982); J. Chem. Phys. 76, 445 (1982).
(14) E. J. Weniger and E. O. Steinborn, Phys. Rev. A 29, 2268 (1984).
(15) H. Shull and P. O. Löwdin, J. Chem. Phys. 23, 1362 (1955).
(16) J. P. Gazeau and A. Maquet, J. Chem. Phys. 73, 5147 (1980).
(17) S.V. Khristenko and S. I. Vetchinkin, Opt. Spectrosc. 31, 269 (1971); see also : G. S. Soloveva, P. F. Gruzdev and A. I. Sherstyuk, Opt. Spectrosc. 57, 473 (1985).
(18) H.Shull and P. O. Löwdin, J. Chem. Phys. 30, 617 (1959).
(19) B. G. Adams, J. Cizek and J. Paldus, Int. J. Quant. Chem. 21, 153 (1982); see also : E. R. Vrscay, Phys. Rev. Lett. 53, 2521 (1984); Phys. Rev. A 31 2054 (1985).
(20) A. R. Edmonds, J. Phys. B 6,1603 (1973).
(21) C. W. Clark and K. T. Taylor, J. Phys. B 13, L737 (1980);ibid 15, 1175 (1982).
(22) D. Delande and J. C. Gay, J. Phys. B 19, L 173 (1986); Phys. Rev. Lett. 59, 1809 (1987).
(23) L. Hostler, J. Math. Phys. 11, 2966 (1970); this expansion is the position space version of Schwinger's O(4) representation : J. Schwinger, J. Math. Phys. 5, 1606 (1964)
(24) A. Maquet, Phys. Rev. A 15, 1088 (1977) and references therein; see also : S. Klarsfeld and A. Maquet, Phys. Lett. 78A, 40 (1980).
(25) S. Jetzke, J. Broad and A. Maquet, J. Phys. B 20, 2887 (1987).
(26) C. J. Joachain, P. Francken, A. Maquet, P. Martin and V. Véniard, to be published.

AN ANALYTICAL L^2 METHOD FOR CONTINUUM AND AUTOIONIZING STATES

R. Moccia
Dipartimento di Chimica e Chimica Industriale
dell' Universita' di Pisa
P. Spizzo
Istituto di Chimica Quantistica ed Energetica
Molecolare del C.N.R.
Via Risorgimento 35 - 56100 PISA - ITALY

INTRODUCTION

In previous works[1-6] we have developed an efficient L^2 method for calculating both the discrete and the continuum spectrum of atomic and molecular systems. It allows to represent all the states by a convenient expansion over manageable analytical L^2 functions and therefore makes very easy to perform all the subsequent calculations of several interesting observables. Moreover, it appears to present computational advantages over the more usual close-coupling methods for the highly correlated continuum states and for the narrow resonances. In this paper we discuss our previous results for atomic Magnesium[4] and extend them by including the phenomenological polarization potential[7] for the intershell correlation effects.

Atomic Magnesium appears to be a system of great experimental and theoretical interest, since many states are strongly correlated and the continuum is highly structured; a very careful numerical treatment is therefore required to obtain reliable results. In addition, the error sources intrinsic to a frozen-core approach (with or without phenomenological intershell corrections) and in the use of the electrostatic Hamiltonian are still small in Magnesium, so that the results may be useful for interpretation purposes. Indeed, the lowest part of the Magnesium spectrum (i.e. that below the 3p $^2P^o$ ionization

M. Defranceschi and J. Delhalle (eds.), Numerical Determination of the Electronic Structure of Atoms, Diatomic and Polyatomic Molecules, 301–310.

threshold) displays very interesting features which correlate, in the independent particle limit, to doubly excited bound states 3pnl. These configurations form the so-called "displaced term system" and may have natural parity $(-1)^L$ (e.g. $3p^2\ ^1S^e$) or non-natural one $(-1)^{L+1}$ (e.g. $3p^2\ ^3P^e$); in this latter case they correlate to true bound states of the electrostatic Hamiltonian, while in the former one to autoionizing states. In this region some assignements are rather controversial[4] and the present results should be helpful in clarifying part of the problem.

Unless specified, all quantities are given in atomic units.

METHOD OF CALCULATION

The states are represented by

$$\Phi_i(X_c,X_v) = \hat{A}\ G_c(X_c)\ G_{vi}(X_v)$$

where $\hat{A}$ is the antisymmetrizer, X_c (X_v) denotes the space-spin coordinates for the ten core (two valence) electrons, G_c is an SCF group function for the core electrons and G_{vi} is a CI group function for the valence electrons. A proper choice of the basis configurations and orbitals employed to build the valence group functions G_{vi} allows to obtain very reliable results, without leading to untractable redundancies or extremely large eigenvalue problems. The lowest 3snl and 3pnl bound states are highly correlated and localized, so that their valence group functions require many configurations and high-l orbitals; simple STO are quite adequate for building up these configurations. This is also true for the localized components of the lowest 3pnl autoionizing states. The higher Rydberg levels and the ionization continuum of the 3s $^2S^e$ channel, on the contrary, are very delocalized and reduce asymptotically to the ion ground state coupled to a very diffuse orbital; therefore a proper description of their long-range behaviour requires valence basis configurations of the form $(\varphi_{3s}\chi_{nl})$ where φ_{3s} is the ion SCF 3s orbital while χ_{nl} are diffuse orbitals to be discussed later. The short-range components of these states are described by the many localized configurations included to obtain the lowest levels. It should be evident that the present method is quite efficient for recovering the valence-shell correlation effects, since it is possible to employ many localized basis configurations with a modest computational effort. An equivalent description of the correlation effects by a close-coupling approach would require many closed-channel target states and pseudostates, and hence the very cumbersome numerical solution of many coupled differential equations for the radial waves.

The diffuse orbitals for representing the long-range behaviour of the higher Rydberg states and of the continuum ones include Hydrogenic orbitals and STOC (STO times a Cosine) whose radial dependence is

$$R_{zkl}(r) = N_{zkl}\ r^{l} \exp(-zr) \cos(kr)$$

These orbitals lead to close analytical expressions for all the integrals needed in a one-center expansion and keep the basis redundancy into tractable limits. If the k-values are chosen according to

$$k_j = (j+1/2)\pi/R_0$$

it is possible to reproduce correctly up to about R_0 all the continuum states with a node in $r=R_0$ and with a wave energy below $k^2_{max}/2$. By varying R_0, it is then possible to sweep finely the continuum spectrum. For typical R_0 values (400-500 in this work), the closed channel components die out well before R_0 and therefore there is a large region of space in which the charge density of the variational states Φ_i reduces to that of a shifted Coulomb wave. From the analysis of this region it is then possible to obtain the phaseshift and to compute a normalization constant C_i such that the C^2-normalized state $C_i\ \Phi_i$ has up to R_0 the energy-normalized amplitude (fig. 1). The very good agreement between these two charge density is more clearly shown in fig. 2, where the low-r region of fig. 1 is expanded.

The wavefunctions of the autoionizing states present large-amplitude localized components on the corresponding quasistationary states. Their unit-normalized variational representations will be therefore almost identical to the quasistationary states and have only a small long-range component (fig. 3).

The presence of an autoionizing state produces an additional term in the phaseshift of the form

$$\tan^{-1}\left[\Gamma/2(E_0-E)\right] + \pi\Theta(E-E_0)$$

which therefore increases of π in an energy range of about Γ. This means that for every R_0 there occurs a state with a node at $r=R_0$ and which lies in the resonance energy neighbourhood; in other words the resonance is a persisting eigenvalue with respect to R_0 (fig. 4). This

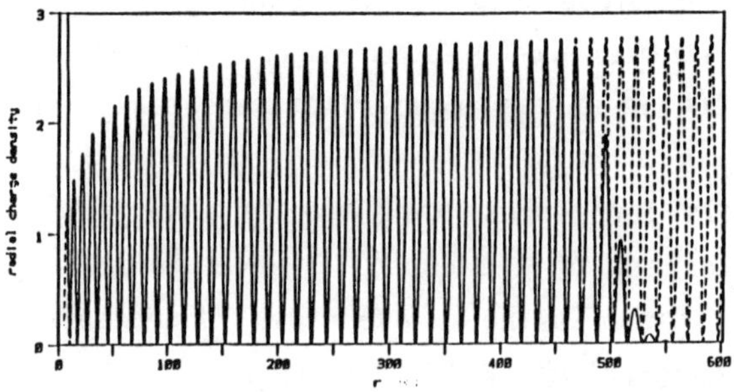

Figure 1. Radial charge density for the $^1P^o$ C^2-normalized variational state of wave energy 0.02432 obtained with R_0=500 (full line) and for the energy-normalized Coulomb wave of the same phaseshift and energy (broken line).

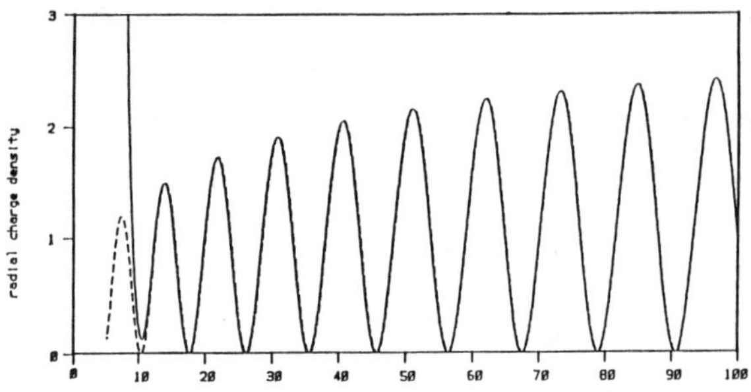

Figure 2. Magnification of the low-r region of figure 1.

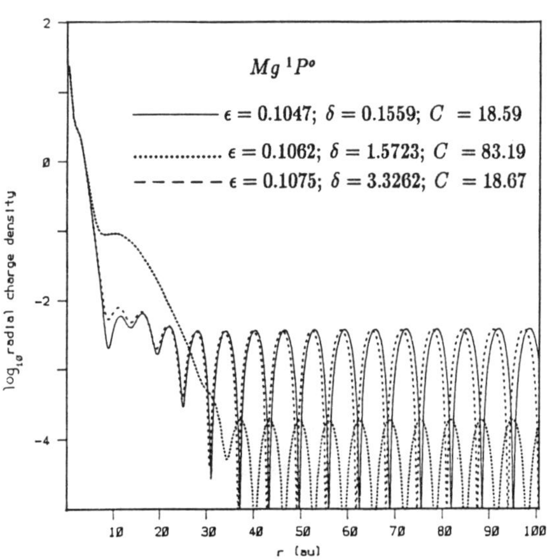

Figure 3.
Radial charge densities for three consecutive unit-normalized states obtained with R_0=506 around the 3p3d $^1P^o$ resonance.
The lowest two states correspond to the points a and h in figure 4.

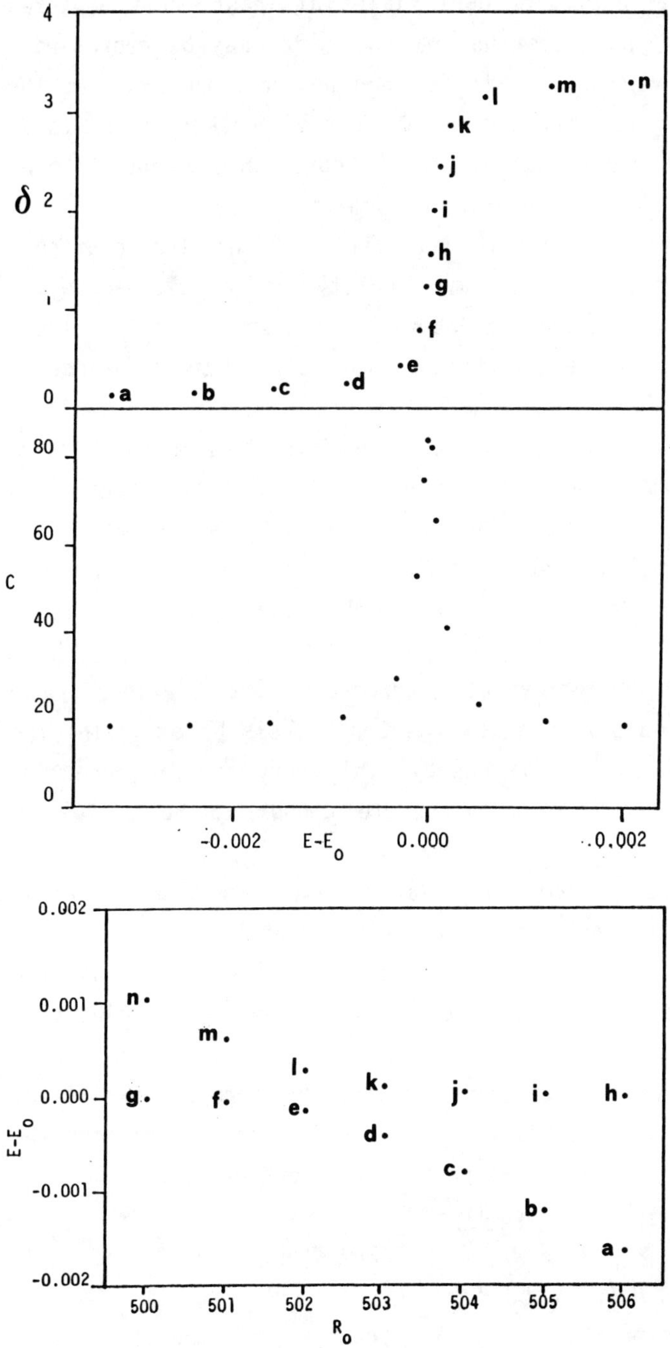

Figure 4
Behaviour of the phaseshift and of the normalization constant around the 3p3d $^1P^o$ resonance.

top:
phaseshift vs. the eigenvalues
middle:
normalization constant vs. the eigenvalues
bottom:
eigenvalues vs. R_o

The lettering denotes the corresponding points.

ensures that the variational procedure yields at least one wavepacket for all the resonances, no matter how narrow, which may be represented by the basis set. By varying R_0, it is then possible to sweep at the same time all the resonances reproduced. The set of phaseshift points so obtained may be then employed to compute the position and the width of each resonance by the usual interpolation formula[4].

In conclusion, the analytical representation of the continuum electron wave may achieve the same quality of a close-coupling calculation and does not require the laborious (and sometimes fruitless) searching required by these last techniques for very narrow resonances.

Apart for the lowest n states of the form 3snl and 3pnl are expected to have the same non-valence (core plus intershell) correlation energy of their parent ion levels, 3s $^2S^e$ and 3p $^2P^o$ respectively. This means that accurate valence shell group functions should reproduce correctly the positions of the higher levels with respect to their parent ions.

Accordingly, in our previous frozen-core calculations[4], we corrected the calculated positions of the 3pnl levels by employing the experimental 3s-3p ion excitation energy; this very simple correction yielded already energy differences accurate enough to be useful in assignement problems.

These calculations yielded oscillator strengths and radiative lifetimes in good agreement with the experimental data; for the lifetimes part of the differences are due to the errors in the transition energies. Only for the 3snd $^1D^e$ levels we found relatively larger discrepancies.

To reduce the error sources due to the frozen-core approximation, we have reinvestigated in this paper the discrete spectrum including the phenomenological correction known as polarization potential[7]:

$$V_p(\underline{r},\underline{r}') = V(\underline{r}) + V(\underline{r}') + V_{de}(\underline{r},\underline{r}')$$

$$V(\underline{r}) = -\alpha \sum_{lm} |lm\rangle\, W_6(r/r_l)\, 1/2r^4 \langle lm|$$

$$V_{de}(\underline{r},\underline{r}') = -\alpha \left[W_6(r/R)\, W_6(r'/R)\right]^{1/2} \cos\widehat{rr'} \;/\; (rr')^2$$

$$W_6(x) = 1 - \exp(-x^6) \qquad R = (r_0+r_1+r_2)/3$$

where $\langle\hat{r}|lm\rangle = Y_{lm}(\hat{r})$, α is the core dipole polarizability and the semiempirical parameters r_l are chosen to reproduce the position of the lowest ^2l level of Mg^+ relative to Mg^{++}.

It is very gratifying that, on the whole, this correction modifies significantly only those results which presented some disagreements with the experimental data, leaving the other pratically unchanged. It appears that the polarization potential causes the largest alterations on the states (and hence on the other observables) in presence of large configuration mixings. The lowest $^1D^e$ states have been therefore the most affected by this correction, because the $3p^2\ ^1D^e$ configuration is completely smeared out in the lowest part of this manifold.

RESULTS

The frozen-core results have been presented, together with the interpretation of the experimental data, in our previous work[4], to which we refer the reader for a detailed discussion. In this paper we deal mainly with the effects of the polarization potential.

The introduction of the polarization potential reproduces exactly the 3s-3p ion excitation energy and reduces the discrepancies in the term energies with respect to the parent ionization threshold. This is shown in table 1 for the lowest 3snl levels and in table 2 for the 3pnl ones. For this last table, it should be noted that the experimental assignements of the $3p^2$ and 3p3d are well estab lished, but those of the 3p4p ones are tentative and controversial[4], so that the present calculations should help to settle the question.

Table 1: Term energies of the lowest 3snl levels.

level	frozen	pol.	exp.	level	frozen	pol.	exp.
$3s^2\ ^1S^e$	-0.27684	-0.28067	-0.28099	$3s4s\ ^3S^e$	-0.09250	-0.09320	-0.09328
$3s3p\ ^1P^o$	-0.11973	-0.12108	-0.12129	$3s3p\ ^3P^o$	-0.18142	-0.18123	-0.18125
$3s3d\ ^1D^e$	-0.07107	-0.06947	-0.06957	$3s3d\ ^3D^e$	-0.06257	-0.06231	-0.06247

Table 2: Term energies of the lowest 3pnl bound levels. The frozen-core results are corrected as discussed in the text.

level	frozen	pol.	exp.	level	frozen	pol.	exp.
				$3p^2\ ^3P^e$	-0.01547	-0.01746	-0.01739
$3p4p\ ^1P^e$	0.09297	0.09206	0.09204	$3p4p\ ^3P^e$	0.10117	0.10084	0.10082
$3p3d\ ^1D^o$	0.08645	0.08662	0.08667	$3p3d\ ^3D^o$	0.10001	0.09963	0.09958

The $3p^2\ ^1S^e$ resonance of Magnesium is one of the best characterized autoionizing states and its theoretical study appears a very interesting topic. While the frozen-core calculations placed this state at the wave energy 0.0273+0.054=0.0327, the inclusion of the polarization potential shifted it to 0.0306, in very good agreement with the experimental value 0.0301. The agreement is not so good for the width, for which the phaseshift analysis yielded the results $1.51 \cdot 10^{-3}$ and $1.72 \cdot 10^{-3}$ respectively without and with the polarization potential, while the two available experimental widths are $1.26 \cdot 10^{-3}$ and $1.27 \cdot 10^{-3}$. It should be noted that a recent close-coupling calculation with polarization potential[8] yields 0.0323 and $1.74 \cdot 10^{-3}$ for the position and the width.

As mentioned above, the $^1D^e$ states are the most affected by the polarization potential. For instance, the oscillator strength for the $3s3p\ ^1P^o$-$3s3d\ ^1D^e$ transition changes from 0.207-0.209 (Length and Velocity Gauge) to 0.240-0.243, in excellent agreement with the experimental value 0.24±0.02. The correction to the radiative lifetime of the $3s3d\ ^1D^e$ level is even more striking, due to a large percentual error in the frozen-core transition energy. The lifetime is indeed reduced from 106-105 ns (LG-VG) to 81-80 ns, again in excellent agreement with the recent experimental values 81±6 and 72±4 ns.

The other oscillator strengths and radiative lifetimes, which are in better agreement with experience, undergo only minor variations. For instance, the oscillator strength for the $3s^2\ ^1S^e$-$3s3p\ ^1P^o$ transition changes from 1.761-1.746 to 1.765-1.745; but the greater transition energy (0.15959 vs. 0.15711) reduces the radiative lifetime of the upper level from 2.15-2.17 ns to 2.08-2.10 ns, against the experimental values 2.09±0.1, 2.00±0.04 and 2.0±0.1 ns.

The variational states so obtained allowed to build a localized representation of the resolvent and hence to compute the two-photon processes. For the bound-bound transition probabilities we employ the transition rate parameter K

$$K_{fi} = (2\pi)^3\ \alpha^2\ (\omega_1 \omega_2)^q\ |M_{fi}|^2\ w_0$$

$$M_{fi} = (1+P_{12})\ \langle f O_1 G(E_i+\omega_2) O_2 i \rangle$$

$$= (1+P_{12})\ \Sigma_j\ \langle f O_1 j \rangle \left[\ \omega_2-(E_j-E_i)\right]^{-1} \langle j O_2 i \rangle$$

where P_{12} exchanges the photon interactions, q=+1 and $O_i=\hat{e}_i \cdot \underline{r}$ in the LG or q=-1 and $O_i=\hat{e}_i \cdot \underline{p}$ in the VG and w_0 is the state density (lineshape

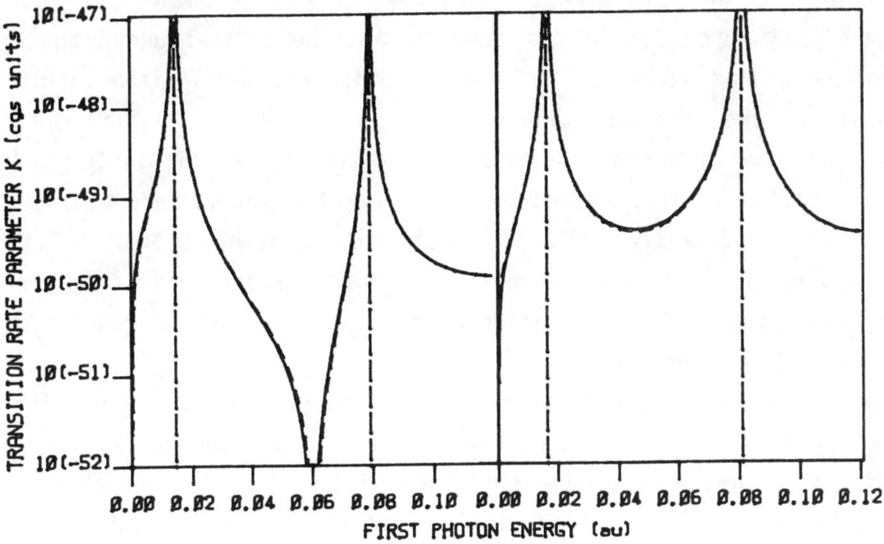

Figure 5. Two-photon transition rate parameter K (frozen-core results) for the transitions from the ground state $3s^2\ {}^1S^e$ to the 3s5s ${}^1S^e$ (left) and 3s4d ${}^1D^e$ levels (right). Full line: LG results; broken line: VG ones.

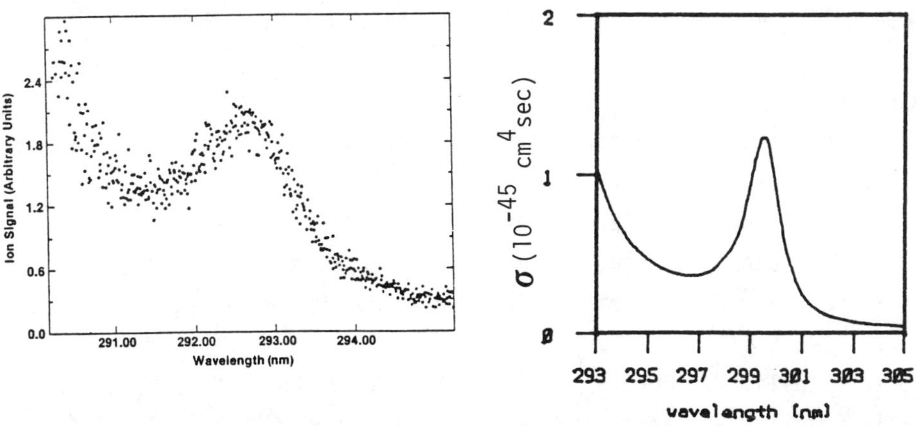

Figure 6. Single-beam two-photon ionization cross section from the ground state with linearly z-polarized light in the region of the $3p^2\ {}^1S^e$ autoionizing state. Experimental data (left) and frozen-core results (right). LG and VG results agree within 2% and cannot be distinguished on this scale.

function) for which we assume an unitary cgs value (i.e. w_0=1/erg).

Fig. 5 reports some our results for the bound-bound transitions from the ground state for two beams of linearly z-polarized light. The quality of the representation of the resolvent G may be inferred from the very good agreement of the LG and VG results, particularly when one of the photon energies tends to zero. In this limit, indeed, the zero limit of K is easily achieved in the LG, where q=+1, but in the VG it requires a noteworthy cancellation in the transition matrix element M_{fi}, since now q=-1. As a matter of fact, the present VG results begin to diverge only for $\omega_1<10^{-4}$.

In fig. 6 we present our frozen-core results and the available experimental data for the single-beam two-photon ionization. The qualitative agreement may be considered satisfactory in view of the fact that for these photon energies the process depends critically on the close relative position of the $3p^2\ ^1S^e$ autoionizing state and of the $3s3p\ ^1P^o$ intermediate resonance.

In conclusion, it appears that the present analytical L^2 method is capable of dealing with highly correlated continua and narrow resonances with performances which, since it does not suffer from searching problems or limitations in the target state number, may be superior to those of the more standard numerical close-coupling methods.

REFERENCES

1 R. Moccia and P. Spizzo, J.Phys.B: At.Mol.Phys. 18, 3537 & 3555 (85)
2 R. Moccia and P. Spizzo, J.Phys.B: At Mol.Phys. 20, 1423 (87)
3 R. Moccia and P. Spizzo, Photons and Continuum States of Atoms and Molecules, Springer Proceedings in Physics 16, pag.47 (87)
4 R. Moccia and P. Spizzo, J.Phys.B: At.Mol.Opt.Phys. 21, 1121, 1133 & 1145 (88)
5 I. Cacelli, V. Carravetta and R. Moccia, J. Chem. Phys. 85, 7038 (86)
6 I. Cacelli, V. Carravetta and R. Moccia, Chem. Phys. 120, 51 (88)
7 C. Bottcher and A. Dalgarno, Proc. R. Soc. Lond. A340, 187 (74)
8 C. Mendoza and C.J. Zeippen, Astron. Astrophys. 179, 339 & 346 (87)

Application of the Two-dimensional Fully-Numerical RHF Method to Open-Shell Hydrides

Florian Müller-Plathe and Geerd H. F. Diercksen
Max-Planck-Institut für Astrophysik, Karl-Schwarzschild-Str. 1
D-8046 Garching, Federal Republic of Germany

Leif Laaksonen
Finnish State Computer Centre, University Support Department
POB 40, SF-02101 Espoo, Finland

Abstract. Total energies and one-electron properties of BeH, BeH*, CH, NH, OH, MgH and SiH are calculated by the recently implemented fully-numerical RHF method for open-shell diatomics and are compared to previously obtained results of SCF calculations in large STO basis sets. The comparison shows that the basis-set truncation error depends systematically on the number of electrons in the molecule, as well as on the internuclear distance.

1. Introduction

The two-dimensional fully-numerical RHF method [1] has been extended to open-shell diatomics [2]. It is used to calculate the Hartree-Fock (HF) total and orbital energies, the multipole moments and other one-electron properties of the electronic ground states of BeH, BH, CH, NH, OH, MgH and SiH, as well as of the $A\,^2\Pi_r$ excited state of BeH at their respective experimental equilibrium internuclear distances, $R_e(exp)$ [3]. Calculations of the potential energy curves are reported for CH and OH.

The SCF total energies obtained by Cade and Huo [4] in extensively optimized STO basis sets have served long as benchmark data for the calibration of other basis sets of much lower quality. In this contribution the errors of their basis sets (which are very small anyway) for the above molecules are quantitatively established. Special attention is focused on the issues of

(i) systematic changes of the basis-set truncation error along potential energy curves, and

(ii) basis-set error dependence on the number of electrons in the molecule, i.e. on the changes as one proceeds from LiH to BeH and so on.

M. Defranceschi and J. Delhalle (eds.), Numerical Determination of the Electronic Structure of Atoms, Diatomic and Polyatomic Molecules, 311–315.

Table 1. Fully-numerical Hartree-Fock total energies and dipole moments (in a.u.).

Molecule	$R_e(exp)^{a)}$	Total Energy	Dipole Moment
BeH($X^2\Sigma^+$)	2.538	-15.153240	-0.112584
		-15.15312[b)]	-0.10898[b)]
BeH($A^2\Pi_r$)	2.519	-15.050809	-1.001556
		-15.05062[b)]	
CH($X^2\Pi_r$)	2.124	-38.279666	0.615799
		-38.27935[b)]	0.63027[b)]
NH($X^3\Sigma^-$)	1.9614	-54.978429	0.633427
		-54.97806[b)]	0.64876[b)]
OH($X^2\Pi_i$)	1.8342	-75.421304	0.691811
		-75.42083[b)]	0.69558[b)]
MgH($X^2\Sigma^+$)	3.271	-200.16456	-0.585469
		-200.1566[b)]	
SiH($X^2\Pi_r$)	2.874	-289.43779	0.09247
		-289.4362[b)]	
		-289.437142[c)]	

[a)] Ref. [3,4]. [b)] Ref. [4]. [c)] Ref. [5].

2. Results and Discussion

The details of the calculations and the numerical results have been reported elsewhere [2]. Here, Table 1 presents the total energies and the dipole moments of the investigated molecules at their respective equilibrium bond lengths, together with the corresponding STO-SCF data of Cade and Huo [4]. The numerical HF energies agree with the STO-SCF data to within 1 mHartree, but are always lower. This reflects the careful selection and validation of the basis sets of Cade and Huo. The numerical HF dipole moments deviate from the previously reported STO-SCF values by 0.5-3.0 %.

Figure 1 shows the basis-set errors of all neutral first and second-row diatomic hydrides treated so far in the fully-numerical HF scheme. (The energies of the closed-shell hydrides are taken from [1], the energy of HCl is interpolated from the data of Laaksonen et al. [6].) A monotonic increase of the basis-set error ΔE is noted for the first-row hydrides, even though Cade and Huo subjected all their basis sets to the same optimization procedure. This behaviour is caused by accommodating more and more electrons within basis sets of the same size. The errors for the second-row hydrides are considerably larger than for the first-row hydrides, probably because the inner shells are not described adequately. A recent calculation on SiH,

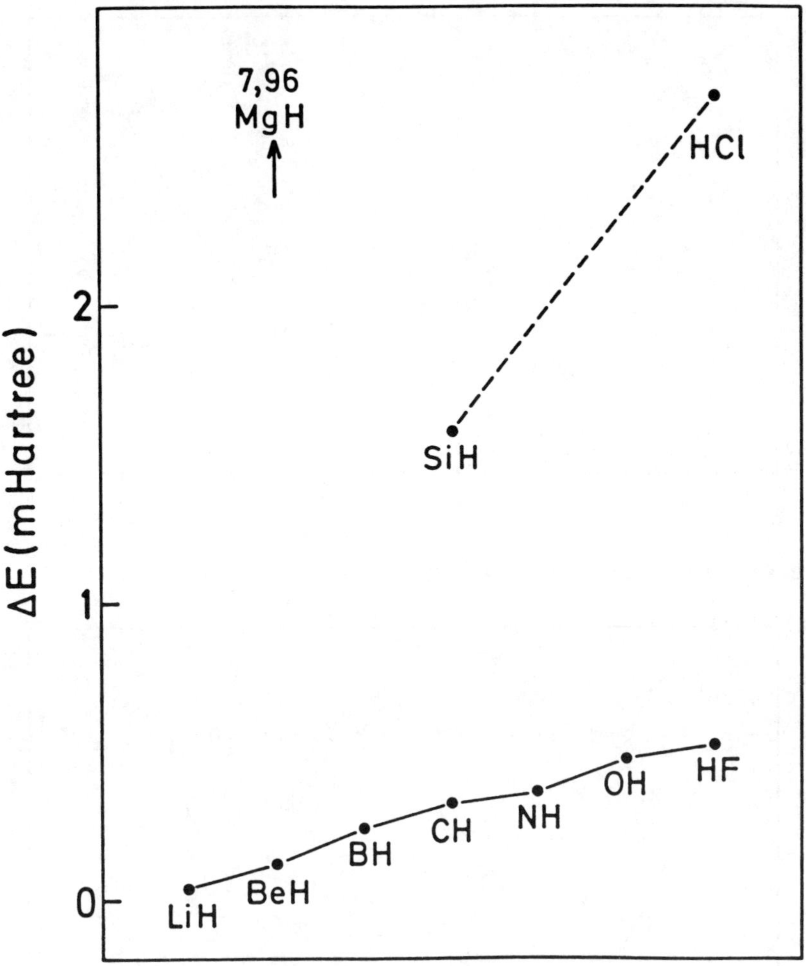

Fig. 1. Basis-set errors of various first and second-row hydrides.

in a (8s,6p,3d,2f/4s3p2d) STO basis, has a basis-set error of only 0.65 mHartree [5] which is comparable to those of the first-row hydrides. A steady increase of the basis-set error with the atomic number is also expected for the second-row hydrides, and is preliminarily verified by comparing the errors for SiH and HCl. However, there is an irregularly large basis-set error for MgH, for which no explanation can be provided at present.

Figures 2 and 3 show the dependence of the basis-set error on the internuclear separation for CH and OH, respectively, calculated using the STO-SCF total energies of Cade and Huo [7]. For both molecules the basis-set error appears to be a smooth function of the bond distance with a minimum near $R_e(exp)$, where the STO basis has

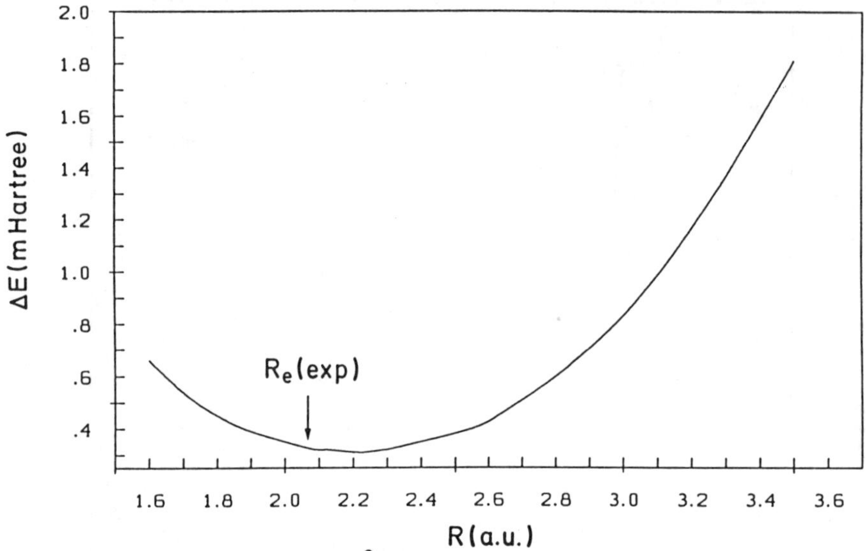

Fig. 2. Basis-set error of CH ($X^2\Pi$) along the potential curve.

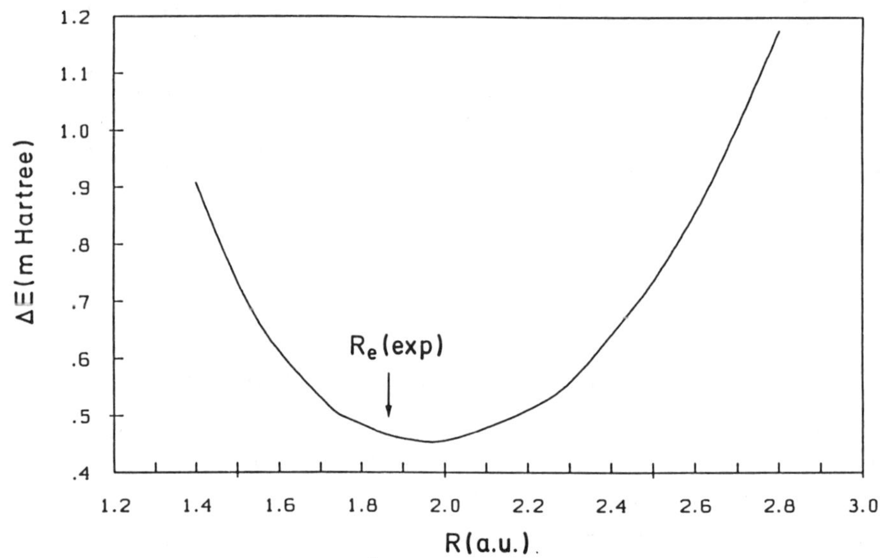

Fig. 3. Basis-set error of OH ($X^2\Pi$) along the potential curve.

been extensively optimized. However, this minimum is not located exactly at $R_e(exp)$, but is shifted by about 0.1 a.u. towards larger bond lengths for both molecules. The basis-set error grows larger toward shorter and longer bond lengths, because the flexibility of the linear expansion coefficients is insufficient to account for the changes of the wave functions with the bond distances. Instead, more compact basis functions are needed at short separations while more diffuse functions are necessary for larger distances. Indeed, Cade and Huo have shown that a modest reoptimization of the orbital exponents of CH at $R = 1.60$ a.u. lowered the basis-set error from 0.66 to 0.40 mHartree, which is similar to the error at $R_e(exp)$ of 0.32 mHartree. In contrast, a similar reoptimization for OH at $R = 1.40$ a.u. yielded only a negligible decrease of the basis-set error which remains almost twice as large as the error at $R_e(exp)$.

For SiH a second calculation, performed at $R = 2.0$ a.u., results in a basis-set error of 3.76 mHartree, in contrast to the one at $R_e(exp)$ of 1.59 mHartree. This increase is much larger than that of either CH or OH, and indicates that not just the basis functions for the valence orbitals need reoptimization. If so, the change of the basis-set error would be of an order similar to the first-row hydrides. The inner shells are also affected by the change of the bond length, and the corresponding basis function exponents should be readjusted as well.

3. Conclusions

There is a systematic dependence of the basis-set error at the HF level of theory on both the internuclear separation and the number of electrons in the molecule. It is found to be rather small for the thoroughly selected and optimized basis sets of Cade and Huo, but it is expected to be much more apparent in basis sets of lower quality that are commonly employed in standard calculations. The basis-set error changes should affect calculated properties that depend on subtle features of potential energy surfaces such as reaction barriers or rovibrational parameters. It can only be removed partly and at the cost of an extensive reoptimization of orbital exponents. Because basis-set errors persist at the correlated level these conclusions should apply accordingly.

4. References

[1] L. Laaksonen, P. Pyykkö and D. Sundholm, Comp. Phys. Rep. **4**, 313 (1986).

[2] L. Laaksonen, F. Müller-Plathe and G.H.F. Diercksen, J. Chem. Phys. (submitted).

[3] G. Herzberg, Spectra of Diatomic Molecules (van Nostrand, New York, 1950).

[4] P.E. Cade and W.M. Huo, At. Data Nucl. Data Tables **13**, 415 (1973).

[5] M. Larsson, J. Chem. Phys. **86**, 5018 (1987).

[6] L. Laaksonen, S. Wilson and I.P. Grant, J. Phys. B (submitted).

[7] P.E. Cade and W.M. Huo, J. Chem. Phys. **47**, 614 (1967); ibid. 649.

INTERPOLATION OF NUMERICAL ORBITALS IN MOMENTUM SPACE

Jorge NAVAZA,
Laboratoire de Physique, Tour B,
Centre Pharmaceutique,
92290 Chatenay-Malabry, France.

ABSTRACT. Numerical Orbitals defined in a 3-dimensional momentum space are interpolated by reconstruction from the values they take on a set of quadrature abscissas.

The Hartree-Fock equations in momentum space are integral equations. The numerical algorithms so far developed to solve them use gaussian quadrature formulas in order to evaluate the integrals. The orbitals and other related quantities are thus sampled at the points of a spherical grid defined by the quadrature abscissas [1-2]. However, most of the integrals are in fact convolutions whose calculation requires grid interpolation. It turns out that the shift in the argument involved in the convolutions can always be performed on the orbitals, which are smooth functions of the momentum. Here we propose a method of interpolation based on the reconstruction of the orbitals from the values they take on the set of quadrature abscissas.

We first state the numerical integration rule over 3-dimensional regions, which is a modification of the PEIRCE method [3-4], to allow for a more general radial dependence of the integrand: if the integration points $\mathbf{p}_{nij} = (p_n, \theta_i, \phi_j)$ are taken at all intersections of the cones θ_i ($i=1,\ldots,L+1$), the half planes ϕ_j ($j=1,\ldots,2L+2$) and the spheres p_n ($n=1,\ldots,N$), then the formula

$$\int f(p,\theta,\phi)\, d\mathbf{p} = \sum_{n=1}^{N} \sum_{i=1}^{L+1} \sum_{j=1}^{2L+2} f(p_n,\theta_i,\phi_j)\, \omega_{nij} \qquad (1)$$

M. Defranceschi and J. Delhalle (eds.), Numerical Determination of the Electronic Structure of Atoms, Diatomic and Polyatomic Molecules, 317–319.

gives the correct result whenever f is a limited expansion in spherical harmonics of orders at most 2L+1, whose coefficients are polynomials in ρ, of degree at most 2N-1, times any integrable nonnegative function $w(\rho)$. The weights are of the product form $\omega_{nij} = A_n B_i C_j$; ρ_n and A_n are the abscissas and weights of a N-points Gaussian quadrature relative to the positive function w, $\cos(\theta_i)$ and B_i are the abscissas and weights of the (L+1)-points Legendre-Gauss quadrature formula, finally $\phi_j = 2\pi j/(2L+2)$ and $C_j = 2\pi/(2L+2)$.

We will then assume that the orbital $\phi(\mathbf{p})$ can be faithfully represented by a limited expansion in spherical harmonics,

$$\phi(\mathbf{p}) = \sum_{l=0}^{L} \sum_{m=-l}^{l} A_{lm}(\rho)\ Y_{lm}(\boldsymbol{\Omega}), \tag{2}$$

with $\rho = |\mathbf{p}|$ and $\boldsymbol{\Omega} = \mathbf{p}/\rho$. The maximum order L may eventually depend on the particular value of the radial coordinate ρ. The functions A_{lm} are given by integrals over the surface of the unit sphere $\boldsymbol{\sigma}$,

$$A_{lm}(\rho) = \int_{\sigma} \phi(\rho.\boldsymbol{\Omega})\ Y_{lm}{}^*(\boldsymbol{\Omega})\ d\boldsymbol{\Omega}, \qquad l \le L. \tag{3}$$

These integrals are correctly evaluated with the quadrature formula

$$\int_{\sigma} \phi(\rho.\boldsymbol{\Omega})\ Y_{lm}{}^*(\boldsymbol{\Omega})\ d\boldsymbol{\Omega} = \sum_{q=1}^{Q} \phi(\rho.\boldsymbol{\Omega}_q)\ Y_{lm}{}^*(\boldsymbol{\Omega}_q)\ \omega_q, \qquad l \le L, \tag{4}$$

if the angular integration points $\boldsymbol{\Omega}_q = (\theta_i, \phi_j)$ and the corresponding weights $\omega_q = B_i C_j$ are chosen according to the above stated rule. Substituting for A_{lm} into the limited expansion (2) and using the addition theorem for the spherical harmonics, we obtain

$$\phi(\rho.\boldsymbol{\Omega}) = \sum_{q=1}^{Q} \phi(\rho.\boldsymbol{\Omega}_q) \sum_{l=0}^{L} \sum_{m=-l}^{l} Y_{lm}(\boldsymbol{\Omega})\, Y_{lm}{}^*(\boldsymbol{\Omega}_q)\, \omega_q$$

$$= \sum_{q=1}^{Q} \phi(\rho.\boldsymbol{\Omega}_q) \left[\sum_{l=0}^{L} (2l+1) \; P_l(\boldsymbol{\Omega}_q.\boldsymbol{\Omega}) \right] \frac{\omega_q}{4\pi}, \qquad (5)$$

where P_l is the LEGENDRE polynomial of order l. We see that the values taken by the function on the set $[\boldsymbol{\Omega}_q$, $q=1...Q=2(L+1)^2]$ are multiplied by expressions which are independent of the radial variable ρ. Moreover, the expression between brackets in (5) is a polynomial of a single variable, of order L. The angular and the radial interpolations are then performed independently of each other.

Similar equations (now based on the Darboux-Christoffel identity) can be obtained for the radial interpolation. They are not developed here because no general "weighting function" $w(\rho)$ has yet been adopted.

References.

[1] J. NAVAZA and G. TSOUCARIS; Phys. Rev., **A24**, 683 (1981)

[2] M. DEFRANCESCHI, M. SUARD and G. BERTHIER; Int. J. Quan. Chem., **25**, 863 (1984)

[3] J. NAVAZA; Acta Cryst., **A43**, 645 (1987)

[4] W.H. PEIRCE; Math. Tables Aid Comput., **11**, 244 (1957).

ON THE ACCURACY OF THE ALGEBRAIC APPROXIMATION IN RELATIVISTIC ELECTRONIC STRUCTURE CALCULATIONS

H.M. QUINEY,* I.P. GRANT* and S. WILSON,+

*Theoretical Chemistry Department, University of Oxford, 1, South Parks Road, Oxford, OX1 3TG, England,
+Rutherford Appleton Laboratory, Chilton, Oxfordshire, OX11 0QX, England.

ABSTRACT. The accuracy of relativistic electronic structure calculations for atomic systems made within the algebraic approximation, that is by making a finite basis set expansion, is assessed by comparison with results obtained by using finite-difference methods. It is demonstrated that the accuracy achieved in numerical studies can be matched by calculations performed within the algebraic approximation. Some advantages of the basis set expansion approach are emphasized.

1. Introduction.

In this note, we consider the accuracy that be achieved in relativistic electronic structure formulated within the algebraic approximation. Section 2 provides the non-relativistic background whilst in section 3 the new features which arise in the relativistic case are described. Atomic units are employed throughout this paper.

2. The algebraic approximation and the non-relativistic electronic structure problem.

In 1963, Schwartz [1] emphasized the need to devize systematic schemes for extending basis sets employed in atomic and molecular calculations stressing the need for "*an orderly plan of attack*" which "*is ... essential for any mathematical analysis of the convergence rate.*". More recently, Huzinaga [2] has underlined this quasi-empirical element of many contemporary electronic structure calculations.

There is a widespread belief (see, for example, [3], pg. 103) that finite difference approximations to atomic and molecular wavefunctions are superior to their finite basis set counterparts [4]. In reality, finite basis set expansions can provide equivalent or superior representations of electronic wave functions. Ruedenberg *et al.* [5,6] constructed systematic sequences of even-tempered basis sets of Gaussian-type functions and obtained atomic self-consistent field energies, often more accurately than finite difference calculations [7]. These sequences of basis sets can also be employed in atomic electron correlation energy calculations [8]. For diatomic molecules, the accuracy achieved in fully numerical calculations [9,10] can be matched by calculations performed within the algebraic approximation [11] when elliptical basis functions are used. The underlying reason for the success of finite difference methods lies in the choice of coordinate system for the special cases in which they have found favour. For general polyatomic molecules, the finite difference method must, in practice, employ a Cartesian grid, which is a manifestly unsuitable coordinate system, particularly in the region of the nuclei. For molecular systems without special symmetry properties, the use of atomic basis functions transfers a local coordinate system to the region where the wave function is most rapidly changing and, more importantly, it yields a square-integrable representation of the *complete* spectrum rather than just a single element. In principle, of course, we could generate finite-difference representations of the complete set but this introduces difficulties associated with the divergence of diagonal interaction matrix elements and is computationally prohibitive for arbitrary molecular systems.

M. Defranceschi and J. Delhalle (eds.), Numerical Determination of the Electronic Structure of Atoms, Diatomic and Polyatomic Molecules, 321–327.

3. The algebraic approximation and the relativistic electronic structure problem.

3.1 BASIS SET EXPANSIONS AND THE FURRY PICTURE.

For more than twenty years relativistic electronic structure calculations employing basis set expansions have been plagued by problems which have now been fully resolved [12-16] and the way is now clear for the relativistic treatment of molecular systems containing heavy atoms, considerably extending the range of applicability of the methods of computational quantum chemistry.

The solution of the Dirac equation within the algebraic approximation presents difficulties which are not present in the corresponding non-relativistic problem [12]. The presence of a 'negative energy' continuum of solutions of the complete Dirac spectrum apparently allows many-electron systems to autoionize, a process which has become known as 'Brown-Ravenhall disease' or 'continuum dissolution'. This difficultly has been resolved by showing [16] that the Dirac-Fock model affords a legitimate zero-order approximation to a relativistic electronic structure theory within the Furry bound state picture [17] of quantum electrodynamics (QED). At the heart of this picture is an expansion of the many-electron wavefunction in a *complete* set of single particle solutions of the Dirac equation. It is assumed that a time-independent potential is present and that the complete set divides cleanly into positive and negative energy branches. The separation of the spectrum is dependent on the strength of the applied potential and on the finite value of c, the speed of light. In practice, the division of the states is achieved by imposition of physically determined boundary conditions. The separation into positive and negative-energy branches may be lost in calculations performed within the algebraic approximation and eigenvalue estimates may fail to be variational upper bounds, which is a problem now known as 'variational collapse' or 'finite basis set disease'. The reasons for unphysical representations of the Dirac spectrum are known to be associated with the failure to satisfy relativistic boundary conditions, particularly in the strong field region near nuclei, where relativistic effects are most pronounced. The possibility of 'variational collapse' in basis set expansions of Dirac wavefunctions has more far-reaching consequences than a loss of accuracy, since the failure to separate into physically meaningful positive and negative energy branches undermines the Furry picture.

3.2 RELATIVISTIC THOMAS-REICHE-KUHN SUM-RULE.

A severe test of the accuracy of basis set approximations is afforded by the relativistic Thomas-Reiche-Kuhn sum-rule [18], which includes contributions from both positive and negative energy branches of the spectrum and is identically zero in the relativistic case. Calculations of this quantity are given in Table 1 as a function of nuclear charge, Z, for 'kinetically balanced' [13,14] and 'spinor' [15,16] even-tempered basis sets of twenty exponential-type functions. The 'kinetically balanced' sets fail to represent the wavefunction near the nucleus correctly, allowing pollution of the branches of the spectrum and a poor value for the sum rule. In contast, the 'spinor' basis set, which satisfies relativistic energy-independent boundary conditions at the point nucleus, provides a smoothly convergent approximation to the relativistic sum-rule as the dimension of the basis set is extended with an accuracy which is only weakly dependent on Z. Note that the ability to evaluate sum-rules accurately is an essential step towards handling *(i)* retarded, transverse Breit interaction in self-consistent field calculations, and *(ii)* the self-energy of a bound electron.

Table 1.

The relativistic Thomas-Reiche-Kuhn sum-rule for 'kinetically balanced' and 'spinor' even-tempered basis sets of exponential-type functions defined by the parameters α=4Z/27, β=3/2, n=20, Powers of ten are given in parenthesis.

Z	'kinetically balanced' set	'spinor' set	Z	'kinetically balanced' set	'spinor' set
1	1.68 (-7)	1.68 (-7)	60	8.35 (-3)	2.09 (-7)
10	7.23 (-6)	1.69 (-7)	70	1.51 (-2)	2.17 (-7)
20	1.12 (-4)	1.74 (-7)	80	2.50 (-2)	2.22 (-7)
30	5.59 (-4)	1.81 (-7)	90	3.90 (-2)	2.27 (-7)
50	4.13 (-3)	2.00 (-7)	92	4.24 (-2)	2.30 (-7)

3.3 RELATIVISTIC SELF-CONSISTENT-FIELD CALCULATIONS.

We report a series of calculations for helium-like species using the matrix Dirac-Hartree-Fock method and systematic sequences of even-tempered 'kinetically balanced' and 'spinor' basis sets of exponential-type functions. We compare the results with those obtained from a relativistic finite difference program of Grant *et al.* [19] in Table 2. In this Table, n is the number of basis functions, α and β define the even-tempered basis set, ϵ is the orbital energy and E is the total energy. The superior convergence properties of the calculations which used 'spinor' sets is evident. Complete agreement with the finite difference calculations was obtained in all cases.

The results presented in Table 2 serve to emphasize the importance of using basis functions which satisfy the appropriate boundary conditions in calculations using point nuclei. In reality, nuclei have a finite size and, use of a uniform charge distribution nuclear model, leads to the requirement that the basis functions be Gaussian at the origin. 'Kinetically balanced' basis sets of Gaussian-type functions with finite nuclei provide the most promising approach to relativistic molecular electronic structure calculations [20,21].

There are advantages to be gained by approximating the solutions of the Dirac-Hartree-Fock equations using finite basis sets which are not shared by finite difference techniques. Firstly, the Breit and Coulomb integrals over the basis functions can be incorporated in matrix Dirac-Hartree-Fock calculations within the self-consistent field iterations [22] rather than, as is usually done in finite difference approaches, as a first order perturbation, presenting no additional computational problems other than storage. Furthermore, this avoids the need to evaluate a complete class of diagrammatic terms, the "bubble" diagrams, when electron correlation effects are considered [22] within the "no pair" approximation.. Chemically significant differences between these two approaches arise for nuclear charges greater than about 50. Secondly, the accuracy which can be achieved in relativistic atomic structure calculations performed within the algebraic approximation promises a new approach to the calculation of QED corrections and PNC (parity non-conserving) terms [23]. Finally, we mention that there is no significant difference between relativistic electronic structure calculations for atoms and molecules once the algebraic approximation has been invoked. Relativistic molecular electronic structure calculations have been performed in a manner analogous to that described here [21,24].

Table 2.

Matrix Dirac-Hartree-Fock calculations for helium-like species using systematic sequences of even-tempered 'kinetically balanced' and 'spinor' basis sets of exponential-type functions. n is the dimension of the basis set, α and β define the even-tempered basis sets, ε is the orbital energy and E is the total energy.

$1s^2$, $Z=2$, kinetically balanced set

n	α	β	ε	E
4	1.401	1.545	-0.91807281	-2.8618106
5	1.309	1.476	-0.91805190	-2.8617996
6	1.239	1.426	-0.91799521	-2.8618078
7	1.184	1.389	-0.91798887	-2.8618121
8	1.139	1.360	-0.91798997	-2.8618132
9	1.100	1.336	-0.91799055	-2.8618133
10	1.068	1.317	-0.91799068	-2.8618133
finite difference			-0.91799068	-2.8618133

$1s^2$, $Z=2$, spinor set

n	α	β	ε	E
4	1.401	1.545	-0.91807405	-2.8618104
5	1.309	1.476	-0.91805161	-2.8617997
6	1.239	1.426	-0.91799524	-2.8618077
7	1.184	1.389	-0.91798889	-2.8618121
8	1.139	1.360	-0.91798995	-2.8618132
9	1.100	1.336	-0.91799055	-2.8618133
10	1.068	1.317	-0.91799068	-2.8618133
finite difference			-0.91799068	-2.8618133

$1s^2$, $Z=92$, kinetically balanced set

n	α	β	ε	E
4	107.992	2.071	-4774.8318	-9622.4121
5	99.970	1.917	-4789.6706	-9649.8237
6	94.032	1.812	-4789.6061	-9650.1950
7	89.393	1.734	-4788.7447	-9648.0997
8	85.625	1.673	-4790.2263	-9651.2851
9	82.451	1.625	-4789.4190	-9649.6128
10	79.790	1.585	-4790.2496	-9651.3178
finite difference			-4790.2724	-9651.3622

$1s^2$, $Z=92$, spinor set

n	α	β	ε	E
4	22.526	1.485	-4785.8319	-9641.7602
5	21.069	1.424	-4790.2373	-9651.2826
6	19.967	1.381	-4790.2703	-9651.3554
7	19.091	1.349	-4790.2724	-9651.3620
8	18.372	1.322	-4790.2723	-9651.3621
9	17.764	1.301	-4790.2724	-9651.3622
10	17.241	1.284	-4790.2724	-9651.3622
finite difference			-4790.2724	-9651.3622

$2p^2$, $Z=2$, kinetically balanced set

n	α	β	ε	E
3	0.378	1.791	-0.19308214	-0.67415834
4	0.346	1.656	-0.19310687	-0.67414082
5	0.323	1.569	-0.19306287	-0.67414272
6	0.305	1.509	-0.19306303	-0.67415885
7	0.291	1.464	-0.19306925	-0.67415931
8	0.280	1.428	-0.19306787	-0.67415945
9	0.270	1.400	-0.19306754	-0.67415956
finite difference			-0.19306763	-0.67415961

$2p^2$, $Z=2$, spinor set

n	α	β	ε	E
3	0.378	1.791	-0.19308308	-0.67415834
4	0.346	1.656	-0.19310649	-0.67414070
5	0.323	1.569	-0.19306287	-0.67414260
6	0.305	1.509	-0.19306307	-0.67415887
7	0.291	1.464	-0.19306925	-0.67415324
8	0.280	1.428	-0.19306787	-0.67415944
9	0.270	1.400	-0.19306755	-0.67415956
finite difference			-0.19306763	-0.67415961

$2p^2$, $Z=92$, kinetically balanced set

n	α	β	ε	E
3	57.629	2.617	-1232.8193	-2487.9673
4	56.062	2.448	-1233.9793	-2490.1830
5	54.902	2.332	-1235.3526	-2492.5162
6	53.990	2.246	-1235.7306	-2493.1938
7	53.242	2.178	-1235.9495	-2493.4584
8	52.610	2.123	-1236.0608	-2493.6290
finite difference			-1235.8786	-2493.2369

$2p^2$, $Z=92$, spinor set

n	α	β	ε	E
3	30.415	1.361	-1235.8204	-2493.0974
4	29.789	1.332	-1235.8763	-2493.2317
5	29.317	1.311	-1235.8787	-2493.2367
6	28.939	1.296	-1235.8789	-2493.2369
7	28.626	1.283	-1235.8790	-2493.2369
8	28.357	1.273	-1235.8790	-2493.2369
finite difference			-1235.8786	-2493.2369

3.4 RELATIVISTIC MANY-BODY PERTURBATION THEORY.

It is often important to go beyond the independent particle picture and take account of electron correlation effects. The diagrammatic many-body perturbation theory affords a powerful and accurate technique for doing this [25-27]. Its linear scaling property with the number of electrons in the system is particularly important in studies of molecules containing heavy atoms since such species also contain large numbers of electrons [24].

In non-relativistic diagrammatic perturbation theory calculations [28] on a model problem for which exact results are known for each of the diagrammatic components of the perturbation series, it has been shown that the use of the algebraic approximation leads to more accurate results than those obtained by using the numerical approach [29] first advocated by Kelly [30]. Difficulties arise in the latter method in the evaluation of integrals over the continuum. The model, which consists of a hydrogenic atom with nuclear charge Z perturbed by an additional charge Z', has been examined within the framework of relativistic quantum mechanics using 'kinetically balanced' basis sets for the ground state with angular symmetry κ=-1 and good agreement between the exact and calculated energy components was obtained. This model problem is a quite sensitive test of basis set quality, since contributions from negative energy states must be included and the energy at each order beyond second results from the near cancellation of diagrams. Here we consider a more difficult case namely states with κ=+1 for which the wavefunctions have a complicated structure near the nucleus. We have performed calculations for systematic sequences of both 'kinetically balanced' and 'spinor' even-tempered basis sets. The calculated energies for the 'spinor' sets converged more rapidly than those for the 'kinetically balanced' sets. The results for the 'spinor' sets are presented in Table 3. Smooth convergence to the exact values is observed for each of the energy components and the importance of the boundary conditions at the point nucleus is again demonstrated.

Table 3.

Relativistic perturbation theory calculations for the model problem with Z=92, Z'=1 using 'spinor' basis sets of exponential-type functions. ϵ_i *denotes the energy through order i and (. .) serves to differentiate the diagrammatic terms according to the conventions established in reference [28]. Powers of ten are given in [...], exact results are given in {...}.*

n	ϵ_0 [+3]	ϵ_1 [0]	ϵ_2 [-1]	ϵ_3(p) [-2]	ϵ_3(h) [-2]	ϵ_3 [-3]	ϵ_4(pp) [-4]	ϵ_4(hp) [-4]
11	-1.25739068	-3.32602454	-3.60572721	-1.39612062	1.03559489	-3.60525726	-5.44921835	3.83367382
12	-1.25739068	-3.32602456	-3.60495902	-1.39561175	1.03496709	-3.60644660	-5.44705832	3.82978511
13	-1.25739068	-3.32602454	-3.60638619	-1.39703570	1.03536182	-3.61673882	-5.45689731	3.83297730
14	-1.25739068	-3.32602428	-3.60583922	-1.39657951	1.03505335	-3.61526162	-5.45430057	3.83086849
15	-1.25739068	-3.32602455	-3.60647508	-1.39723107	1.03525928	-3.61971789	-5.45887424	3.83249546
16	-1.25739068	-3.32602455	-3.60614675	-1.39692499	1.03510510	-3.61819887	-5.45694386	3.83136250
	{-1.25739068}	{-3.32602454}	{-3.60651501}			{-3.61964777}		

n	ϵ_4(hh) [-4]	ϵ_4(R) [-4]	ϵ_4 [-5]	E [+3]
11	-3.84087417	1.12268344	-5.00061441	-1.29101515
12	-3.84091140	1.12176380	-5.06635698	-1.29101508
13	-3.84077785	1.12263591	-5.09084639	-1.29101523
14	-3.84080048	1.12213130	-5.11232773	-1.29101517
15	-3.84075073	1.12255239	-5.12081661	-1.29101524
16	-3.84078390	1.12228302	-5.12719739	-1.29101521
			{-5.12798570}	{-1.29101525}

The high accuracy of the sum-over-states formulation has led us to develop a relativistic many-body perturbation theory [15,24,27,32] without the numerical divergences inherent in the finite difference approach. As long as the positive- and negative-energy branches of the spectrum can be identified, second-quantization can be employed to ensure that all states below the Fermi level are filled according to the Dirac hole theory and the restrictions on the allowable summations involved in the energy expressions corresponding each of the diagrammatic terms in the perturbation series follow naturally. A sequence of calculations of the second-order contributions to the relativistic fine structure of the helium ground state are presented in Table 4 serves to demonstrate the practical implementation of this method. Further details can be found elsewhere [27,32,33].

Table 4.

Relativistic second-order many-body perturbation theory calculations for the ground state of the helium atom. E_2*(relativistic) is given in microhartree with signs reversed and values of* E_2*(relativistic) -* E_2*(non-relativistic) are given in brackets.* n *denotes the dimension of the basis set.*

κ	n=13	n=14	n=15
-1	13496.6 [-1.4]	13497.3 [-1.4]	13497.8 [-1.4]
+1	6322.4 [+2.2]	6322.9 [+2.2]	6323.2 [+2.2]
-2	12649.1 [-0.2]	12649.1 [-0.1]	12651.1 [-0.1]
+2	1273.8 [+0.9]	1274.4 [+0.9]	1274.8 [+0.9]
-3	1911.9 [+0.1]	1912.8 [+0.1]	1913.5 [+0.1]
+3	396.5 [+0.4]	1397.1 [+0.4]	397.5 [+0.4]
-4	529.1 [+0.1]	529.9 [+0.1]	530.5 [+0.1]
+4	157.8 [+0.2]	158.3 [+0.2]	158.7 [+0.2]
-5	197.4 [+0.1]	198.1 [+0.1]	198.6 [+0.1]

References.

[1] Schwartz, C., 1963, Meth. Comput. Phys. **2** 241.

[2] Huzinaga, S., 1985, Comput. Phys. Rept. **2** 279.

[3] Schaefer, H.F., III, 1984, Quantum Chemistry - The development of ab initio methods in molecular electronic structure theory, Clarendon Press, Oxford.

[4] Wilson, S., 1987, Adv. Chem. Phys. **67** 439.

[5] Schmidt, M.W., and Ruedenberg, K., 1979, J. Chem. Phys. **71** 3951.

[6] Feller, D.F., and Ruedenberg, K., 1979, Theor. chim. Acta. **52** 231

[7] Fischer, C.F., 1977, The Hartree-Fock method for atoms, Wiley, New York.

[8] Wilson, S., 1980, Theor. chim. Acta. **57** 53, **58** 31.

[9] Laaksonen, L., Pyykko, P., and Sundholm, D., 1987, Comput. Phys. Rep. **4** 313.

[10] Laaksonen, L., Pyykko, P., and Sundholm, D., 1983, Intern. J. Quantum Chem. **23** 319.

[11] Wells, B.H., and Wilson, S., 1985, J. Phys. B: At. Mol. Phys. **18** L731.

[12] Kutzelnigg, W., 1984, Intern. J. Quantum Chem. **25** 107.

[13] Dyall, K.G., Grant I.P., and Wilson, S., 1984, J. Phys. B: At. Mol. Phys. **17** L45 493 1201.

[14] Stanton, R.E., and Havriliak, S., 1984, J. Chem. Phys. **81** 1910.

[15] Quiney, H.M., Grant, I.P., and Wilson, S., 1987, Physica Scripta **36** 460.

[16] Grant, I.P., and Quiney, H.M., 1988, Adv. At. Mol. Phys. **23** 37

[17] Furry, W.H., 1951, Phys. Rev. **81** 115.

[18] Drake, G.W.F., and Goldman, S.P., 1982, Phys. Rev. A **25** 2877.

[19] Grant, I.P., Mackenzie, B.J., Norrington, P.H., Mayers, D.F., Pyper, N.C., 1980, Comput. Phys. Commun. **21** 207.

[20] Ishikawa, Y., and Quiney, H.M., 1987, Intern. J. Quantum Chem. Symp. **21** 523.

[21] Laaksonen, L., Grant, I.P., and Wilson, S., 1988, J. Phys. B: At. Mol. & Opt. Phys. (in press).

[22] Quiney, H.M., Grant, I.P., and Wilson, S., 1987, J. Phys. B: At. Mol. Phys. **20** 1413.

[23] Grant, I.P., 1988, Meth. Comput. Chem. **2** (in press).

[24] Wilson, S., 1988, Meth. Comput. Chem. **2** (in press).

[25] Wilson, S., 1984, Electron correlation in molecules, Clarendon Press, Oxford.

[26] Wilson, S., 1985, Comput. Phys. Rep. **2** 389.

[27] Wilson, S., 1988, The many-body perturbation theory of atoms and molecules, Adam Hilger, Bristol.

[28] Quiney, H.M., Grant I.P., and Wilson, S., 1985, J. Phys. B: At. Mol. Phys. **18** 577.

[29] Rossky, P., and Karplus, M., 1977, J. Chem. Phys. **67** 5419.

[30] Kelly, H.P., 1969, Adv. Chem. Phys. **14** 129.

[31] Quiney, H.M., Grant, I.P., and Wilson, S., 1985, J. Phys. B: At. Mol. Phys. **18** 2805.

[32] Quiney, H.M., 1988, Meth. Comput. Chem. **2** (in press).

[33] Quiney, H.M., 1987, D. Phil. thesis, University of Oxford.

NUMERICAL MCSCF IN ONE AND TWO DIMENSIONS

Dage Sundholm, Jeppe Olsen, Per-Åke Malmqvist and Björn O. Roos
Theoretical Chemistry, Chemical Center
P.O.B. 124, S-221 00 LUND, Sweden

ABSTRACT. Preliminary results show the feasability of extending modern basis set MCSCF methods to apply also to numerical calculations. Rectangular finite elements are used as one-electron basis. Quadratic convergence is demonstrated for atoms, fast linear convergence for diatomics, with quaranteed convergence to a energy minimum.

1. INTRODUCTION

In the non-relativistic Born-Oppenheimer approximation, spectroscopic and chemical properties of molecules are completely determined by the set of solutions to the time-independent Schrödinger equation:

$$\left(-\frac{1}{2}\sum_i \nabla_i^2 - \sum_{i,A} \frac{Z_a}{r_{iA}} + \frac{1}{2}\sum_{i>j} \frac{1}{r_{ij}}\right) \psi = E\,\psi \qquad (1)$$

In practice, approximate finite representations in numeric or analytic form is obtained only for the ground, and possibly a few excited, states, and some properties are then most conveniently obtained by also calculating perturbed wave functions, i.e. solutions of

$$(\hat{H} - E_o)\,\psi_1 = \hat{A}\,\psi_o \qquad (2)$$

with conventional notation, where $\hat{A}$ is some operator of interest.

The MCSCF method is a well-established procedure for solving eq. (1) in the "analytic" approximation, where the wave function is expanded in a set of configuration state functions

$$\psi = \sum_I C_I\,\phi_I, \qquad (3)$$

which are constructed as spin-coupled products of orbitals. These, in turn, are expanded in a set of analytic functions:

M. Defranceschi and J. Delhalle (eds.), Numerical Determination of the Electronic Structure of Atoms, Diatomic and Polyatomic Molecules, 329–334.

$$\varphi_i(\underline{r}) = \sum_k d_k^i \chi_k(\underline{r}) \qquad (4)$$

There are then two limitations to the accuracy: The truncation of the CI expansion, eq. (3), and of the basis set $\{\chi_k\}$ in eq. (4). Of these, only the first is inherent in the MCSCF method, and the second can be made negligibly small by using a numeric rather than analytic approximation. However, for historical reasons, the numeric approximation has until recently been implemented by difference approximation of the Laplacian in eq. (1). In retrospect, this formulation has some serious disadvantages: The obtained result is not variationally bound by the exact eigenvalues, some powerful numerical tools are not applicable, and numerical implementations have serious convergence control and convergence speed problems, which seem to be solved in modern analytic programs.

However, there are no practical reasons why a basis set approach should not be used also for a numeric representation of the wave function. Such an approach is the Finite Element Method (FEM) [1], which has already been successfully applied to atomic Dirac-Fock [2] and diatomic Hartree-Fock-Slater [3] calculations, although it has not yet been widely recognized to what extent existing, successful analytic methods can be applied also to FEM-based numerical problems.

2. THE FINITE ELEMENT FORMULATION

In our programs, we treat only atomic (1-D) and diatomic (2-D) problems. The 1-D case is simple, and we confine the description to the 2-D case. For this, we use prolate spheroidal coordinates, μ, υ, φ [4], where the φ dependence is predetermined. For μ, we truncate the interval $[0,\infty[$ at some large limit μ_∞. The remaining interval $[\mu_\infty,\infty[$ is easily included, by means of a special element function with exponential tail. However, we have not found this necessary. The last interval can be replaced by suitable boundary conditions.

Each of the intervals $[0,\mu_\infty]$ and $[0,\pi]$ is divided into elements, further subdivided by (m-1) interior and the two boundary grid points. For each grid point, a basis function is defined, which has the value 1 in the defining point and 0 in all other. The function is non-zero within element(s) containing or touching the defining grid point. Within such an element, it is a Lagrange interpolation polynomial (LIP) of degree m. For the 2-D domain $[0,\mu_\infty]\otimes[0,\pi]$, the basis functions are simple products of the LIP's of variables μ and υ. The variation parameters are then simply the values of the approximated function in each grid point.

Basic integrals involving products of two LIP's, and their derivatives, are precalculated and stored. Strictly speaking, these basic integrals are insufficient for calculating the interaction of an electrostatic field with a density exactly, since that involves integrals which are products of three LIP's. The error is small, and is presently ignored for simplicity. It can be removed at the price of small complications in the code.

3. THE POISSON SOLVER

In the numerical calculations, up to 10^4 basis functions are used, and precalculation and storage of electrostatic integrals is impossible. Instead, a number of numerically represented electrostatic fields are recomputed at each MCSCF iteration. In FEM form, the method is still variational if formulated as a minimization of electrostatic field and interaction energy.

The resulting Poisson equations take the form

$$\underline{\underline{A}}\,\underline{v} = \underline{\underline{S}}\,\underline{\rho} \qquad (5)$$

where $\underline{\underline{A}}$ and $\underline{\underline{S}}$ are positive definite. Traditional overrelaxation methods were found to be too slow, and we developed a fast Poisson solver based on the preconditioned conjugate gradient (PCG) method [5]. A requisite for success in the PCG method is the ability to solve efficiently $\tilde{\underline{\underline{A}}}\,\underline{x} = \underline{y}$ for any $\underline{y}$, where $\tilde{\underline{\underline{A}}}$ is an approximation to $\underline{\underline{A}}$. It was chosen as an incomplete block LU factorization, in which inverted matrix block were replaced by a (2j+1)-diagonal matrix, with j in the range 5--10. Due to the properties of the LIP representation, the same preconditioner could be used for any order of the LIP's. The performance of the method is summarized in Table I.

Table I. Number of iterations and CPU-time in seconds to solve the Poisson equation.

Grid size	Order of Lagrange interpolation functions			
	m = 1	m = 2	m = 4	m = 6
13 x 13	4 0.03	13 0.09	31 0.20	74 0.46
25 x 25	6 0.11	12 0.18	34 0.45	97 1.26
49 x 49	8 0.33	13 0.45	32 1.01	108 3.38
97 x 97	14 1.54	20 1.91	44 3.97	129 11.59

On an IBM 3090/150 VF, $\|r\|_\infty \leq 10^{-6}$

4. THE WAVE FUNCTION OPTIMIZATION

In each iteration of the MCSCF procedure, a number of Poisson equations are solved for the necessary electrostatic fields. From these, and using current orbitals, and density matrices from the CI package, the energy gradient with respect to orbital variations is obtained. Applying an inverse Hessian yields orbital updates. However, this inverse is not exact, and the update is not immediately applied. Instead, the orbital space is increased to contain both the current orbitals and the updates, all necessary electrostatic and 1-electron integrals are evaluated, and MO and CI coefficients are fully optimized by a second-order conventional basis-set MCSCF method in the enlarged orbital space. The approximate inverse Hessian is updated by a modified BFGS technique, and is block-diagonal. Since it is positive definite, energy decrease is guaranteed in each step. Meaningful orbital updates are produced as long as the gradient remains non-zero, i.e., convergence is guaranteed.

Some preliminary results can be seen in Tables II - IV. True second-order and true sesquilinear convergence can be obtained for atoms (1-D). The outline above applies to the more difficult 2-D case, and convergence is then linear and its speed depends on the quality of the inverse Hessian approximation.

Note that the 2-D energies are slightly too low, in spite of the fact that they are variationally obtained minima. This error results from the approximation of the electron-field interaction. As mentioned earlier, this approximation is not necessary.

Table II. Convergence of 2-D HF calculation on BH (97 x 97 grid points and 6th degree LIP's).

Iteration	Energy	Energy Change	Gradient norm
0	-21.5311522	-	1.0873846
1	-24.3425373	-2.8113851	0.6080568
2	-24.8932233	-0.5506860	0.5429001
3	-25.0464332	-0.1532099	0.4896432
4	-25.1173350	-0.0709018	0.4601838
5	-25.1273911	-0.0100561	0.2968516
6	-25.1301310	-0.0027399	0.1297089
7	-25.1313443	-0.0011833	0.0425374
8	-25.1315440	-0.0002298	0.0278797
9	-25.1316133	-0.0000692	0.0146867
10	-25.1316262	-0.0000130	0.0093496
11	-25.1316284	-0.0000022	0.0051942
exact	-25.131609	-	0.0

Table III. Second-order convergence of 1-D HF calculation on He (401 grid points and 6th degree LIP's).

Iteration	Energy	Gradient norm
0	-2.84750000000	0.488
1	-2.86157164411	$0.545\ 10^{-2}$
2	-2.86167999557	$0.235\ 10^{-4}$
3	-2.86167999560	$0.300\ 10^{-9}$

Table IV. 2-D Hartree Fock results. (97 x 97 grid points and 6th degree LIP's)

System	Iterations	Energy	Energy (exact)
He	6	-2.8616835	-2.8616800
H_2	5	-1.1336295	-1.13362957
HeH^+	5	-2.9331032	-2.93310325
Be	11	-14.5730402	-14.5730232
LiH	11	-7.9873552	-7.9873524
BH	11	-25.131628	-25.131609

5. THE CI ALGORITHM

Our numerical MCSCF has been equipped with a new CI algorithm, based on Slater determinants (SD's) [6]. The loss of permutation symmetry in two-electron integral indices, due to the complex φ-dependence, is treated without loss of computational efficiency. For most full CI calculations (CASSCF) it is probably the fastest existing program. More general MCSCF is done by subdividing the active space into subspaces with occupation restrictions. The CI part is not yet operative in the MCSCF program. As already seen, orbital optimization has been tested for HF calculations, where no CI is needed. The CI package has been tested in a conventional CASSCF program.

Correlating all 10 electrons of Ne in 16 orbitals produced $2.4 \cdot 10^6$ SD's. The CPU time was 171 seconds per CI iteration on our IBM 3090/150VF. An MR-SDCI from a reference space of 420000 SD's yielded $1.1 \cdot 10^6$ SD's and took 140 seconds. Correlating the 6 electrons of CH^+ in 26 orbitals gave $1.7 \cdot 10^6$ SD's, and the iteration time was 290 seconds.

6. CONCLUSIONS

In this preliminary report, we have demonstrated the feasability and advantages of extending modern basis-set MCSCF methods to apply also to numerical calculations. This was possible by using high-order Lagrange interpolation functions within rectangular elements in a finite element method. Quadratic convergence has been demonstrated for atoms, fast linear convergence for diatomics, with guaranteed convergence. The resulting energies are variationally bound by the exact MCSCF solutions, apart from a small and removable approximation of the electrostatic interaction energy. The method relies on three new major algorithms: A fast FEM Poisson solver, a non-linear optimization algorithm for orbitals and CI coefficients, and a fast configuration interaction algorithm.

REFERENCES

1. R. Gruber (ed.), Finite Elements in Physics, Proceedings of the 1st European Graduate Summer Course on Computational Physics, Elsevier (1987).

2. W.R. Johnson, S.A. Blundell and J. Sapirstein, Phys. Rev. **A37**, 307 (1988).

3. D. Heinemann, B. Fricke and D. Kolb, Chem. Phys. Letters **145**, 125 (1988).

4. L. Laaksonen, P. Pyykkö and D. Sundholm, Comp. Phys. Reports **4**, 313 (1986).

5. D. Sundholm, Comp. Phys. Commun. **49**, xxx (1988).

6. J. Olsen, B.O. Roos, P. Jørgensen and H.J.Aa. Jensen, J. Chem. Phys. **89**, xxx (1988).

NUMERICAL METHODS FOR CALCULATING MULTICENTER INTEGRALS FOR ARBITRARY ORBITALS

JAMES D. TALMAN
Departments of Applied Mathematics and Physics and Centre for Chemical Physics
University of Western Ontario
London, Ontario, Canada N6A 5B9

ABSTRACT. Techniques for the numerical calculation of convolution integrals, nuclear attraction three-center integrals and four-center integrals that are important in molecular Hartree-Fock calculations are discussed. The methods are based on an algorithm for computing spherical Bessel transforms that is very accurate and efficient.

1. Introduction

In this article the problem of the numerical calculation of integrals of the form

$$I_1 = \int f(\mathbf{r}-\mathbf{a}_1)\, g(\mathbf{r}-\mathbf{a}_2)\, d\mathbf{r} \tag{1.1}$$

$$I_2 = \int f(\mathbf{r}-\mathbf{a}_1)\, |\mathbf{r}-\mathbf{R}|^{-1}\, g(\mathbf{r}-\mathbf{a}_2)\, d\mathbf{r} \tag{1.2}$$

$$I_3 = \iint f_1(\mathbf{r}-\mathbf{a}_1)\, f_2(\mathbf{r}-\mathbf{a}_2)\, |\mathbf{r}-\mathbf{r}'|^{-1}\, f_3(\mathbf{r}'-\mathbf{a}_3)\, f_4(\mathbf{r}'-\mathbf{a}_4)\, d\mathbf{r}\, d\mathbf{r}' \tag{1.3}$$

where the functions are angular momentum wave functions of the form

$$f_{\ell m}(\mathbf{r}) = f_\ell(r)\, Y_{\ell m}(\theta,\phi) \tag{1.4}$$

will be discussed. The radial functions f_ℓ can describe arbitrary bound state wave functions. Integrals of this form are central to quantum chemistry theory. The three forms of integrals are known respectively as overlap integrals, three-center nuclear attraction integrals and four-center integrals. They are readily calculated for so-called Gaussian type orbitals (GTOs) with the consequence that practically all quantum chemistry calculations are based on GTOs. Nonetheless, GTOs may not be suitable for all applications, and there remains interest in the evaluation of such integrals for more general basis functions [1].

The methods that will be discussed are based on a novel method of computing spherical Bessel (or Hankel) transforms [2]. This method will be described briefly in Section 2. The central problem of the translation of angular momentum wave functions from one center to another will be discussed in Section 3. The calculation of nuclear attraction three-center integrals, and of four-center integrals will be discussed in Sections 4 and 5 and the application of the methods to a molecular calculation will be described in Section 6.

M. Defranceschi and J. Delhalle (eds.), Numerical Determination of the Electronic Structure of Atoms, Diatomic and Polyatomic Molecules, 335–339.

2. Numerical Spherical Bessel Transforms

It is well known that the calculation of integrals such as those in eqs. (1.1), (1.2), (1.3) can be facilitated by transforming to momentum space. The Fourier transform of a function f(**r**) will be defined by

$$\tilde{f}(\mathbf{k}) = \int e^{i\mathbf{k}\cdot\mathbf{r}} f(\mathbf{r})\, d\mathbf{r} \tag{2.1}$$

with

$$f(\mathbf{r}) = (2\pi)^{-3} \int e^{-i\mathbf{k}\cdot\mathbf{r}}\, \tilde{f}(\mathbf{k})\, d\mathbf{k}. \tag{2.2}$$

Then for example

$$I_1 = (2\pi)^{-3} \int e^{i\mathbf{k}\cdot(\mathbf{a}_1 - \mathbf{a}_2)}\, \tilde{f}_1(\mathbf{k})\, \tilde{f}_2(-\mathbf{k})\, d\mathbf{k} \tag{2.3}$$

and the latter integral may perhaps be more readily calculated than the integral in eq.(1.1).

The Fourier transforms of angular momentum wave functions of the form given in eq.(1.4) are expressed in terms of spherical Bessel or Hankel transforms and are given explicitly by

$$\tilde{f}_{\ell m}(\mathbf{k}) = 4\pi\, i^{\ell}\, Y_{\ell m}(k)\, \tilde{f}_{\ell}(k) \tag{2.4}$$

where

$$\tilde{f}_{\ell}(k) = \int_0^{\infty} j_{\ell}(kr)\, f_{\ell}(r)\, r^2\, dr. \tag{2.5}$$

The numerical calculation of integrals of the form given in eq.(2.5) is difficult at large values of the momentum k because of the rapid oscillation of the integrand. It is then necessary to use a very fine mesh in the r variable and the numerical calculation may become prohibitively lengthy. Moreover, the result decreases rapidly at large k values and it is difficult to obtain the numerical accuracy required.

It has been found that it is possible to circumvent these difficulties by using a novel method for calculating integrals of the form in eq.(2.5) by using as variable $\rho = \ln r$ and $\kappa = \ln k$ [2]. The integral in eq.(2.5) then becomes a convolution which can be calculated numerically by making a Fourier transform, multiplying by a fixed function and making a second Fourier transforms. These Fourier transforms can be computed very efficiently using the fast Fourier transform algorithm. A subroutine package for this is available from the CPC library and is described in the literature [3]. Normally, the calculations are carried out on meshes in the ρ and κ variables of 64, 128 or 256 points because of the use of the FFT method.

3. Translation of Angular Momentum Wave Functions

The possibility of the numerical calculation of spherical Bessel transforms opens the way to the ability to calculate the expansion of angular momentum wave functions centered at one point in terms of angular momentum wave functions centered at another point. Explicitly, one can write for an angular momentum wave function $f_{\ell m}(\mathbf{r})$

$$f_{\ell m}(\mathbf{r}-\mathbf{a}) = \Sigma_{LM}\, \Sigma_{L'M'}\, f_{LL'MM'}(r)\, Y_{LM}(\theta_a,\phi_a)\, Y_{L'M'}(\theta_r,\phi_r). \tag{3.1}$$

If $f_{\ell m}(\mathbf{r}-\mathbf{a})$ is expressed in terms of the Fourier transform and the factors $e^{-i\mathbf{k}\cdot\mathbf{r}}$ and $e^{i\mathbf{k}\cdot\mathbf{a}}$ expanded in spherical harmonics it is found that

$$f_{LL'MM'}(r,a) = (-1)^m \begin{pmatrix} L & L' & \ell \\ M & M' & -m \end{pmatrix} f_{LL'}(r,a) \tag{3.2}$$

where

$$f_{LL'}(r,a) = 8\left[\frac{(2\ell+1)(2L+1)(2L'+1)}{4\pi}\right]^{1/2} i^{\ell+L-L'} \begin{pmatrix} L & L' & \ell \\ 0 & 0 & 0 \end{pmatrix}$$
$$\times \int j_{L'}(kr)\, j_L(ka)\, \tilde{f}(k)\, k^2\, dk. \tag{3.3}$$

Thus to calculate the expansion of an angular momentum wave function about a new center, it is necessary to calculate a spherical Bessel transform, multiply by $j_L(ka)$ and calculate a new set of spherical Bessel transforms. It should be noted that L and L' are restricted by $|L-L'| \leq \ell$ and $L-L'+\ell$ is even.

The calculation of the integral in eq.(3.3) tends to be inacccurate because of the oscillatory behavior of $j_L(ka)$ for large k. The spacing of the mesh points in k increases proportionally to k, so that a fine mesh is required to deal with the oscillatory behavior properly. In the $\ell = 0$ case, which is the most difficult, $\tilde{f}(k)$ decreases only like k^{-4} for large k if f(r) has a cusp at the origin. This problem can be reduced by subtracting a function of the form $Ae^{-\alpha r}$ from f(r) (when $\ell = 0$) to cancel f '(0) and adding back the known analytic form. This procedure should improve the accuracy in the calculation of $f_{LL}(r,a)$, but a preliminary investigation suggests that it does not make a significant improvement in the calculation of the multicenter integral. The reason for this is that the errors are large only for r close to a and therefore are not so important after the integration.

4. Nuclear Attraction Three-Center Integrals

The translation formula for angular momentum wave functions renders the calculation of the nuclear attraction three-center integral of eq.(1.2) straightforward in principle, if not in practice. Each of the three factors can be translated to a common origin, the angular integral can be calculated in terms of 3j coefficients and the radial integral numerically, and the result obtained as a three-fold sum. There is some advantage to rotating the coordinate system so that the nucleus is on the new z axis [4].

The problem of convergence of the angular momentum sums has been investigated by LaBudde and Sahni in the case that the expansion center is the nucleus or one of the orbital centers [5] and in the general case by the author [4]. It turns out that the series are geometric in the ratios ρ_1/ρ_2, ρ_1/ρ_3 and ρ_2/ρ_3 where ρ_1, ρ_2 and ρ_3 are the distances from the origin to the three centers, arranged in increasing order.

The optimum choice of expansion center is therefore the point that minimizes the largest of these three ratios. This minimum evidently occurs for $\rho_1/\rho_2 = \rho_2/\rho_3$. The geometric problem of finding this point can be solved [4] and it is found that in the worst case, in which the three centers lie on an equilateral triangle the ratio is $(\sqrt{5}-1)/2 \cong 0.618$, the reciprical of the golden mean. In this case accuracy to about 10^{-6} is obtained by retaining L values up to about 10 in the sum. For less symmetric geometries, the maximum ratio is smaller and fewer terms need be retained.

5. Four-Center Integrals

In a molecular calculation involving N orbitals, of the order of $N^4/8$ four-center integrals of the type in eq.(1.3) are required. The method that has been developed is to expand each of the pairs of products of angular momentum wave functions about a point **c** which is the mid-point of the line segment joining the two centers for the orbitals. Since there are of the order of $N^2/2$ such pairs, considerable effort may be expended in this phase of the calculation provided the final calculation of the four-center integral is efficient.

We consider first the calculation of the general double integral

$$I = \int \int f(\mathbf{r}-\mathbf{a}_1)\, |\mathbf{r}-\mathbf{r}'|^{-1}\, f(\mathbf{r}'-\mathbf{a}_2)\, d\mathbf{r}\, d\mathbf{r}' \tag{5.1}$$

It is straightforward to show that in momentum space this is given by

$$I = (2\pi)^{-3} \int (4\pi/k^2)\, e^{i\mathbf{k}\cdot\mathbf{a}}\, \tilde{f}(-\mathbf{k})\, \tilde{g}(\mathbf{k})\, d\mathbf{k} \tag{5.2}$$

where $\mathbf{a} = \mathbf{a}_2 - \mathbf{a}_1$ so that this integral is effectively no more difficult than the convolution integral (2.3).

It can be shown [6] that the expansion of the product of two functions centered at $\mathbf{a}_1$ and $\mathbf{a}_2$ about a point $\mathbf{c}$ on the line segment joining the two centers is

$$f_1(\mathbf{r}-\mathbf{a}_1) f_2(\mathbf{r}-\mathbf{a}_2) = \Sigma_{\lambda\Lambda j\mu\sigma} \begin{pmatrix} \ell_1 & \ell_2 & j \\ m_1 & m_2 & m \end{pmatrix} \begin{pmatrix} j & \lambda & \Lambda \\ m & \mu & \sigma \end{pmatrix} \times F_{\lambda\Lambda j}(r)\, Y_{\lambda\mu}(\mathbf{a})\, Y_{\Lambda\sigma}(\hat{\mathbf{r}}) \tag{5.3}$$

where $\mathbf{a}$ is a unit vector in the direction from $\mathbf{a}_1$ to $\mathbf{a}_2$ and

$$F_{\lambda\Lambda j}(r) = (2j+1)\, [(2\lambda+1)(2\Lambda+1)(2\ell_1+1)(2\ell_2+1)]^{1/2}\, \Sigma_{L_1 L_1'}\, \Sigma_{L_2 L_2'}\, (-1)^{L_2} \times \begin{pmatrix} L_1 & L_2 & \lambda \\ 0 & 0 & 0 \end{pmatrix} \begin{pmatrix} L_1' & L_2' & \Lambda \\ 0 & 0 & 0 \end{pmatrix} \begin{Bmatrix} \ell_1 & \ell_2 & j \\ L_1 & L_2 & \lambda \\ L_1' & L_2' & \Lambda \end{Bmatrix} \times f^{(1)}_{L_1 L_1'}(r,c_1)\, f^{(2)}_{L_2 L_2'}(r,c_2). \tag{5.4}$$

Here c_1 and c_2 are the distances from $\mathbf{c}$ to $\mathbf{a}_1$ and $\mathbf{a}_2$ respectively.

The four-center integrals are calculated by finding the transforms of $\tilde{F}_{\lambda\Lambda j}(k)$ for all pairs of particles. The terms in the sum are restricted by $|\ell_1-\ell_2| \le j \le \ell_1+\ell_2$ and $|\lambda-\Lambda| \le j \le \lambda+\Lambda$. Therefore in the simplest case of s orbitals there is essentially only a single index. In general, the number of functions to be stored is proportional to $(2\ell_1+1)(2\ell_2+1)$. It has been found that retaining terms up to $\Lambda = 4$ seems to give reasonable accuracy (of the order of 10^{-6} au) for the general four-center integral while it may be necessary to retain terms up to $\Lambda = 6$ or 8 in the hybrid integrals in which $\mathbf{a}_1 = \mathbf{a}_2$ or $\mathbf{a}_3 = \mathbf{a}_4$.

6. Discussion

The methods described in the previous sections have been applied in a somewhat preliminary way to the NH_3 molecule. The MO-LCAO Hartree-Fock calculation was carried out using as basis functions the Clementi and Roessi atomic orbitals for the N atom. In this case there are 144 nuclear attraction three-center integrals and 666 four-center integrals to calculate. The calculations were carried out on a mesh of 128 grid points with L_{max} in the three-center integral calculation being 6, and Λ_{max} in the four-center integral calculation being 4 in the general case and 6 in the hybrid integral case. On a CYBER 825 the three- center integrals required 190 seconds, calculating the product functions necessary for the four-center integrals required 90 seconds and calculating the four-center integrals required 60 seconds. The program has also been run on a Cray X-MP 24 and with full optimization of the FFT subroutine these times are reduced by a factor of 100 or better.

The techniques described above are particularly suited to computers with vector or parallel processing capabilities. It may be hoped that with the continuing development of such machines and with improvements in the computational methods, it may be possible to carry out calculation of molecular properties that are not well described by GTOs using more general basis functions.

7. Acknowledgements

This work was carried out in part at the University of Oslo under the support of NORDITA and has also been supported by the Natural Sciences and Engineering Research Council of Canada.

8. References

[1] *ETO Multicenter Molecular Integrals*, edited by C.A.Weatherford and H.W.Jones (Reidel,Dordrecht,1982).
[2] J.D.Talman, J. Comput. Phys.**29**, 35 (1978).
[3] J.D.Talman, Comput. Phys. Comm. **30**, 93 (1983).
[4] J.D.Talman, J. Chem. Phys. **84**, 6879 (1986).
[5] C.D.LaBudde and R.C. Sahni, Phys. Rev. **33**, 1022 (1960).
[6] J.D. Talman, J. Chem. Phys. **80**, 2000 (1984).
[7] E.Clementi and C. Roetti,At. Data. Nucl. Data Tables, **14**, 177 (1974).

NONLINEAR SEQUENCE TRANSFORMATIONS FOR THE EFFICIENT EVALUATION OF AUXILIARY FUNCTIONS FOR GTO MOLECULAR INTEGRALS

E. Joachim Weniger and E. Otto Steinborn
Institut für Physikalische und Theoretische Chemie
Universität Regensburg
D-8400 Regensburg
Federal Republic of Germany

Abstract
It is discussed how the auxiliary function $F_m(z)$ with $m \in \mathbb{N}_0$ and $z \in \mathbb{R}_+$ can be computed efficiently by means of some recently discovered nonlinear sequence transformations. It is shown that the new transformations yield somewhat better results than the already very powerful sequence transformations by D. Levin [Int. J. Comput. Math. B **3**, 371 - 388 (1973)]. The use of this transformation for the evaluation of $F_m(z)$ had been suggested by J. Grotendorst and E. O. Steinborn [J. Chem. Phys. **84**, 5617 - 5623 (1986)].

1. Introduction

In molecular *ab initio* calculations with Gaussian-type basis functions nuclear attraction and interelectronic repulsion integrals are ultimately expressed in terms of the following class of auxiliary functions:

$$F_m(z) = \int_0^1 u^{2m} e^{-zu^2} du, \quad m \in \mathbb{N}_0, \quad z \in \mathbb{R}_+ \tag{1}$$

In molecular calculations these auxiliary functions $F_m(z)$ have to be computed so frequently that this amounts to a significant part of the whole integral evaluation time. Consequently, there is an extensive literature dealing with the efficient evaluation of these functions. Of particular interest here are review articles by Shavitt [1] and Saunders [2,3] in which most of the relevant properties of the $F_m(z)$ can be found and which also contain a fairly complete survey of the literature on this topic.

But although the state of art of evaluating the auxiliary functions $F_m(z)$ has reached a high degree of sophistication, it nevertheless seems to be worthwhile to look for further improvements in computational efficiency since even small gains can possibly add up to significant reductions of the total integral computation time.

In this article we want to discuss a computational approach which is based upon nonlinear sequence transformations. This is not a new idea. Harris [4] discussed the use of various continued fraction representations for $F_m(z)$. Later, Grotendorst

M. Defranceschi and J. Delhalle (eds.), Numerical Determination of the Electronic Structure of Atoms, Diatomic and Polyatomic Molecules, 341–346.

and Steinborn [5] showed that a sequence transformation by Levin [6], which also produces rational approximants, is more efficient than continued fractions or Padé approximants computed by Wynn's epsilon algorithm [7]. However, we shall show that some recently discovered sequence transformations [8] give even better results than Levin's sequence transformation [6] which in extensive numerical tests performed by Smith and Ford [9,10] had turned out to be the currently most powerful and also most versatile convergence acceleration method described so far in the mathematical literature.

2. Mathematical Properties of the Auxiliary Functions

In this section we want to list those mathematical properties of the $F_m(z)$ which will be needed in the sequel. We use the same notation as in the article by Grotendorst and Steinborn [5]. All formulas presented here can be found either in [5] or in the review articles [1–3].

The auxiliary functions $F_m(z)$ possess a Taylor series expansion:

$$F_m(z) \;=\; \frac{1}{2m+1}\, {}_1F_1(m+1/2; m+3/2; -z)\,. \tag{2}$$

Here, the ${}_1F_1$ stands for Kummer's confluent hypergeometric function. Eq. (2) is a special case of the following addition theorem of the $F_m(z)$:

$$F_m(x+y) \;=\; \sum_{k=0}^{\infty} \frac{(-x)^k}{k!}\, F_{m+k}(y)\,. \tag{3}$$

By means of some well-known transformation properties of Kummer's confluent hypergeometric series ${}_1F_1$ a representation for $F_m(z)$ as an exponential multiplied by a power series can be written down immediately:

$$F_m(z) \;=\; \frac{e^{-z}}{2m+1}\, {}_1F_1(1; m+3/2; z)\,. \tag{4}$$

If we compare the two infinite series expansions (2) and (4) we may expect that the alternating signs in the infinite series in Eq. (2) may lead to cancellation errors for larger values of z. In addition, the infinite series in Eq. (4) will converge more rapidly with increasing order m. However, it is quite clear that for larger values of z neither Eq. (2) nor Eq. (4) will be computationally attractive. For very large arguments z suitably truncated partial sums of the following divergent asymptotic series for $F_m(z)$ can be used:

$$F_m(z) \;\sim\; \frac{\Gamma(m+1/2)}{2z^{m+1/2}} \;-\; \frac{e^{-z}}{2z}\, {}_2F_0(1, 1/2-m; -1/z)\,. \tag{5}$$

The auxiliary functions $F_m(z)$ satisfy the following inhomogeneous 2-term recurrence formula:

$$(2m+1)F_m(z) \;=\; 2zF_{m+1}(z) \;+\; e^{-z} \, . \tag{6}$$

For small z, this recurrence formula is numerically stable only if it is used *backwards*. Hence, if a whole string of auxiliary functions $F_0, \cdots, F_m$ with the same argument z will be needed one has to compute the function with maximum value of m and then compute the others. However, for large arguments z, this recurrence formula may safely be used *forwards*, i.e., we may compute F_0 first and then compute the functions with higher orders m.

It is a typical feature of the auxiliary functions $F_m(z)$ as well as of many other special functions that they can be computed accurately and economically for small arguments because Taylor expansions will then converge rapidly and also if the arguments are so large that asymptotic expansions are sufficiently accurate. However, moderately large arguments are troublesome because then Taylor expansions will converge too slowly and asymptotic expansions will be too inaccurate. It is this region of moderately large arguments where sequence transformations can help most.

3. Nonlinear Acceleration Methods

In this section, we shall discuss Levin's u transformation [6] as well as two recently derived nonlinear sequence transformations which are based upon the same remainder estimate as Levin's u transformation [8]. Other variants of Levin's sequence transformation and of the new sequence transformations are not treated here since in this context they give roughly the same results. If s_n stands for the partial sum of an infinite series with terms a_k,

$$s_n \;=\; \sum_{k=0}^{n} a_k \, , \tag{7}$$

then Levin's u transformation may be defined by the following ratio [6]:

$$u_k^{(n)}(\beta, s_n) \;=\; \frac{\displaystyle\sum_{j=0}^{k} (-1)^j \binom{k}{j} \frac{(\beta+n+j)^{k-2}}{(\beta+n+k)^k} \frac{s_{n+j}}{a_{n+j}}}{\displaystyle\sum_{j=0}^{k} (-1)^j \binom{k}{j} \frac{(\beta+n+j)^{k-2}}{(\beta+n+k)^k} \frac{1}{a_{n+j}}} \, . \tag{8}$$

The parameter β has to be chosen in such a way that $n+\beta$ is always positive. Levin's original choice was $\beta = 1$. Both numerators as well as denominators of these transformations can also be computed recursively [8,11]. The following two sequence transformations, which can be derived on the basis of similar assumptions as in the case of Levin's u transformation, are good convergence accelerators and particularly well suited for the summation of violently divergent series [8]:

$$y_k^{(n)}(\beta, s_n) = \frac{\sum_{j=0}^{k} (-1)^j \binom{k}{j} \frac{(\beta+n+j+1)_{k-2}}{(\beta+n+k)_{k-1}} \frac{s_{n+j}}{a_{n+j}}}{\sum_{j=0}^{k} (-1)^j \binom{k}{j} \frac{(\beta+n+j+1)_{k-2}}{(\beta+n+k)_{k-1}} \frac{1}{a_{n+j}}} . \tag{9}$$

The parameter β in Eq. (9) should again be positive. The parameter γ in the following transformation should be negative and satisfy the inequality $-\gamma \geq k-1$.

$$Y_k^{(n)}(\gamma, s_n) = \frac{\sum_{j=0}^{k} (-1)^j \binom{k}{j} \frac{(\gamma-n-j+1)_{k-2}}{(\gamma-n-k)_{k-1}} \frac{s_{n+j}}{a_{n+j}}}{\sum_{j=0}^{k} (-1)^j \binom{k}{j} \frac{(\gamma-n-j+1)_{k-2}}{(\gamma-n-k)_{k-1}} \frac{1}{a_{n+j}}} . \tag{10}$$

Both numerators and denominators of the transformations (9) and (10) can also be computed recursively [8].

4. Numerical Results

Since the orders m of the auxiliary functions $F_m(z)$ depend upon the angular momentum quantum numbers of the GTO's and since the arguments z depend upon the scaling parameters and internuclear distances, a whole range of orders m and of arguments z can occur in a molecular calculation. Hence, let us assume that we have to be able to compute the auxiliary functions $F_m(z)$ for all $m \in \mathbb{N}_0$ with $0 \leq m \leq M$ and all $z \in \mathbb{R}_+$ with $0 \leq z \leq z_{max}$ efficiently and also sufficiently accurate.

We applied the nonlinear sequence transformations (8), (9), and (10) to the partial sums s_n of the infinite series representations (2), (3), (4), and (5) and tested their performances over a wide range of orders and arguments. Since we also wanted to find out whether and how the sequence transformations discussed in this article are affected by rounding errors we did all computations both in DOUBLE PRECISION (15 - 16 decimal digits) and in QUADRUPLE PRECISION (31 - 32 digits). According to our results the sequence transformations are remarkably stable and in most cases it is possible to abtain 14 decimal digits in DOUBLE PRECISION.

When we applied the sequence transformations to the partial sums of the Taylor series (2) we found that the rate of convergence did not depend much upon the order. It decreased slightly with increasing m. In the case of Eq. (2), Levin's sequence transformation (8) was marginally better than the other two. For instance, for $m = 10$ and $z = 2.5$, the partial sums $s_0, \cdots, s_{17}$ were needed to produce an accuracy of 14 decimal digits.

We found that the infinite series representation (4) is computationally clearly superior, in particular since its rate of convergence increases significantly with increasing order m and since the new sequence transformation (10) is remarkably effective in the case of large m. For instance, for $m = 16$, to produce an accuracy of 14 decimal digits the transformation (10) only needs the partial sums $s_0, \cdots, s_7$ for $z = 5$, or the partial sums $s_0, \cdots, s_9$ for $z = 8$. The transformation (9) needs under the same conditions the partial sums $s_0, \cdots, s_{10}$ and $s_0, \cdots, s_{11}$, respectively, whereas Levin's transformation requires the partials sums $s_0, \cdots, s_{12}$ and $s_0, \cdots, s_{13}$,respectively.

The numerical properies of the asymptotic expansion (5) depend very much upon the magnitude of the order m of the auxiliary function. If m is small, the sequence transformations yield good results already for relatively small arguments z. For instance, for $m = 0$ and $z = 2.5$ Levin's transformation (8) required the partial sums $s_0, \cdots, s_{16}$ to produce an accuracy of 14 decimal digits, whereas the other two transformations (9) and (10) only needed the partial sums $s_0, \cdots, s_{13}$. However, for large values of m convergence becomes quite bad even for larger arguments. For instance for $m = 16$ and $z = 5$ none of the transformations was able to extract more than 11 digits out of the partial sums $s_0, \cdots, s_{21}$.

Finally, we also investigated the convergence of the addition theorem (3). Since for fixed argument z the function $F_m(z)$ is strictly decreasing with increasing order m, we may immediately conclude that the infinite series in Eq. (3) converges at least as fast as the Taylor series for e^{-x}. Hence, for small values of x convergence will be rapid and not much can be gained by applying a convergence accelerator. For instance, for $m = 0$, $x = 1$, and $y = 3$ Levin's transformation yielded 14 decimal digits after 9 terms whereas the infinite series needed 14 terms to produce the same accuracy.

Hence, on the basis of our numerical tests we would suggest the following approach for the computation of all $F_m(z)$ with $0 \leq m \leq M$ and $0 \leq z \leq z_{max}$:

(i) Choose a sequence of arguments $z_0 < z_1 < \cdots < z_k$ which are all contained in the interval $0 < z \leq z_{max}$.

(ii) If $z_j \in \{z_0, \cdots, z_k\}$ is small compute $F_M(z_j)$ using the infinite series (4) and compute the remaining $F_m(z_j)$ by backward recursion using Eq. (6) and store them.

(iii) If $z_j \in \{z_0, \cdots, z_k\}$ is moderately large compute $F_M(z_j)$ using the infinite series (4) in connection with the sequence transformation (10) and compute the remaining $F_m(z_j)$ by backward recursion using Eq. (6) and store them.

(iv) For very large $z_j \in \{z_0, \cdots, z_k\}$ compute $F_0(z_j)$ using the asymptotic expansion (5) in connection with either sequence transformation (9) or (10) and compute the remaining $F_m(z_j)$ by upward recursion using Eq. (6) and store them.

(v) For any other z with $0 \leq z \leq z_{max}$ compute $F_m(z)$ with the help of the addition theorem (6). Since one would like to use the precomputed values $F_m(z_j)$ choose $x = z - z_j$ and $y = z_j$ in such a way that the magnitude of x becomes minimal. If the differences $z_j - z_{j+1}$ are small enough no sequence transformation will be needed here.

5. References

[1] I. Shavitt in B. Alder, S. Fernbach, and M. Rotenberg (eds.), *Methods in Computational Physics* **2**, pp. 1 - 45 , Academic Press, New York, 1963.

[2] V. R. Saunders in G. H. F. Diercksen, B. T. Sutcliffe, and A. Veillard (eds.), *Computational Techniques in Quantum Chemistry and Molecular Physics*, pp. 347 - 424, Reidel, Dordrecht, 1975.

[3] V. R. Saunders in G. H. F. Diercksen and S. Wilson (eds.), *Methods in Computational Molecular Physics*, pp. 1 - 36 , Reidel, Dordrecht, 1983.

[4] F. E. Harris, Int. J. Quantum Chem. **23**, 1469 - 1478 (1983).

[5] J. Grotendorst and E. O. Steinborn, J. Chem. Phys. **84**, 5617 - 5623 (1986).

[6] D. Levin, Int. J. Comput. Math. B **3**, 371 - 388 (1973).

[7] P. Wynn, Math. Tables Aids Comput. **10**, 91 - 96 (1956).

[8] E. J. Weniger, *'Nonlinear Sequence Transformations for the Acceleration of Convergence and the Summation of Divergent Series'*, in preparation.

[9] D. A. Smith and W. F. Ford, SIAM J. Numer. Anal. **16**, 223 - 240 (1979).

[10] D. A. Smith and W. F. Ford, Math. Comput. **38**, 481 - 499 (1982).

[11] T. Fessler, W. F. Ford, and D. A. Smith, ACM Trans. Math. Software **9**, 346 - 354 (1983).

Acknowledgment

Support from the staff of the Rechenzentrum der Universität Regensburg is gratefully acknowledged. E. O. S. thanks the Fond der Chemischen Industrie for financial support.

CONCLUDING REMARKS

This meeting has, to our knowledge, for the first time gathered scientists from various horizons and interested and/or working in the field of numerical evaluations of the electronic structure of atoms and molecules to share their knowledge and computational experiences acquired over the years.

Beyond the definite interest of having a timely and up to date picture of the present status of the field, as provided by the lectures, this NATO-ARW had also the purpose of establishing bridges between the various working directions and to identify the major problems to be addressed for future progress.

An obvious benefit of the meeting was to give a chance to the participants to actually meet. They had the opportunity to present their specific results during specially arranged sessions for short communications and discuss in more details the associated problems in working groups informally arranged during the free time left.

It appeared that configuration space approaches are now capable of treating, fully numerically and with high accuracy, diatomic molecules, not speaking of atoms, including elements up to Ar ($1 \leq Z \leq 18$). The algorithmic developments have reached such a level that it is now safe to state that numerical calculations on diatomics are truly competitive with the existing "algebraic" approaches. Prospects for molecules including heavier systems are quite promising, but consideration of polyatomic systems with the existing methodologies still faces the problem of finding "priviledged" coordinate systems.

The newly emerging momentum space method is quite promising in that respect, since calculations on polyatomic molecules have been reported and even a simple model polymer has been partly considered. However, being in its infancy, this approach requires further theoretical and computational elaborations before its promise can suitably be characterized.

Recent progress made by Monte Carlo methods is impressive. Their versatility is particularly well adapted to treat small polyatomic molecules, electron-positron systems, vibrational problems, finite temperature effects, etc. However, difficulties remain important when treating many-fermion systems with high nuclear charges ($Z>10$). This is especially true for accuracy questions.

Of great value were the exchanges with mathematical physicists to identify ways to improve the convergence (reduce the number of integration points) of the numerical calculations. It was generally recognized that, for all methods considered in during the NATO-ARW, removing singularities occuring in the problems was a major problem to address. Concrete suggestions have been made. Similarly, numerical integration techniques should be chosen and in some cases tailored to take advantage of the basic mathematical features of a particular approach, *e.g.* configuration versus momentum space methods.

Impressions spontaneously given by the participants tend to indicate that this NATO-ARW was successful, and hopes that meetings on this subject could be organized in the future were expressed.

We would like to thank all the participants for their crucial cooperation in making the atmosphere of the meeting highly stimulating scientifically and very friendly.

Mireille Defranceschi
Joseph Delhalle
November, 1988

LIST of PARTICIPANTS

ADAMOWICZ Ludwik
College of Arts and Sciences, Faculty of Sciences, Department of Chemistry
The University of Arizona, Tucson, Arizona 85721 - USA

AHLRICHS Reinhart
Institut für Physikalische und Elektrochemie, Lehrstuhl für Theoretische Chemie
Universität Karlsruhe
Kaiserstr. 12, D-7500 Karlsruhe - WEST GERMANY

ALEXANDER Steven A.
Quantum Theory Project, Williamson Hall
University of Florida, Gainesville, Florida 32611 - USA

ANDERSON James B.
Department of Chemistry, 152 Davey Laboratory
The Pennsylvania State University, University Park, Pennsylvania 16802 - USA

AUBERT-FRECON Monique
Laboratoire de Spectrométrie Ionique et Moléculaire - Bât. 205
Université Lyon I, Campus de la Doua
43, boulevard du 11 novembre 1918, F-69622 Villeurbanne Cedex - FRANCE

BAK Keld
Kemisk Institut
Aarhus Universitet, DK-8000 Aarhus C - DENMARK

BERTHIER Gaston
Institut de Biologie, Université Pierre et Marie Curie
13, rue Pierre et Marie Curie, F-75005 Paris - FRANCE

CAFFAREL Michel
Laboratoire de Dynamique des Interactions Moléculaires
Université Pierre et Marie Curie, Tour 22, Couloir 22-23
4, place Jussieu, F-75230 Paris Cedex - FRANCE

DEFRANCESCHI Mireille
Ecole Normale Supérieure, Laboratoire de Chimie
1, rue Maurice Arnoux, F-92120 Montrouge - FRANCE

DELHALLE Joseph
Laboratoire de Chimie Théorique Appliquée
Facultés Universitaires Notre-Dame de la Paix
61, rue de Bruxelles, B-5000 Namur - BELGIUM

DINER Simon
Institut de Biologie, Université Pierre et Marie Curie
13, rue Pierre et Marie Curie, F-75005 Paris - FRANCE

FONTE Giacomo
Universita di Catania, Dipartimento di Fisica
Corso Italia, 57, I-95129 Catania - ITALY

FORTUNELLI Alessandro
Laboratorio di Chimica Quantistica ed Energetica Molecolare del C.N.R.
Via Risorgimento, 35, I-56100 Pisa - ITALY

FRIPIAT Joseph
Laboratoire de Chimie Théorique Appliquée
Facultés Universitaires Notre-Dame de la Paix
61, rue de Bruxelles, B-5000 Namur - BELGIUM

GODEFROID Michel
Laboratoire de Chimie Physique Moléculaire, C.P. 160
Université Libre de Bruxelles
50, avenue F.D. Roosevelt, B-1050 Bruxelles - BELGIUM

HEINEMANN Dirk
Farchbereich Physik der Universität Kassel
Heinrich-Plett Strasse, 40, D-3500 Kassel - WEST GERMANY

ISHIKAWA Yasuyuki
Department of Chemistry, University of Puerto Rico
Rio Piedras Campus, Rio Piedras, Puerto Rico 00931 - USA

KOCH Henrik
Kemisk Institut
Aarhus Universitet, DK-8000 Aarhus C - DENMARK

KOLB Dietmar
Fachbereich Physik der Universität Kassel
Heinrich Plett Strasse, 40, D-3500 Kassel - WEST GERMANY

KUTZELNIGG Werner
Theoretische Chemie, Ruhr Universität Bochum
Postfach 102148, D-4630 Bochum 1 - WEST GERMANY

LASSETTRE Edwin N.
Department of Chemistry, Carnegie Mellon University
4400 Fifth Avenue, Pittsburgh, Pennsylvania 15213 - USA

MAQUET Alfred
Laboratoire de Chimie Physique, Université Pierre et Marie Curie
11, rue Pierre et Marie Curie, F-75231 Paris Cedex 05 - FRANCE

MÅRTENSSON-PENDRILL Ann Marie
Chalmers University of Technology
Department of Physics, S-43144 Göteborg - SUEDE

MOCCIA Roberto
Laboratorio di Chimica Quantistica ed Energetica Moleculare del CNR
Via Risorgimento, 35, I-56100 Pisa - ITALY

MONKHORST Hendrik J.
Quantum Theory Project, Williamson Hall
University of Florida, Gainesville, Florida 32611 - USA

MORGAN John D. III
Department of Physics
University of Delaware, Newark, Delaware 19716 - USA

MÜLLER-PLATHE Florian
Max-Planck-Institut für Astrophysik
Karl Schwarschild Strasse 1, D-8046 Garching - WEST GERMANY

NAVAZA Jorge
Laboratoire de Physique, Tour B, Faculté de Pharmacie
Avenue J.B. Clément, F-92290 Châtenay-Malabry - FRANCE

PYYKKÖ Pekka
Department of Chemistry, University of Helsinki
Et. Hesperiankatu 4, SF-00100 Helsinki - FINLAND

SUARD Michèle
Laboratoire de Chimie, Ecole Normale Supérieure
1, rue Maurice Arnoux, F-92120 Montrouge - FRANCE

SUNDHOLM Dage
Department of Theoretical Chemistry, Chemical Center
University of Lund, P.O.B. 124, SF-22100 Lund - FINLAND

SUREAU Alain
Laboratoire de Spectroscopie Atomique et Ionique
Faculté des Sciences d'Orsay, Bât. 350, F-91405 Orsay Cedex - FRANCE

TSOUCARIS Georges
Laboratoire de Physique- Tour B, Faculté de Pharmacie
Avenue J.B. Clément, F-92290 Châtenay-Malabry - FRANCE

TALMAN James D.
Faculty of Sciences, Department of Applied Mathematics
University of Western Ontario, CD-N6A 5B9 London - CANADA

WEDIG Ulrich
Max-Planck-Institut für Festkörperforschung
Heisenbergstrasse 1, Postfach 80.06.65
D-7000 Stuttgart 80 - WEST GERMANY

WELLS Bryan
Physical Chemistry Laboratory, South Parks Road
UK-0X1 3QZ Oxford - GREAT BRITAIN

WENIGER Ernst
Institüt für Physik und Theoretische Chemie
Universität Regensburg, D-8400 Regensburg - WEST GERMANY

WILSON Stephen
Rutherford Appleton Laboratory
UK-0X11 OQZ Chilton - GREAT BRITAIN

SUBJECT INDEX